W9-DGL-799

DEFINING VISION

FLORIDA STATE
UNIVERSITY LIBRARIES

AUG 21 1998

TALLAHASSEE, FLORIDA

Also by the author

The Circus Master's Mission
(A novel)

DEFINING VISION

The Battle for the Future of Television

JOEL BRINKLEY

A HARVEST BOOK
HARCOURT BRACE & COMPANY
San Diego New York London

HE
8700.74
.U6
B75
1998

Copyright © 1997 by Joel Brinkley

All rights reserved. No part of this publication may be reproduced or transmitted in any form or by any means, electronic or mechanical, including photocopy, recording, or any information storage and retrieval system, without permission in writing from the publisher.

Requests for permission to make copies of any part of the work should be mailed to: Permissions Department, Harcourt Brace & Company, 6277 Sea Harbor Drive, Orlando, Florida 32887-6777.

Library of Congress Cataloging-in-Publication Data
Brinkley, Joel, 1952–
Defining vision: the battle for the future of television/Joel Brinkley.
—Rev., enl. ed., with new material added.
p. cm.
Includes index.
ISBN 0-15-100087-5 0-15-600597-2 (pbk)
1. High definition television—United States. 2. High definition television—Government policy—United States.
3. Television—United States. 4. Competition, International. I. Title.
HE8700.74.U6B75 1998 384.55'2—dc21 98-17794

Designed by Ivan Holmes
Printed in the United States of America
First Harvest edition 1998
E D C B A

For Sabra

Contents

Introduction

In the sultry summer days of 1988, America was seized by a panic, and it rose from the most unlikely of threats: a new TV that offered startlingly sharp pictures on a wide, movie-style screen. High-definition television, it was called. Japan had invented this new wonder, and Japanese manufacturers were promising to market it soon. America didn't have anything like it, and the *New York Times* editorial page decried the nation's failure: "For America not to compete in HDTV would be tantamount to abandoning a wide range of markets. But it might already be too late."

Congress was writing bills and holding heated hearings. U.S. companies were forming multibillion-dollar HDTV consortiums while columnists published funereal predictions. Almost overnight, high-definition television became a powerful icon, a symbol of everything that was wrong with the nation. "This is the same kind of thing as Sputnik in 1957," as Ervin Duggan, a commissioner of the Federal Communications Commission, saw it. "It's causing a broad national reassessment."

Out of that tumult an extraordinary idea was born, a course of action unprecedented in the history of this nation or probably any other. The United States government started a race, a titanic competition among the major corporations of the world for a prize so rich no one could even comprehend its worth. Whoever created the cleverest, most advanced new television would be crowned the winner. The government would anoint that company; it would hold licensing rights for every next-generation television set sold in America and probably much of the world.

Millions of TV sets. Billions of dollars.

When the competition began, the Japanese were the uncontested front-runners. They already had high-definition television, and everyone agreed that it looked awfully good. In Tokyo, the engineers and corporate chieftains were confident they would win. Nonetheless, twenty-three other players filled out their entry forms—huge corporations and backyard tinkerers, all driven by their own dreams, ambitions, vanities, and fears.

As long as the panic held Washington's attention, high-definition television remained a visible public issue, though not without backroom intrigue. But as soon as the race began in earnest, the policymakers and pundits quieted down. The focus of attention turned to the contestants' laboratories and boardrooms. There for eight long years, largely out of public view, they remained locked in desperate battle. Several of the contending companies threw everything else aside and put their best minds into the quest. As fast-paced contest deadlines loomed, the race grew to be a reckless, fevered scramble. In labs across the land, engineers abandoned families and slept on cots between shifts that ran around the clock.

But inventiveness and dedication weren't necessarily enough to assure victory. So some contestants cheated. Others lied and disparaged their competitors. Still others tried espionage—even bribery. Then, just as the contestants were finally nearing the finish line, the government again grew interested in the race it had started almost a decade before. Many of the same government leaders who had issued panicky complaints about Japan years earlier suddenly realized that they might be able to auction the airwaves set aside for HDTV and raise tens of billions of dollars. Congressmen talked of using all the new money to pay for tax cuts or to salvage endangered government programs. Dazzled by the allure of this, the government threatened to scuttle the race just as it was about to produce a winner.

"This is really a *saga,*" one of the contestants lamented near the end, as the race took one more tortured turn. "It just won't quit."

Today, manufacturers are finally readying the new televisions for sale, and with both piety and pride, politicians and marketeers are promoting America's triumph. The new sets, vastly superior to those that started the Sputnik-like panic almost a decade ago, were invented and designed in America. They will offer wide-screen images of striking brilliance and clarity. For the first time, TV will seem almost three-dimensional, real.

High-definition TV doesn't feel like television. Watching it is almost like looking out a window.

But that's not all. These new machines are wondrous indeed because, unlike the earlier Japanese models, these are *digital* televisions. A TV that receives its signal digitally is no longer just a dumb box passively displaying pictures and sound. Digital televisions, properly equipped, can be powerful, interactive computers, hardly different from desktop PCs. With those capabilities, suddenly television comes alive.

Instead of leafing through a TV guide, viewers will be able to program their computer-televisions to search through on-line schedules for the sorts of programs they like. The TV will then lay out menus of suggested viewing options, including shows that first aired days or weeks ago. Some TV stations will allow viewers to customize their local news programs—watch an expanded sports report, say, or get a traffic report tailored to show just their own route to work.

With digital television, viewers will be able to download an infinite range of movies, music, video games, or anything else. They can page through interactive home-shopping catalogs and direct an on-screen model to show the outfit with the suit coat on, or off—then press a few buttons to order the product. Viewers will be able to take interactive college courses at home, search electronic Yellow Pages, summon menus from local restaurants onto the screen and then make a reservation or place an order to go.

Viewers who don't regularly use computers will be able to converse with their friends through on-line chat groups or pick up their E-mail simply by punching a button on the remote control. And finally, almost everyone will gain full access to the Internet, right on the television screen. Viewers watching a football game, say, will be able to click on the World Wide Web address for the home team's Web page and go there instantly. The Web site might provide player stats or highlights from previous games; they appear in a window in a corner of the TV screen. Manufacturers are beginning to offer limited access to the Internet on present-generation televisions. But on conventional, *analog* TVs, these new services will forever be jury-rigged, awkward, and slow.

The digital television frontier is limitless; even the engineers who designed it admit they cannot even begin to dream of all the ways that inventors and marketeers will mold their new machine. But it does seem certain that digital television will serve as a powerful engine for the vast communications revolution rushing to capture America. These new computer-TVs will be display devices for all the programming and advanced services that the cable, telephone, satellite, computer, and broadcast industries are scrambling to bring into each of our homes.

The computer industry argues that computers, not televisions, will remain the primary point of access to the Internet and all that follows. For millions of Americans, no doubt, that will continue to be true. But fifteen years after the personal computer was introduced, fewer than half of American homes had one, and not everyone in all of those homes felt comfortable using it. Even with all the ease-of-use improvements of recent years, personal computers remain finicky, difficult appliances.

At the same time, more than 99 percent of American homes have televisions, and it's a rare soul who does not know how to use one. So for many Americans, digital television will offer an easy, natural entryway to the "information superhighway"—the World Wide Web, electronic mail, and everything else.

All of this comes with another change that may not be so welcome. Soon the clear distinctions between televisions and computers will blur. And with that, a disease of the computer world will probably infect the television industry, too. Until now, a family could buy a new TV with confidence that it would last eight or ten years, maybe longer. After all, the technology behind televisions has barely changed in three decades. But now TVs will be like computers, and a personal computer bought just two years ago is a technological midget today. Soon TVs are likely to grow obsolete almost as quickly. Each new model will come with ever faster chips, larger memories, better modems, more expansive options.

Sure, three to five years from now, owners of the first digital HDTVs will still be able to get personalized stock quotes on screen, select from vast movie libraries, shop at home, or simply watch the network news or the NCAA basketball championship in stunningly sharp detail—complete with six-channel surround sound. But will that be enough when manufacturers start offering new models with built-in videophones, or interactive pornographic channels?

And how will the owners of *those* cutting-edge sets feel a few years later, when the newer models offer flat-panel displays you can hang on the wall and a sex or violence dial: Twist it up or down to get the program version you prefer.

A few years after that, will the owners of even those models be content when the next generation comes along with the capability to offer holographic video games, or other marvels as yet unthinkable?

The creation of digital, high-definition television is an American triumph, no question. At the start of this decade, television engineers around the world, without exception, believed that digital TV defied the laws of physics. But one contestant in the HDTV race proved them wrong. Soon, wide-screen digital televisions will be commonplace features in almost every home, and today's square-screen sets will seem as archaic as a record player. There may be little choice because under the government's present plan everyone in the nation will have to buy a digital television sometime in the next ten or fifteen years. On a date certain sometime in the next decade, conventional television broadcasting will end. The transmitters will go dark. The entire nation will have switched to digital TV.

But don't believe all the boosterism that will accompany the introduction of digital HDTV. In truth, the origins of this remarkable new technology could hardly have been more cynical, the triumphs more serendipitous.

The high-definition television race did spawn exceptional creative genius. But it also pointed up with astonishing clarity how willing government leaders are to use American business for their own opportunistic ends—promoting it at one moment, betraying it at the next. And it displayed individual behavior—rank hypocrisy, conniving duplicity, selfish disregard—of which few people would be proud.

This is the story of the race and its players.

1

The Scheme

1 HDTV...Maybe that's it!

John Abel hadn't quite known what to expect, but certainly not this. He and his colleague Tom Keller had flown down to Fort Lauderdale for a visit with Bill Glenn, arguably the nation's foremost broadcast television engineer. They wanted to see Glenn's high-definition television system, said to be the most advanced America had to offer, and Glenn had promised to take them for a visit to his lab. But here they were, standing under the scorching sun at Pier 66, waiting for Glenn to pick them up in a boat. Abel already had a bad feeling about this trip, and now he was beside himself. What kind of research lab do you visit by boat?

A little outboard was plowing toward them through the swells, and Abel could see a white-haired man behind the wheel. That must be Glenn, he thought, shaking his head. A couple of minutes later, Glenn docked the boat and stepped onto the pier. As everybody shook hands, Abel looked him over. Glenn looked to be about sixty years old, an affable man with a full head of wavy white hair. Now this college professor, the former director of research for CBS Labs, was directing them into his boat. Before Abel could catch his breath, they were plowing across the harbor, and Abel was wiping spray off his face, thinking, This is crazy, just plain crazy. What the hell am I doing here?

Abel was a vice president of the National Association of Broadcasters, a Washington-based trade group that represented radio and television stations nationwide. Trade organizations like the NAB lobbied Congress and the federal agencies to promote, or stop, bills and regulatory initiatives of concern to their membership. And now, in the autumn of 1986, the NAB was locked in a desperate lobbying battle. The broadcasters were trying to hold on to the extra TV channels allotted for television service in every city, even though they weren't actually being used. In Washington, for example, networks and independent stations broadcast on channels 4, 5, 7, and 9 on the VHF dial and 20, 26, 32, and 50 on the UHF side. The rest of the designated broadcast television channels—2 through 69, the ones viewers picked up with an antenna if they did not have cable TV—sat vacant, and the situation was similar in most other cities.

The mobile-communications industry, led by Motorola, wanted to take some of those vacant channels away and use them for two-way radio transmissions. The broadcasters didn't like this one bit. These were *their* channels, after all, and as they saw it the industry's very future depended on the fight to keep them. They had to show Congress and the Federal Communications Commission that they needed the channels for something, *anything*. So Abel had come up with a scheme: Let's tell everyone we need those channels for high-definition television.

Broadcasters had liked the idea. But for this pitch to work, they needed to put on a demonstration. They had to show that HDTV was real, not just one more fuzzy fantasy of the far distant future. The NAB had been giving Glenn a little money for his HDTV research, and now his system was said to be the only one under development in America that could actually show a picture. Abel wanted to have a look.

They bounced across Port Everglades for a few minutes, then Glenn maneuvered into a tiny harbor and docked between a couple of small pleasure boats. The three of them clambered ashore. Abel had expected to see a multibuilding research complex by the sea; instead they walked past a trailer and a small aluminum shed. Just in front of them was a rusting chain-link fence, and on the other side lay a dozen large spools of heavy marine cable scattered in the sand and underbrush. Glenn led them down a dirt track past a garage, and when they turned the corner he pointed to his lab just ahead—a small two-story structure.

That's it? Abel thought, hardly able to contain himself now. It's cinder-block. The damned thing looks like a garage. And look at that; it's sitting there in a vacant lot surrounded by weeds! Inside, at least the building was air-conditioned, and Glenn led them into a small office, where he and Keller, the NAB's technology officer, chattered between themselves. Abel half listened, impatient again.

As you know, Glenn was saying, we have only a "hard-wire" system. It's closed circuit: a camera, a signal processor, and a monitor. We aren't ready to actually transmit anything over the air. We can't afford to take that on just yet. Still, Glenn added with a smile and a lift of the chin:

"We can aim that camera at anything we can find and get high-definition images." Then he led them across the hall to his lab.

Abel was no engineer, but he'd been in more than a few broadcast labs. He could tell a shoestring operation when he saw it. Glenn didn't have much equipment, and most of it was several generations old. Abel half expected the supply cabinets to be stocked with Wrigley's gum and baling wire.

Unaware of Abel's ruminations, Glenn showed off his invention with obvious pleasure. He pointed to the camera, such as it was. It wasn't mounted on a tripod. Without any casing, the camera lay splayed in the middle of a table. A couple of circuit boards stuck out horizontally from behind the lens, wires dangling from their sides. The damned thing looked like a gutted fish!

The camera was aimed at another table a few feet away, this one washed in bright light. Two spotlights shone down on the tabletop from poles planted on either side. In the middle of this table a tiny toy ballerina in a bright crinoline dress stood frozen in midpirouette, rotating slowly on a plastic turntable. Clearly somebody had picked this up at Toys "R" Us.

Finally the professor directed them to the monitor, a 19-inch Sony high-definition television set. The NAB had given Glenn the money for this, and he'd bought the Sony model because no one in America made anything like it. When Abel looked, sure enough, the picture was pretty good. As the ballerina twirled, you could see all the individual folds of her little dress. The colors were vivid, the resolution was crystal clear.

"We've actually got better resolution than the Japanese do," Glenn was saying with a smile, but Abel had heard enough.

This is 1986, he was thinking, and the Japanese have *two hundred engineers* working on HDTV. They've got cameras and recorders and TV sets *in production.* They're actually broadcasting. And here we are; this is the best America has to offer, and we're looking at it *in a garage!*

That's it, Abel said to himself. We're in trouble. Time to call the Japanese.

That autumn afternoon, Abel was a relative newcomer to the NAB. The seeds of the organization's crisis had actually been sown more than a decade before, while Abel was still teaching classes at Michigan State University.

In the 1960s and 1970s, the leaders of the three television networks stood confidently in positions of unparalleled importance. Did the president of the United States wield greater influence over public opinion than the men who decided whether *Gunsmoke,* or *60 Minutes,* would return for another season? Even their office towers in midtown Manhattan were American landmarks: Rockefeller Center, Broadcast House. And the top-floor executive suites were the regally appointed homes for the Broadcast Barons, the royalty of America's new, electronic age.

Across the country, meanwhile, the families that owned the television stations carrying those network shows had long ago learned that they had only to sit back and smile as the profits poured in. These people had grown to be community leaders of the first rank—heads of the arts commissions, directors of the United Way campaigns. They held forth from offices that were proud downtown monuments, just around the corner from city hall. Their satellite-TV trucks sallied forth across town, greeted everywhere they ventured almost as if they were official city vehicles.

Out of all that, their appointed representatives in Washington, the lobbyists at the NAB, had grown fat and comfortable. Their royal clients held official licenses to print money—permits that were well protected by their own special agency of government, the Federal Communications Commission. Everybody was getting rich; everyone was happy.

But then the *Mongols* began pounding at the gates. Cable TV.

In 1976, barely 15 percent of American homes were wired for cable, and to the broadcasters then, the cable operators were irritating, oafish figures. They seemed to favor polyester. They worked out of buildings with corrugated metal walls somewhere out there by the warehouse district. Bluntly put, these people were parasites. They offered no shows of their own; they simply sucked up the network programs and sent them out over a wire. To the local broadcasters, these were people of a decidedly lower caste. Nonetheless, as the 1980s dawned, Americans began falling in love with cable and its promise of interference-free TV pictures and vast new selections of programming. Cable offered an escape from the tyranny of the networks. As more and more homes hooked up, ever greater numbers of viewers were choosing ESPN or the Movie Channel during prime time, instead of *Love Boat* or *LA Law.* Marilyn Chambers or Linda Lovelace late at night, instead of Johnny Carson or *Matlock.* The Broadcast Barons watched, helpless, as their profits and power began to slip away. Do something! they pleaded. But it wasn't so easy.

With every new cable hookup the broadcasting industry lost a few more drops of its lifeblood. Some of the broadcaster lobbyists in Washington began to imagine one of those giant plywood thermometers out in front of the cable television headquarters just across town, but this one wasn't there to show how well the United Way campaign was going. No, in the lobbyists' minds, the bright red fever line inching up day by day showed how many American homes were unhooking their TV antennas and plugging in the cable instead: 19 percent in 1979, 28 percent in 1981. *Forty-three* percent in 1984.

Then in 1985, the Cable Mongols won a decisive battle. A federal appeals court ruled that cable systems were no longer required to carry broadcast television programming at all. That meant the cable companies could simply choose not to transmit all those programs the Broadcast Barons worked so hard to produce—along with all those commercials they struggled even harder to sell. Right away several cable systems threatened to drop some of the weaker stations in their communities. Who was next? Forty million households had already unhooked their TV antennas. The Mongols had guns to the broadcasters' heads.

With that the NAB seemed to be at its nadir, as everyone in Washington could easily see. Senator Bob Packwood had said it first—and while he was a guest at the NAB's own convention. Now his statement had become common wisdom in Washington: "The NAB can't lobby its way out of a paper bag."

But darker days were yet to come. Even as they lost every skirmish with the Cable Mongols, a new and even more dangerous enemy appeared.

Land Mobile.

The Cable Mongols were stealing their audience. But the way the Broadcast Barons saw it, Land Mobile was challenging their very existence. To make matters worse, the FCC was on Land Mobile's side.

The FCC's most difficult task was allotting uses for the crowded airways. Everyone wanted to transmit on this invisible highway—ham radio operators, air traffic controllers, county rescue squads, the military, radio-dispatched taxicabs, pizza delivery trucks. The list went on and on, but there were only so many lanes. Of all these users, however, the broadcasters had the choicest space—or *spectrum,* as it was called—because these particular channels allowed long-distance transmissions with the greatest clarity. And it's no wonder: wireless, and then radio, were the first users of the electromagnetic spectrum, a hundred years ago. That's why, even today, when the airways are crowded with varied services, all of it is still known as the radio frequencies.

Now, in the mid-1980s, a new group was clamoring for space—the manufacturers and users of two-way radios. Police departments, ambulance services, commercial delivery companies. Motorola made most of these radios and led this lobby, which was known as Land Mobile. And Motorola's lobbyists were trying to convince the FCC that broadcasters had no real use for much of the choice spectrum they controlled. After all, most cities had only eight or ten TV stations at most, so fifty or more of the channels set aside for television broadcasting lay fallow. Some of those were left unoccupied on purpose, to reduce interference between adjacent channels. Still, more than half of the channels allotted for TV service in most cities were sitting idle. Why not give some of those channels to us? Land Mobile asked. By 1986, the FCC had pretty much decided to do just that. Several vacant UHF channels in ten big cities were to be taken away from the broadcasters and given to Land Mobile.

Nothing was more certain to rouse the broadcasters. Above all else, they held sacred the eleventh commandment: Thou Shalt Not Give Up Spectrum. Their assigned channels were precious electronic real estate—beachfront property, they liked to call it. They argued that mobile-radio transmissions would cause static, noise, and other irritating interference on the broadcast stations, driving even more viewers to cable.

Here was a big potential problem. And broadcasters did not like the principle of it, either. Like cattle ranchers on the western plains, the broadcasters saw themselves as the descendants of heroic frontiersmen. Their forebears—Marconi, de Forest, Armstrong, Sarnoff—had tamed this spectrum, cultivated it, and then passed it on to them. By god, it was *theirs,* and they were not about to give up any of it. Land Mobile was just the camel's nose under the tent. Let them in, and others would stream through right behind them. Pretty soon the broadcasters would have no spectrum left. And would anyone really care, once most of the nation was hooked to cable? Truly, the broadcasters felt, their very survival was at stake.

John Abel came to be the NAB's point man for this battle, and he loved a good fight. Abel looked unimposing; he had a bushy mustache, a head of thick black hair, and a sad-sack expression when his face was at rest. But with little provocation he broke into a conspiratorial grin, for Abel was an exceedingly clever man. Some might say conniving, though he was also likable and generally honest about his intentions. Abel's special qualities certainly had not been lost on Eddie Fritts, the association's president. When he met Abel on an airplane in 1983, Fritts promptly offered him a job.

Abel, an Indiana native, had been chairman of the Department of Telecommunications at Michigan State University, but not even the intrigue of big-campus politics had been enough to keep him challenged. University life was too parochial. He had spent a sabbatical year working as a consultant at the FCC, and Washington power politics was more exciting than he had ever imagined. All of a sudden his old position seemed irrelevant. So he accepted the job at the NAB.

Abel took on Land Mobile with relish. First he did some research, found a few potentially malignant spots on Motorola's record: questions about the company's government contracts or its environmental policies. He spread that around, but the FCC didn't seem to care. When that tactic failed, Abel and the other lobbyists tried logical arguments. What about the interference? Viewers didn't want a flash of static on their TV sets every time a pizza delivery truck drove past. Land Mobile shot back that the discussion wasn't about pizza trucks. It was about ambulances and police cars. In any case, just what was this sacred programming that couldn't stand any interference? *Laverne and Shirley?* Ex-Lax commercials? So much for that pitch.

Abel was losing. Then Land Mobile repeated its most telling argument: You broadcasters aren't *doing* anything with those vacant UHF channels, and you have no plans for them. Tell us what you're going to use them for. *Tell* us! Abel and the others had puzzled and worried over that. But they hadn't been able to think of any answer at all. Now there seemed no way to stop Land Mobile. Around Abel the Broadcast Barons were asking, If we lose this one, can our industry survive?

One afternoon in the summer of 1986, Abel led a meeting of officers from the NAB and allied groups in one of the association's wood-paneled conference rooms on the first floor of NAB headquarters on R Street in downtown Washington. Their mission: Devise one last, desperate lobbying strategy. But as Abel looked around the table he grew depressed. The mood could not have been grimmer. The lawyers and lobbyists mumbled, sighed, fidgeted in their seats, and looked down at the table. No one had a good idea. To Abel, it seemed as if everyone had already given up.

But then, out of the corner of his eye Abel happened to notice Tom Keller, the NAB's technology officer. Abel could hardly miss Keller because he was sound asleep, as he sometimes was during meetings, chin resting on his chest. Abel was Keller's boss, and as he glared at him suddenly a thought struck.

Through Keller, the NAB had been giving money to Bill Glenn, that college professor down in Florida. The sums weren't large because the broadcasters didn't really care much about high-definition television. Glenn's work was like a school science project, really. And HDTV—that was a technology of the far distant future that few of them had ever seen or really even thought much about. But as Abel stared at Keller, who was drawing long breaths as he dozed, a realization dawned: Bill Glenn's high-definition system wouldn't fit on a single television channel. If broadcast, it would fill up all of channel 3, say, and half of channel 4 as well.

Wait a minute, Abel thought, sitting up straight as an idea began racing through his head. Here's an argument. HDTV takes more channels. Land Mobile wants to know what we need those extra channels for? Well, we need them for high-definition television.

HDTV . . . Maybe that's it!

"I've got an idea," he told the group as a grin slowly spread. A dozen dour faces looked up. "What about high-definition television? Why don't we tell them we need all that extra spectrum for high-definition television?" The others didn't say anything at first. Quizzical expressions crossed several faces, as Abel's listeners tried to recall exactly what HDTV was. To Abel, that puzzlement was exactly the point. Land Mobile probably hadn't ever considered HDTV either.

"They'll *never* have thought of this," he said, looking around the table with that conspiratorial grin. "It'll really take them by surprise, put them on the defense. And it's a positive argument, not negative." The broadcasters could offer the lofty idea that they needed all that extra spectrum so they could *bring HDTV to America.*

Slowly, cautiously, some of the others began to nod. It just might work. They certainly had nothing else. Then, as Abel recounted it, one of them stopped short as he thought ahead.

"Yeah, but what if we actually *get* it?" he asked. It was obvious that broadcasters would have to spend quite a lot of money if HDTV ever became a reality. They would have to buy new high-definition cameras and recorders. They would need monitors, transmitters . . . the works. Some of them might have to shop for a new TV tower. All of that would cost millions, and where would they get the money? Stations wouldn't be able to sell more advertising just because the ads were broadcast in high definition. Ad rates couldn't be raised just because the detergent boxes showed up better. All that and more ran quickly through some people's minds, but they brushed it off. They had a problem *now*. HDTV . . . that was years away, maybe even decades. Hell, some of them figured they'd probably have retired or died by the time high-definition television was on the air. Finally Abel said, "Finding something that works *right now* is more important than where we end up." The discussion moved ahead.

"We've gotta have a strategy," Abel said. "We can't just tell them; we've got to show them." He turned to Keller; with all the commotion, he was awake now. Tom, he asked, can we get Bill Glenn up here for a demonstration?

Keller shook his head. "I don't think his system's ready yet."

Abel wasn't really surprised. From what he'd heard about Bill Glenn's machine, it sounded like "a baling-wire kind of thing." Still, he said, maybe we'd better go down and have a look. But if Glenn's system wouldn't do, where else could they get an HDTV system to put on display?

Across the table, Greg DePriest was listening with a bemused smile on his face. DePriest was a vice president of the Association for Maximum Service Television. This group had a small office suite a few blocks away, where six employees worked on behalf of their broadcaster members to keep ahead of technological developments in television. DePriest had been watching the progress of HDTV with more interest than the others at the table; this fell squarely within his organization's charter. He knew full well that the Japanese already had an HDTV system up and running. They had cameras, transmitters, TV sets, and VCRs—everything. In fact, DePriest had been talking with NHK, Japan's public broadcasting network, about putting on a demonstration in the United

States. The idea had been to make sure that broadcasters remained competitive. What would happen if the Cable Mongols started offering high-definition television before broadcasters were able? Another disaster. Cable would need all the new equipment, too. But unlike the broadcasters they wouldn't have to get permission from the government to broadcast HDTV.

The talks with the Japanese had been largely theoretical; DePriest's organization didn't have the money or wherewithal to stage a big public demonstration. DePriest had been thinking about asking the NAB for help, but the two groups didn't get along very well. In DePriest's view, Abel and the others looked down their noses at his little organization. He found it so unpleasant to work with them, and he hadn't quite gotten around to asking. Nonetheless, he spoke up now: "What about NHK? We've been talking to NHK about doing a demo. They could do it." NHK's system needed more than one TV channel, too.

That's a possibility, Abel said.

They talked through the idea some more and finally concluded: We have a strategy. The meeting adjourned, and now Abel had to decide on the best way to carry it out. A short time later, he and Keller took their trip down to Fort Lauderdale.

Even before Abel left Glenn's cinder-block lab, he had concluded that the professor's system simply would not do for the extravagant show he was planning. It didn't take much to imagine the senators, congressmen, and FCC commissioners seated expectantly in front of that little TV monitor, noting the fat cable that snaked away from the back of the set. As the picture came on, the honored guests would follow the cable with their eyes until they spotted that prototype camera, the gutted fish with the wires hanging out, focused on the Toys "R" Us ballerina—or maybe on the traffic outside the window.

Abel turned to Keller and said, "Call the Japanese."

Stand up close to any television set and you'll see that the picture is actually striped. Sent from the tower, the television signal is a series of lines that, laid down side by side, make up a complete TV picture. When the U.S. standard for broadcast television was set, by the National Television Standards Committee in 1941, the engineers of the time had decreed that a television picture should be made up of 525 lines. This had seemed an extravagant number at the time, but a few years later the Europeans adopted their own standards and settled on 625 lines.

But television came late to Japan as the nation recovered from World War II. By the mid-1950s hardly anybody had a set, even though the American occupiers had left the nation with the 525-line television system used in the United States. Still, the Japanese people were enthralled by the televised marriage of their crown prince in 1959. Many viewers stood transfixed before the tiny screens in department store windows. They were captivated again by the television coverage of the Tokyo Olympic Games in 1964. Television sales boomed after that, and soon engineers at NHK, Japan's public broadcasting network, began thinking about television's future. These engineers chafed at the limitations of the television system the Americans had given them. "Conventional TV systems fail to attain the level at which the functions of the human visual system can be effectively utilized," a company monograph argued. "As a result, TV cannot be compared with movies or printing in terms of picture clarity, impact, or immediacy."

Many of Japan's best engineering minds had been dedicated to the nation's military effort while TV was being developed in America, so the network's engineers had little investment in the status quo. They wanted to create a new television—one that could, as the monograph put it, "appeal to a higher level of psychological sensation and emotion by transmitting highly intellectual information with detailed characters and graphics." Toward that end, NHK began its research by asking the most basic questions: What does the human eye really see? How can we get our viewers even more involved in the programming?

One conclusion came quickly: the closer viewers sat to a picture, the more it filled their field of vision and the greater their emotional involvement. This was a well-established visual principle. Even a family-vacation photograph, for example, is more effective when the subjects fill the foreground.

But sitting too close to a television set made viewers dizzy. Too far, and the picture seemed remote. There had to be an ideal spot. To find it, NHK engineers asked volunteers to watch movies and TV shows from various distances. After several months of study, the engineers found that ideally the viewer should sit at a distance four times the height of the screen. If the screen is three feet high, say, viewers should sit twelve feet away. There they will feel involved without getting dizzy.

But that ideal ratio didn't work with television. The picture was too crude to be viewed close up. Years before, as a result, American engineers had found the best viewing distance to be about seven times the picture height. "Watching the screen at a distance nearer than this," the Japanese wrote, "the viewer would find the picture coarse and blurred." At the recommended distance, however, the TV picture seemed remote, uninvolving.

That led the researchers to an immediate, obvious conclusion: Improve the picture quality so that it doesn't seem coarse when viewed close up.

The Japanese engineers reached a second conclusion, too. TV screens are nearly square, but the human field of vision is wider. That's why movie screens are almost twice as wide as they are high. Television's picture, too, should be wider. It should fill the entire field of vision.

Once they finished their audience research and set their goals, the Japanese quickly realized that the surest way to raise the quality of a television picture was to increase the number of lines. They settled on 1,125 lines, doubling the current standard. With twice as many lines on the screen, each line was only half as wide. That meant the picture could show twice as much detail. The higher quality easily allowed comfortable viewing for people sitting closer to the TV.

The problem was that a television channel was only so wide—not wide enough to carry all those additional lines. Back in 1941, the American standard setters had decided that each TV channel could occupy only 6 megahertz on the airwaves. A channel only 6 megahertz wide is roughly analogous to a water pipe that is only 6 inches in diameter; the wider the pipe, the more water can flow through it. In 1941, 6 megahertz seemed lavish, yet still narrow enough so that lots of channels could fit on the airways.

The Japanese got around this 6-megahertz limitation by using various strategies to make the signal more compact. But most important, they used two TV channels; they supplemented the signal that was being broadcast on one TV channel with additional electronic information broadcast on another. Two-thirds of the signal would be sent on channel 6, say, and the rest on channel 8. The television receiver would combine the two signals to create a unified, high-resolution picture with 1,125 lines, viewed on channel 6.

Now they had two broad goals and strategies for achieving them: Improve the picture quality by doubling the number of lines, and widen the screen. And as the Japanese began the actual engineering work in the 1970s, they were driven by more than one motivation. Sure, NHK's scientists and engineers hoped to create something new for the world.

They wanted recognition, and they hoped to provide their country with an important new export product. But as the years went by and the research progressed, the project also seemed as if it might save the network.

Most of NHK's operating revenues, about $2 billion a year, came from receiver fees set by the Diet, Japan's parliament. Everyone who owned a TV set had to pay the fee for the right to receive the public network's programming. During the 1960s and 1970s, NHK's receipts kept rising and rising as more and more Japanese bought sets. But by the late 1970s almost every home had a TV, and NHK's income from the fees began to level off—even though the network's costs continued to rise. Japanese companies couldn't easily lay off workers or reduce benefits, so NHK began running annual deficits.

Beginning in the early 1980s, the network raised the receiver fees every few years, but the increases were always greeted with rancor from both the Diet and the public. At the same time, however, the new TV research began to produce results, and NHK's leaders got an idea: Maybe offering viewers a dramatic new service might give the network a better excuse to raise the fees or justify other new charges. High-definition television, as it came to be called, might be the solution to the network's budget crisis. So NHK began trying to make deals with Sony, Ikegami, and other Japanese companies to manufacture high-definition cameras, recorders, and TV sets for the Japanese market.

At first, most of the manufacturers were unenthusiastic. These people at NHK were broadcasters, after all. What did they know about marketing consumer electronics? But over time, more of them began to see the product's potential. Soon the Japanese government got involved and made long-range plans to broadcast HDTV by satellite to the entire nation within the next decade. Everyone in Japan would buy a new HDTV. That would be great for the manufacturers, and NHK might solve its financial problems, too.

With fourteen thousand employees in the early 1980s, NHK was the second largest television network in the world, just behind the BBC. The work on HDTV had been the biggest engineering project in the network's history and a great source of pride—not just for NHK but for much of Japan. After all, how often had the Japanese created an important technological breakthrough on their own rather than simply improving an invention that originated someplace else?

In 1981, they were ready to go public. So they demonstrated their system outside Japan for the first time, at a television engineers' convention in San Francisco. The Americans were impressed. At that time, however, NHK had only an experimental, closed-circuit system, like the one Bill Glenn developed years later. And NHK's signal was so complex that it would have filled up five entire television channels. Sure, the picture's great, the Americans said after the demonstration. With all that spectrum, *anybody* could produce a high-definition picture. In the end, the demonstration caused a ripple of interest among American engineers, but it didn't get much notice otherwise. It seemed little more than a novelty.

But as work continued, NHK took up a new strategy. Maybe Japan could sell the world on the *technical* standards behind the system, even if the hardware wasn't quite ready. That idea held some inherent appeal: as it was, different regions of the world used different standards. Japan formally proposed its standard—1,125 lines and 60 images, or fields, per second—as a world standard, and the Consultive Committee of International Radio, a world body set up to consider such matters, put the proposal on its agenda. The Japanese seldom brought it up, and in the West few people talked about it. But the commercial potential behind this innocuous-seeming proposal was immense. If the Consultive Committee agreed to adopt 1125/60 as the world standard, then Sony, Toshiba, JVC, and other manufacturers would begin setting up dozens of new production lines. Very soon they would start pumping out millions, *tens of millions*, of 1125/60 televisions, cameras, recorders, transmitters, film converters, and every other manner of equipment. Japan would capture an entire new consumer electronics market—the biggest one of them all.

Nonetheless, the U.S. State Department liked the idea. The United States exported movies, TV shows, and other recorded entertainment around the world. Even with differing TV standards, the market was worth $5 billion. If one standard were adopted for every nation on the globe, then people everywhere would be able to slip American-made videotapes into their VCRs and, U.S. officials figured, the American trade surplus in this market would soar even higher. State Department officials—pinstriped diplomats afflicted with the myopia that blinds so many residents of Foggy Bottom—decided that it made sense to support the Japanese request. This certainly would do wonders for U.S.-Japanese relations, ever the diplomats' goal.

A few people elsewhere in government opposed the State Department on this issue; oddly, though, the policy was not especially controversial. By then, most American manufacturers had dropped out of the TV business. The United States was already heavily dependent on the Japanese for televisions and related equipment. To the few people who noticed, the prevailing thought was, What's the harm?

But the Europeans, particularly the French, found the whole idea appalling. They didn't want their people easily slipping videotapes of *Rambo* or *Dallas* into their VCRs. And Europe, unlike the United States, still had a consumer electronics industry to protect. So the Europeans launched a ferocious counterattack. The European Community began talking about setting up a consortium to create an HDTV system of its own. The Americans saw this as only a paper enterprise. Soon, however, the question of a world TV standard was overcome by bitter nationalist bickering. And when the Japanese proposal was debated at the Consultive Committee meeting in Dubrovnik in May 1986, it plummeted to the ground in flames. When the NAB called a few months later, asking NHK to put on a demonstration in Washington, the Japanese were still sifting through the ashes. As a result, the Japanese engineers and executives were delighted, even ecstatic. Yozo Ono, NHK's chief scientist based in New York, could hardly contain himself. "To be invited to put on a demonstration in Washington, in the capital," he marveled, "and to begin technical cooperation with the United States"—well, NHK was "extremely happy to do that."

Warm visions of "close cooperation between us and American broadcasters" began filling Masao Sugimoto's head, too. From the beginning, he had directed NHK's program to develop high-definition television. In the West, the system had been given the name "Muse," a loose acronym for the English-language technical description: "multiple subnyquist sampling encoding." Sugimoto was known as "the father of Muse." He happily arranged to send twenty engineers—virtually his entire team—to Washington. So much rode on how well their new invention was received in America. If they could just sell the United States on Muse, the European obstinacy would hardly matter. HDTV would certainly spread around the world. This was their last, best chance.

Much rode on this for the American broadcasters, too. The NHK demonstration stood as their last chance to stave off Land Mobile and keep their extra channels. And as preparations proceeded, the Japanese and Americans fixated on their own agendas, mostly blind to the other's motivations and goals. Few American broadcasters had been following the debate over the Japanese broadcast standard. It was a remote issue for John Abel and for most others at the NAB as well. They had their own problems—the Cable Mongols and now the Land Mobile war. NAB officers weren't necessarily trying to hide their concerns about Land Mobile, but they never explicitly explained to the Japanese that the demonstration was a lobbying tactic, nothing more. NHK's engineers, for their part, saw the NAB as a group of fellow broadcasters—sympathetic colleagues who were going to give them a boost. So both camps plunged into the planning, each of them convinced that the demonstration could save them.

The Japanese engineers first had to make some modifications so that Muse could be sent over the air from a TV tower instead of from a satellite, as originally designed. The NAB enlisted the help of WUSA, the CBS affiliate in Washington. The Japanese would set up at the station and use WUSA's tower for the broadcast. The NAB, meanwhile, filed an application with the FCC to use two UHF channels that were vacant in the Washington area, 58 and 59, for the experimental broadcast—a routine request that was quickly granted. As the application noted, the Muse signal was 8 megahertz wide, meaning that it would fill all of channel 58 and part of channel 59—exactly the kind of space on the airwaves that Land Mobile wanted to take. The FCC also agreed to let the NAB stage the first demonstration at FCC headquarters, in the commissioners' eighth-floor meeting room.

That done, the NAB formally scheduled the event for just after the New Year—on January 7, 1987. Abel and the others had already prepared the lobbying strategy. Eddie Fritts, the NAB president, opened the campaign in a speech to the Annenberg school of communications on December 15. Suddenly, as everyone could see, Fritts was an HDTV convert.

"HDTV is a vital development on the global television scene," he told the startled group—few of whom had ever heard of high-definition television before. But Fritts went on: "We all know that implementation of broadcast HDTV will require more spectrum. Where will that spectrum come from? We propose that it be drawn from the existing UHF broadcast allocation.

"But ladies and gentlemen," he added ominously, "the Federal Communications Commission appears predisposed to give Land Mobile users the available UHF-TV frequencies we need to transmit HDTV to all of the American public." If that were to happen, what of America's cherished tradition of free, over-the-air television for everyone? Would the American people be offered high-definition television only if they were willing to pay for it on cable? What would happen to broadcast television then? Would it go the way of AM radio, an irrelevant, forgotten service? Then came the broadcasters' new rallying cry, a mantra repeated over and over again in the following months: If America's broadcasters "are precluded from offering HDTV as a free, over-the-air service to the nation," that will bring *the death of local broadcasting as we know it!*

The broadcasters sent invitations to everyone who mattered: Reagan administration officials, diplomats, Federal Communications Commission officers, leaders of the television broadcasting industry. On a crisp January morning, everyone gathered in the FCC's big meeting room, coats draped over their arms, ready for a show. Many of them gawked at the odd-shaped television sitting on a low platform at the front of the room. The picture tube was almost twice as wide as it was high, just like a movie screen. A conventional, square-screened TV sat alongside. Some guests wandered to the front of the room so they could look over the supporting electronic equipment laid out on a long table against the right-hand wall. As they did, Japanese engineers glanced up momentarily, then resumed urgent whispering as they made last-minute adjustments to racks of experimental gear that hummed and glowed—fed by fat cables that trailed off behind the table to 240-volt outlets specially installed for this event.

As the room filled up, James McKinney wandered in and pushed his way up front to have his own look. McKinney was the head of the FCC's Mass Media Bureau, the division that regulated television and radio, and he could see that NHK had brought along Sony's very best, studio-quality conventional receiver. He was impressed. The Japanese were ready and willing to compare their new, high-definition system with the most sophisticated regular set they had.

Finally, chattering among themselves, McKinney and the others settled into their chairs. More people arrived, and soon two dozen spectators were crowded together in the aisles. As show time neared, the room was tense with anticipation, and people jockeyed to get a clear view of the two screens. Hardly anyone here had ever seen high-definition television before, and the audience exchanged excited whispers as they waited. Still, with the possible exception of the Japanese engineers manning the equipment, everyone knew exactly what this event was really about. That's why, with minutes to go, only one of the five FCC commissioners had appeared, and she was there just to be polite. The broadcasters, Commissioner Mimi Dawson realized, "aren't known for rushing headlong into expensive new technology." Clearly this was "just a political strategy shift."

John Abel stood off to one side anxiously watching the door, waiting for the other four commissioners to arrive. How could they not come? Abel wondered. Their offices were just down the hall. McKinney and some other members of the FCC's senior staff were in the audience, but only the FCC commissioners had the votes to turn back Land Mobile. They were the ones the broadcasters wanted to impress this morning. If they didn't show up, it could only mean that they were deliberately choosing to stay away, and the broadcasters' last, desperate gambit would certainly fail.

Just then FCC Chairman Mark Fowler stepped into the room—through the back door, behind most of the audience. Abel and everyone else looked up. Fowler waved, then stood for a moment, looking at the two television sets at the front of the room, sixty feet away. Abel watched to see if he would walk up front and take his seat. Fowler was a classic Reagan administration appointee who believed that market forces, not the government, should direct events. To him, HDTV was "just a speck on the horizon, not yet perfected." The thing he liked least about being chairman was "constantly being bombarded with high-powered lobbyists from one industry or another." And by holding this demonstration, right here in the FCC's own headquarters, Fowler believed, the NAB was trying to flex "raw political power" in a demonstration that was "more part of a lobbying battle than what the public interest required." Fowler was not at all afraid of confrontation. Long ago his opponents had nicknamed him "Mad Mark." After a couple of minutes, Fowler waved again, then turned around and left the room. Abel let out a long sigh. The demonstration seemed pointless now.

A moment later, however, several NAB officers addressed the crowd. Then the engineers down at WUSA turned on the transmitters, and at the FCC both screens clicked on, carrying the station identifier: "This is Experimental Station WWHD, Channel 58-59, Washington D.C. Transmitting HDTV in MUSE. National Association of Broadcasters." After that another NAB officer came on the screen with a prerecorded pitch: If the FCC gives away those vacant UHF channels, broadcasters will be unable to deliver HDTV. And that will lead to *the death of local broadcasting as we know it!*

With the commercial over, the show began. Action scenes from the 1984 Olympic games in Los Angeles filled both screens. The wide screen showed the picture in high definition, while the square monitor showed the same picture in conventional 525-line television. The demonstration moved from Olympic swimmers plunging through laps to sweeping aerial views of the Grand Canyon, Mount Rushmore, and Washington, D.C. Clips from the recent movie *Top Gun* were next; F-18 fighters screamed off carrier decks, and the engine noise roared with full realism through the two high-quality speakers.

The HDTV images were so vivid, the improvements over conventional TV so striking, that even some of the jaded professionals couldn't help but exclaim out loud. People sitting in the front rows could see each leaf on every tree. Faces in the Olympic stands were clear and recognizable, while on the other TV they were little better than a blur. On the high-definition set, the audience could pick out individual footprints on the Olympic track. On the regular television the track looked like mud. Looking at the high-definition set, they felt as if they were being carried right into the action. And when the demonstration ended with the fireworks that closed the Olympic games, surprised gasps and delighted murmurs swept the room.

"It's just a completely different viewing experience," the FCC's chief engineer leaned over and whispered to McKinney. This was high praise, coming from a jaded elder statesman. "All of a sudden you're looking at things you've never seen on TV before. Remarkable!"

On their way out the door a few minutes later, the broadcasters, the FCC staffers, and the rest glowed as they shook hands with Abel, Fritts, and the Japanese. The guests' eyes sparkled, and their grins were lustful, as one after another declared, "I want one for my den—right now!" But Abel's smile was hollow. Not one of these people had a vote on the FCC. What good was their enthusiasm? He knew he had failed.

Actually, Abel had dropped a match on some leaves in the forest that morning. Even now, as he walked away, the leaves smoldered. Before long Abel's match would set off a blaze in the brush, then a raging fire.

Minutes after the NHK demonstration, a camera crew from CBS showed up in Jim McKinney's FCC office. On McKinney's desk were phone messages from NBC, the Associated Press, the *Washington Post.* All of them wanted to know what he thought of this new-generation television. "A landmark," he told them. "Impressive." But the bottom line was the sound bite shown on the networks that night: "There is no chance the broadcasters are going to get a second channel for this." In fact, Chairman Fowler said he was positively scornful of the "Kabuki dance" put on by the NAB. Fowler wanted to take those UHF channels away, and that was that.

The thinking behind all this wasn't just political. As Lex Felker, another senior FCC official, considered high-definition TV, he remembered the time, a few years earlier, when he visited citizens-band-radio enthusiasts around the country. Felker was doing research as the commission considered new CB rules, and as he sat in these people's living rooms he couldn't help noticing that they were picking up the broadcasters' carefully calibrated television signals on bent and broken rabbit ears or coat hangers wrapped in aluminum foil. Give me a break, Felker was thinking now. These people are going to buy HDTVs?

Only one of the five commissioners, James Quello, a seventy-two-year-old former broadcasting executive who had been a commissioner since 1974, was on the broadcasters' side. The other three commissioners all wanted to give Land Mobile the extra channels. So did Fowler. And the chairman of the FCC generally can do whatever he wants.

To broadcasters and others in the telecommunications industry, the FCC is an extraordinarily powerful agency. Outside that world, however, it's almost invisible. Yes, the commission is in Washington, but the offices are in a nondescript commercial office building in the city's business district, across the street from Burger King and The Gap. The rest of the federal government barely knows it's there.

"You could put the FCC in Indianapolis, and it really wouldn't make any difference," said Kenneth Robinson, assistant to the chairman for several years. Within this cloistered world, the chairman holds unusual power. Proposals aren't even put on the agenda for a vote unless he approves.

After the demonstration at the FCC, the NAB and the Japanese moved their road show over to the Capitol and set up in the Senate Caucus Room, the decorous high-ceilinged chamber where the Army-McCarthy hearings were staged in 1954. Many senators and congressmen were curious about this new form of television, though few of them cared about the Land Mobile debate—if they knew about it at all. To most of Washington this particular issue was a parochial squabble that had merited barely a mention in the *Washington Post*. Still, broadcasters had often found that one good way to sway the FCC was to get congressmen on their side. Usually that wasn't especially difficult. When these people went home to visit their districts, they just *loved* to be on TV.

Through the day, senators, representatives, and their staffs filtered into the Caucus Room and took seats for the show, repeated every hour or so. The choreography for this demonstration was the same as the last. The NAB officers gave their pitch about the death of local broadcasting, then came the action scenes from the 1984 Olympic games and the rest. The program was the same, but the audience response could not have been more different—and in a way that neither Abel nor anybody else at the NAB had anticipated. When the congressmen looked at the sparkling high-definition pictures, their eyes widened. Their political pulses quickened—but not because any of them saw a mortal threat to the broadcasting industry. No, for the senators, representatives, and their aides, this show demonstrated only one salient fact.

This stunning new television was *Japanese!*

By that time in 1987, the Japanese were already manufacturing one-third of the television sets sold in America, and enthusiasm for videocassette recorders was reaching its apex. Almost everyone had a VCR, and video rental stores were opening in even the most isolated, rural areas. As politicians could not fail to note, nearly all these VCRs were made in Japan. An American company, Ampex, had invented the video recorder, but Ampex had been interested only in the selling the larger, professional models. The company hadn't tried to design a smaller version for consumers, and when Ampex approached other American manufacturers, they hadn't shown any interest, either. Finally, Japanese companies had asked to license manufacturing rights, and Ampex agreed; by 1987 Japan and Korea had sold more than 100 million VCRs around the world. That story was a dark legend in the consumer electronics industry.

When Representative Mel Levine of California got up from his seat in the Senate Caucus Room, dumbstruck by the power of this new television, just one sharp question filled his head: "Are we going to let the next major development in consumer electronics go the way of the VCR?" Other congressmen started to grumble, too. Soon news stories began to appear carrying a thinly veiled Yellow Peril tone, and some of the Japanese made things worse. Hikehiko Yoshita, a Toshiba vice president who had helped to arrange the NAB demonstrations, bubbled in one interview that he was "truly convinced of the successful penetration of HDTV receivers into almost every home in the world in the not too distant future." And within a short time, a ringing cry was heard across town: "The Japs are coming, the Japs are coming!"

With that, Chairman Fowler suddenly realized he had "this political problem." A month after the Capitol Hill demonstration, Fowler was testifying before a House subcommittee, and the congressmen peppered him with questions about HDTV. Fowler told them, "I think the broadcasters are overreacting, frankly." Still, before he got up from the witness table Fowler was forced to offer the vague promise that the broadcasters would not be precluded from offering HDTV. What choice did he have? These were the people who set the FCC's budget. A few days later, two letters landed on his desk signed by two dozen senators and congressmen. "We are concerned that the commission is acting prematurely," the representatives warned. The Land Mobile rule "could seriously hamper" American development of HDTV, wrote the senators.

Over at the NAB, John Abel began to realize that his strategy was producing results he hadn't expected. Until this moment, Abel hadn't "fully grasped the true, big political picture." Now he was excited. Maybe he hadn't failed after all. Maybe, just maybe, by playing this Japanese card the broadcasters could turn things around.

The TV industry immediately petitioned the FCC to open an official inquiry to see what effect high-definition television might have on the broadcasting business. And by the way, it said, the Land Mobile decision will have to be postponed until this study is finished. Fifty-eight broadcasting organizations signed the document—the first time these normally competitive, fractious companies had spoken in one voice. "We're fighting for the future of HDTV!" a broadcast industry lobbyist exclaimed to a *Broadcasting* magazine reporter. A few months earlier, this lobbyist probably couldn't have said what those initials stood for.

In early March 1987, the FCC was scheduled to vote on the Land Mobile rule. Even in February the Land Mobile decision had seemed, as Ann Hagemann, another broadcasting lobbyist, put it, "as much a done deal as anything I'd ever seen at the FCC—signed, sealed, and delivered." Then on Wednesday, March 11, the commission released its agenda. Across town, lawyers and lobbyists grabbed the paper as soon as it landed on their desks. Maybe the commissioners would schedule a discussion of HDTV, giving the broadcasters one last shot. But some could not believe what they saw. A few looked the agenda over twice just to be sure, but it was true: the Land Mobile decision had been pulled from the schedule. The commissioners were postponing the vote. Right now the issue was just too hot. Abel was almost giddy. What a lobbying coup! "We've moved mountains," he said.

But Land Mobile wasn't dead. Far from it.

A few weeks later, Chairman Fowler resigned to go into business, as FCC chairmen often do. Commissioner Dennis Patrick was chosen to replace him, and he seemed the broadcasters' worst nightmare. Patrick's father had been a Los Angeles police officer for thirty years, and the common wisdom, as McKinney put it, was that before Patrick came to Washington his father had told him, "I want you to get more channels for Land Mobile." In fact, Patrick got no such instruction, even though his father wasn't the only policeman in the family. His brother and uncle were law enforcement officers, too, and Patrick knew that all three of them had at times relied on mobile radios for their lives. As a result, "my exposure to law enforcement did sensitize me to the importance of mobile communications," he said.

At thirty-six, Patrick was the second youngest person ever to serve as FCC chairman. (William Henry, appointed by President John F. Kennedy, was eleven weeks younger.) But he carried himself as if he were far older. That wasn't surprising; Patrick had been in difficult political fights since his earliest days. In the early '70s, at the height of the Vietnam War, he was chairman of the Young Republicans at Occidental College—a rather small group. ("I was not alone," he quipped later. "There were three or four of us.") After law school, he clerked for a judge who was a friend of Ronald Reagan, and through that connection he landed a junior position in the White House personnel office. His job was to review candidates for positions in "the cat-and-dog agencies," as the FCC and similar low-profile commissions and bureaus were known. When a position came open on the FCC in 1983, at Mark Fowler's urging Patrick accepted his own nomination.

Now, as chairman, his swept-back hair was prematurely silver, with just a shiny hint of mousse. His shirts were always stiff white, his necktie knots as tight as they could be. He asked the FCC's press officers not to tell reporters that he had been a surfer as a young man. And when he spoke, the words came out in a slow, carefully measured cadence. He liked to offer self-important sounding political aphorisms. "I am interested in notions of optimality," he would say. "The perfect can be the enemy of the good."

Early in the spring, Jonathan Blake, a Washington lawyer representing the broadcasters, decided it was time to have a talk with the new chairman. The broadcasters knew Patrick's reputation; he had been a commissioner since 1983. But he was the chairman now. They had to deal with him.

Blake showed up with a couple of his broadcaster clients, and they arranged themselves around the conference table in the chairman's office. After they exchanged pleasantries, Patrick and the broadcasters traded well-worn arguments: HDTV is the wave of the future, a broadcaster said. If Land Mobile needs spectrum more than you, Patrick countered, then they should get it. Blake listened and considered the situation. They were just hurling statements at each other—the least effective kind of lobbying. Then everyone turned to him. It was Blake's turn, and as he looked back on it later, he said, "I guess I was, as the athletes say, 'in my zone.' "

Blake reminded Patrick of the importance to America of free, over-the-air broadcasting for everyone, rich and poor. For the price of a TV set, every citizen could get local and national news, weather bulletins, and a wide array of programming at no additional cost. But HDTV was coming, no question about it. And all the other services—cable, satellite, and the rest—would be able to transmit it without asking permission from anybody. Only the broadcasters were constrained by FCC regulation. Without help from the commission, only the broadcasters would be left behind. Could the industry survive if it was prevented from providing HDTV? Could America's tradition of free, local broadcasting survive?

Patrick was listening, so Blake closed with a punch: "You simply can't take the risk that we are right, and find years from now that the bulk of the United States will be precluded from getting HDTV over the air."

Patrick sighed. To Blake he seemed "kind of resigned." Blake remembers that after a moment the chairman told him, "I guess I have to do it."

Blake heaved a relieved sigh of his own. Now it was finally clear: Abel's strategy had worked. The Land Mobile decision would be overturned. The broadcasters had won.

In April, the FCC formally announced that it was reversing itself: no UHF channels would go to Land Mobile until the commission could determine what should be done about HDTV. Then in August the commission opened a special three-month HDTV inquiry. Only after that would the commission decide.

Meanwhile, the grumbling on Capitol Hill had been growing incrementally louder, until finally, in October 1987, it broke the sound barrier: the House scheduled its first HDTV hearing. Congressman Edward J. Markey, the new chairman of the House Telecommunications and Finance Subcommittee, had missed the NHK demonstration early in the year, but he'd heard from others that "HDTV cuts through an audience like a hot knife through butter." Markey represented a district just outside Boston that included the Route 128 corridor, a thick concentration of high-tech enterprises second in size only to Silicon Valley. Technology was one of his most important constituent issues. And Markey was also a "TV junkie," as his aide, Larry Irving, put it. He wanted to see this new Japanese wonder for himself. The chairman also wanted others in Congress to have a look, because to him "the promise of this new technology is astounding." So Markey's staff asked American companies that were working on high-definition television to show their systems in the Capitol, or at least talk about what they had planned. As it turned out, NHK had scheduled another Muse demonstration in Canada at about the same time, so Markey's staff also arranged to have that demonstration beamed to the Capitol by satellite. The event came to be billed as "the first international transmission of high-definition television." Markey told every member of his subcommittee, "You've just *got* to come to this hearing so you can have your own look at this new technology."

On the appointed day, Bill Glenn and half-a-dozen people working on systems less advanced than his set up their equipment in a large chamber adjacent to the subcommittee hearing room. Glenn was showing his Toys "R" Us ballerina, and if Glenn's system was "a baling-wire kind of thing," as Abel had put it, then some of the other systems were laughably amateur.

Markey and others filtered in throughout the day, stopping at each booth. The congressmen nodded and smiled politely, but after looking all of it over Markey couldn't help but say, "I'm puzzled as to why the American consumer electronics manufacturers are so far behind their Japanese counterparts in developing competitive HDTV systems."

The Japanese, meanwhile, had set up their monitor in the hearing room itself. Their show went off without a hitch, leaving Markey and the other committee members stunned. An enthusiastic story in the *Washington Post* the next morning described the heady feeling that filled the room: HDTV, the paper said, "is an advance so startling that few people exposed to it are likely ever to feel satisfied with the set in the family room again." Congress had been grumbling for months, but now some of the committee members started to shriek.

"The Japanese have absolutely leapfrogged the rest of the world!" cried Representative Don Ritter of Pennsylvania. "We have been watching from the sidelines as the Japanese have taken this technology and run with it so many laps around this course. And here we are, so far behind." Ritter was the only engineer in Congress, and his view of Japan was already well known. A few weeks before, he had joined a select group of congressmen who gathered on the Capitol steps with sledgehammers in hand. As soon as the news photographers had finished setting up, Representative Ritter and the others pulverized a Toshiba radio, affecting outrage because, it had just been disclosed, a Toshiba subsidiary had sold some military technology to the Soviet Union.

After lunch, Chairman Markey called a panel of engineers who were working on HDTV systems up to the witness table, and asked them, one by one, how far along they were in their work. Steven Bonica, vice president for engineering at NBC, said the network, working with the

Sarnoff Research Center (the former RCA labs), had "designed a system using computer emulation" and "the circuit boards are now being designed." In other words, they had nothing to show. William Schreiber, of MIT's Center for Advanced Television Studies, said he, too, was doing work "primarily by computer simulation." Larry French, vice president for technology and manufacturing for North American Philips, said his company had put on an "in-house, prototype demonstration in our labs" last spring that had also included "a simulated cable TV presentation of these signals." Philips, he added, did not yet have "a productworthy system." Only Bill Glenn had demonstrated his system "with hardware," he noted. But even a year after John Abel's visit, Glenn still had no more than a closed-circuit version, able to transmit pictures only through a wire.

Also at the witness table was Masao Sugimoto of NHK, "the father of Muse." As the congressmen had just seen, NHK had televisions, cameras, recorders, transmitters, and everything else in full production. Not only that, they'd just beamed a perfect HDTV signal across half a continent! Ritter wanted to know how Japan had accomplished this, so he asked Sugimoto, "How much money has NHK invested in HDTV over the last seventeen years?"

"In seventeen years, by rough measure, maybe between $10 million and $12 million," Sugimoto responded.

"That isn't much," Markey said with a shrug.

But then all of a sudden, Schreiber, the MIT professor, blurted out, "NHK was in charge of this program, but a lot of money was spent by Sony, Matsushita, Hitachi, and other Japanese companies."

"What would you estimate?" Markey asked, his interest picking up now.

"The numbers I have heard are between $200 million and $300 million."

Listening to that, cold rage began flooding into Sugimoto's head. Schreiber has never even visited NHK, he was thinking. He's never studied this himself. He's just playing back *rumors.*

But the Japanese engineer didn't say anything. That was not his style. Meanwhile, up on the dais, eyes opened wide. For just a moment, the congressmen sat frozen. At last Markey spoke, but slowly, with an air of astonished disbelief: "The Japanese companies have invested some $200 million to $300 million?"

"That was the dynamite charge," he said later. The HDTV problem had his full attention now. So he began asking the American witnesses, "What would you recommend that the Congress or the FCC do?"

After moments like that one, FCC Chairman Patrick knew he couldn't just let this problem slide. He'd have to do something significant when the FCC's three-month inquiry ended, or the arguments would land right back in his office, even louder. This was no longer just another of those parochial, inter-industry disputes. No, the FCC had to "get this off our plate," as Commissioner Mimi Dawson told Patrick. The other commissioners agreed.

"We need to ship it out of here," said Commissioner Patricia Diaz-Dennis.

So Patrick did the natural thing. He appointed an advisory committee to study the matter for a while, the government's time-honored solution to thorny dilemmas. Not only that, he stacked the committee with broadcast industry officials. That would keep them quiet. They'd consider the issue for a while—a year, maybe two. By the time they came back with their report, Dawson believed, maybe interest will have flagged, technological difficulties will have come along. Maybe the problem will have solved itself, and the HDTV crisis will have simply faded away.

John Abel loved the idea. "Advisory committees typically are zoos," he said. "They can be a mess. There are so many ways to slow things down." With a little behind-the-scenes manipulation, the Land Mobile decision could be delayed for *years*!

Patrick named his new group the Advisory Committee on Advanced Television Service and, like the name, the charter was vague enough to cover almost anything: "The Committee will advise the Federal Communications Commission on the facts and circumstances regarding advanced television systems for Commission consideration of the technical and public policy issues." The debate over those extra channels would be delayed until this new advisory committee finished its work.

Appointment of an advisory committee was hardly a momentous event in Washington. Dozens were formed every year. The leaders of almost every agency in government established them anytime they had a thorny problem they could not solve—and also could not ignore. Still, all of a sudden an official government body was in charge of HDTV. High-definition television was no longer just a lobbyist's mantra. With appointment of an advisory committee, the issue had taken clear form. It had assumed life.

Patrick had to choose a chairman for his new body, and he knew full well that no lightweight would do. The issue was too hot; Congress was watching too closely. Still, it also had to be somebody loyal. He didn't want someone with an agenda who would run the committee off in wild directions—riling Congress, obligating Patrick to do things he didn't want to do.

Patrick talked it over with Mimi Dawson, who suggested Richard Wiley. He was the FCC chairman during the Ford administration and was seen as a friend of the broadcasters. As chairman he had seemed to take their side in some of the early skirmishes with the Cable Mongols. As Abel put it, Wiley was "clearly one of us."

Now Wiley headed a large and powerful law firm that specialized in telecommunications issues. Lawyers in his office represented a host of broadcasters. Wiley, in fact, was general counsel for CBS, and he'd become Washington's most influential lawyer-lobbyist in the field. At the same time, the former chairman still loved the FCC and missed his days in government service. Almost every day, he walked over to the commission and stopped into offices to chat with the commissioners or the staff. Inside that building, he didn't just know everyone's name, he really *knew* almost everybody there, from the chairman to the janitors. So Patrick was confident Wiley would be loyal. Wiley wanted to be liked, *needed* to be liked, by everyone in the building. His business and his personal happiness both depended on it.

Patrick took Dawson's suggestion and appointed him. Wiley was one man who wasn't going to freelance. And it was true; Wiley was loyal. But as Patrick quickly learned, Dick Wiley was no patsy. Under him, Patrick's little Advisory Committee was not going to backpedal and stall until the issue withered and died.

2 *Washington to the rescue*

Alfred Sikes, a radioman from Missouri, wanted to be the head of the Federal Trade Commission in Ronald Reagan's second term. His good friend John Danforth, the Republican senator from Missouri, told Sikes he would help him get the job. So one day toward the end of 1985 Sikes arrived at the White House to see Robert Tuttle, Reagan's personnel director. There's an awful lot of competition for that FTC job, Tuttle told him. You're in radio. How about the Federal Communications Commission? We're going to have an opening there. Sikes knew that in situations like these, "you don't blink." In his slow and careful Missouri cadence, he told Tuttle, "I didn't come here to be on the F-*C*-C. I came here to be on the F-*T*-C."

Next thing Sikes knew, he had been appointed an assistant secretary of commerce, head of the National Telecommunications and Information Administration, an obscure little office charged with promoting effective telecommunications policies. This was not what he had asked for, and Sikes had barely heard of the NTIA. But he was not unhappy to be there. Coming from a small town in Missouri, he was impressed just to be in Washington at all. He'd held a couple of midlevel Missouri state government positions in the 1970s. After that, he bought several small radio stations, built up their value, and then sold them for a profit of about $1 million. Sikes was financially secure, and he had some background in communications issues—at least as they related to radio.

He was tall and unassuming in appearance; large steel-rimmed glasses framed an ever earnest face. He liked to describe himself as "a glass-half-full kind of guy." And he was always a careful listener; he made a practice of looking people straight in the eye as they spoke. Sikes was bright—a determined, independent thinker who was willing to fight convention at times. His small-town, Midwestern roots were never far from the front of his mind.

The NTIA was one of those out-of-the-way agencies that could do whatever it wanted within its special field. Or it could do nothing at all. Either way, people seldom noticed. More often than not, the assistant secretary chose to focus on one subject at a time, and when Sikes arrived the agency had just finished up some work on telephone issues. If Sikes wanted to make a mark, he would have to find something new to do. During his first weeks in the job, he was struck by the dire tone of the official discussions with Commerce Secretary Malcolm Baldridge. Back in Missouri just a few weeks earlier, all the talk had centered around selling advertising time on his little radio stations. But here in the secretary's office, the officials were wringing their hands about the trade deficit, foreign competition, the continuing loss of American manufacturing jobs. Sikes was impressed, not to say awed.

Then in April of 1986, Sikes found himself at the National Association of Broadcasters' annual convention in Las Vegas, a huge extravaganza attracting fifty thousand or more television executives, managers, and technicians. And on the opening day, Jim McKinney of the FCC took Sikes over to see the exhibit hall in the Las Vegas convention center. This cavernous chamber was filled with elaborate displays of the very latest in television, radio, and related broadcast equipment. As Sikes and McKinney stood in the doorway, they looked over a room so vast they couldn't see the far wall. Almost to the horizon, the manufacturers' booths looked like expensive and exclusive showrooms—carpeted and carefully lit. Above each, massive neon signs reached toward the rafters announcing the names of the world's largest electronics companies: Sony and Toshiba. Thomson, Philips, Goldstar.

Sikes turned to McKinney. A note of astonishment crept into his normally even voice as he said, "I don't see a sign for a single American company out there."

A few weeks after that, in May, Sikes was getting ready for one of his first major responsibilities as a federal government official. As head of the NTIA, he was the official American representative to the world body charged with setting international broadcast standards. And this May, the committee was gathering in Dubrovnik to decide whether to adopt a world standard for producing high-definition television shows.

Sikes knew little if anything about HDTV. Back in Missouri, he had seen an article or two about it in *Broadcasting* magazine, but generally he'd glanced quickly past them. His interest was radio. So he asked one of his assistants, "What's this about? What's our goal here?" The United States position, the aide told him, is to support the Japanese proposal to have NHK's system—1125 lines and 60 fields per second—adopted as a world standard.

Sikes couldn't believe what he was hearing. He thought about all those troubling meetings in Baldridge's office and recalled all those neon signs at the NAB convention. "It strikes me as rather odd that one of our key objectives is to promote Japanese hegemony over advanced television," he said. "How did we get in this crazy position?" Al Sikes had found his cause.

Of course, the Japanese proposal was defeated in Dubrovnik, largely because of Europe's vehement objections. But when John Abel put the issue back on the table, with the NHK demonstration in Washington in January 1987, Sikes took an interest again. In the spring, he wrote an official letter to FCC Chairman Patrick urging him not to give those vacant UHF channels to Land Mobile, saying that the FCC should be pushing for "state-of-the-art use" of the spectrum and therefore "must resist the short-term cost savings to the detriment of future, innovative uses." Patrick didn't respond, so Sikes wrote other letters, gave interviews, testified before Congress—always urging that the FCC step up its promotion of HDTV. "This is not a small, inconsequential public policy issue," he told the *Washington Post*. "If you start adding up the total TV equipment investment in this country, you find that this is potentially a $100 billion problem."

To Patrick and others at the FCC, Sikes's remonstrations were "just satellite stuff," as Lex Felker put it with a dismissive wave. The FCC, not the Commerce Department, had responsibility for the airwaves. Patrick and the others considered Sikes just a gnat making a bit of irritating noise. But Sikes was undaunted. The man from Missouri had a mission, and to him the debate wasn't just about the next generation of television, as important as that might be. It was about "a whole new world of technology products that would be affected by this work—defense, medicine, printing, advertising. Huge revenue streams can come from this. And the American manufacturing community *has* to have an opportunity to participate in this." European companies were forming a well-funded, government-backed consortium to create a high-definition television system of their own. Japan continued pushing Muse. As all this went on, Sikes saw, America was "sitting it out, fat and dumb."

Sure, Dick Wiley was organizing an Advisory Committee. In fact, Sikes was on it. But the first months were consumed by squabbling between the Broadcast Barons and the Cable Mongols. Cable people were furious that Patrick had stacked the committee with broadcasters. But the broadcasters insisted, as John Abel put it with sanctimonious certainty, that "the Advisory Committee was established to create a standard for *broadcast* television. The charter doesn't say a thing about cable." The argument seemed intractable; when the two groups sat together for the first time, it was almost as if North and South Koreans had been thrown into the same room. Wiley, a master of consensus and inclusion, was spending all of his time creating an elaborate structure of subcommittees and working parties so that he could appoint chairmen and vice chairmen from both groups. He had dutifully told the FCC about the delay, but the commissioners hadn't seemed particularly concerned. "This is going to take another year?" Commissioner Dawson shrugged. "That's great."

It all seemed so hopeless. Sikes coaxed and complained, but hardly anyone was paying attention. We need to give the debate some teeth, he concluded. So Sikes called a meeting with his friend Larry Darby, an economist and former FCC official who was now a private consultant, and asked him: Can you do a study that will show how important HDTV might be for the American economy? To Darby, this sounded like a singularly bad idea. He looked at Sikes and said, "You can't predict a market for a product that doesn't even exist yet." Sikes and his aides were undeterred. Do the best you can, they said. Though Darby wanted the work, he had to tell them, "Anything I come up with is going to be a really iffy kind of thing." Fine, fine, they said. And a short time later Commerce issued Darby's contract.

Working from his Northwest Washington home, Darby began pulling together statistics that showed how quickly sales had taken off for previous new consumer electronics products: color TVs, VCRs, compact-disc players, personal computers. Not surprisingly, he found that more expensive items sold more slowly. Then he asked experts to help him guess what a high-definition television set would cost. Though the product didn't exist yet—at least not in the United States—it was already clear that an HDTV would cost several thousand dollars. But the biggest problem was that neither Darby nor anybody else had any idea whether consumers would actually buy these new TVs. All the politicians were certain they would, but who really knew? The damned things were going to be expensive. How could he predict the market?

At the same time, Darby could hardly ignore the political environment. His sponsors at the Commerce Department and many others in town were already convinced that "HDTV is very, very important," he said. Nobody at Commerce had tried to tell him how his study should come out, but Darby also knew that if his study "purported to show that this thing is a turkey," he'd be hooted right out of town. After all, "the Japanese had already poured hundreds of millions of dollars into this. So a three-month Commerce Department study is going to say the Japanese wasted all that money?" No way!

Darby knew the best he could do was "try to pass the straight-face test"—produce a document he could offer without apology or excuse. So when he sat down to write the report, he cautioned that the market for HDTV "may simply not materialize, as was the case with the picturephone, videotext, and a host of other failed consumer products. The high current visibility of advanced television products could, in short, turn out to be much ado about nothing."

That bit of analysis wound up on page 31. But the executive summary at the front, the only part of the report that most people ever read, offered only the barest reference to this idea. More prominently, it forecast that manufacturers could take in as much as $144 billion from HDTV sales in the next twenty years, noting that "over 100,000 U.S. jobs may be at stake by the turn of the century." HDTV-related products could have "immense potential impacts on both the balance of trade and the federal budget." If HDTVs were made in America, that new market might "lead to significant contributions toward restoring balance" in both the federal budget deficit and the U.S. trade deficit.

The Commerce Department quickly published the Darby Report with its official imprimatur. For weeks, months, it was quoted over and over again. Nobody bothered to mention anything but Darby's most optimistic scenarios. If the debate needed a focus, here it was: High-definition television would bring back America's failing consumer electronics industry. It would create tens of thousands of jobs, maybe hundreds of thousands—the number seemed to grow with each telling. HDTV would eliminate the trade deficit, *pay off the budget deficit*! It would lead America to *Nirvana!* And with that, in very short order, Washington fell into an HDTV frenzy.

After George Bush's election in the fall of 1988, a warm feeling settled over moderate Republicans, senior civil servants, even some Democrats in Congress. They figured that the government, while still Republican, would loosen up a bit after the ever-doctrinaire Reagan years. Out of this new glow stepped Craig Fields, a senior officer with the Defense Advanced Research Projects Agency, known in Washington as Darpa. This Pentagon office was founded in the late 1950s in response to the Soviet Union's launch of Sputnik, the first man-made satellite. Darpa's mission was to make sure nothing like that ever happened again. The agency gave much-sought-after grants to companies working on advanced research of interest to the military. And late in 1988, Fields turned the office's spotlight onto HDTV.

What interest did the Department of Defense have in improved-resolution television? None, really. But high-resolution imaging would certainly be of value for tactical displays in warplanes and battlefield computers, and in assorted other areas. Besides, Fields explained, "the loss of our consumer electronics industry may be undermining our electronics industry. The implications for national security could not be greater." So in December 1988 Fields announced that Darpa would award grants totaling $30 million to firms working on high-definition display systems. With that, all of a sudden the government formally joined the fight for HDTV.

A few weeks later, Robert Mosbacher, Bush's new commerce secretary-designate, came before the Senate Finance Committee for confirmation. And in the first question put to him at the hearing, Committee Chairman Lloyd Bentsen asked what he intended to do about high-definition television. Without hesitation, Mosbacher promised that the new administration would step in to promote the development of HDTV. Normally, a millionaire Republican oilman who is the president's good friend might expect more than a few squeaks and bumps as he faced confirmation by a Democratic Congress. But Mosbacher's views on HDTV were just the grease needed to slide his nomination through.

"The United States cannot afford to see its traditional technological and commercial leadership further eroded," he averred. HDTV "is a high-priority item. It's late in the game, but not too late. I intend to work directly and personally with the assistant secretary"—Al Sikes—"on these policy priorities." Most important, Mosbacher suggested, the time for government support of private industry might now be at hand.

Those were just the words congressmen dreamed of hearing: A new spending program! A rich opportunity to send jobs and largesse to their home districts after years of Reagan-era deprivation. Across Capitol Hill, lips trembled, hearts raced. Grown men and women salivated. Larry Darby's report had already said HDTV could solve the budget crisis, eliminate the trade deficit. Now the Bush administration was saying it was willing to spend some money.

All the ingredients were in place, and congressmen started pushing and shoving, elbowing their way to the trough. HDTV hearings began showing up on the docket in both the Senate and the House. Two senior House members, Mel Levine and Don Ritter, the radio-bashing Republican from Pennsylvania, formed an HDTV Caucus, and forty members joined right up. HDTV was going to create "millions of jobs," Levine crooned, and "cut through so many industries and technologies. If we don't get our act together, those jobs are going to go someplace else." He and Ritter offered a bill that would award tax benefits for industrial research and development and provide a guaranteed government market for American-made HDTVs. This idea proved modest compared to the legislation that followed.

In the Senate, Commerce Committee Chairman Ernest Hollings called for a government-backed manufacturing consortium for which Washington would provide $30 million. Representative George Brown of California, a senior member of the Science, Space, and Technology Committee and due to become chairman soon, proposed a plan that would go one better. *His* HDTV research consortium would get $100 million. That still wasn't enough for Senator John Glenn of Ohio, chairman of the Government Affairs Committee. His bill proposed a "civilian Darpa" that would promote development of HDTV. It would get $500 million in government money. And the hearings and bills kept coming.

After the ever earnest Reagan years, here was a grand, drunken party, and everyone was invited. Every committee and subcommittee that had even the tiniest connection to communications joined in—Telecommunications and Finance; Science, Space, and Technology; Commerce, Science, and Transportation; International Cooperation; Economic and Commercial Law; Government Affairs—even Armed Services. Over ten months in 1989, eleven Senate and House committees staged more than a dozen hearings on HDTV. On the first day after one congressional recess, the industry newsletter *Communications Daily* noted, "Representative Ritter was outgunned in his apparent effort to become the first in the House to offer HDTV legislation." Representatives Rick Boucher and Campbell got their HDTV bill in first.

As the weeks passed, so many HDTV bills were piling up in the hopper, so many new program pronouncements were pouring out of the bureaucracy, that several congressmen called on the White House to name an "HDTV czar." Some of them suggested Craig Fields. He had grown to be the darling of the congressional committee room; by early 1989, Darpa had received eighty-seven proposals for Pentagon funding—"exciting and cutting-edge" proposals, Fields told a happy House subcommittee in March. Fields then suggested that his office might actually spend more than $30 million. When Fields was finished testifying, Secretary Mosbacher stepped up to say that President Bush had given *him* the mandate to coordinate HDTV policy, and soon he would offer a comprehensive plan.

The partying extended far beyond Capitol Hill. NASA said it wanted to install HDTVs at Mission Control in Houston. The Pentagon's Defense Manufacturing Board formed still another HDTV task force, and its executive secretary proposed to have President Bush give a major speech to the nation about HDTV, presumably during prime time. After all, the defense board said, "the U.S. is in a national crisis of the first magnitude."

Everyone partied on. The revelry grew so loud that TV manufacturers began complaining that Americans were deciding not to buy new televisions until they could get HDTVs. "Our industry is killing itself," howled Bruce Schoenegge of Hitachi.

With all the talk of free government money, leaders of America's manufacturing community began looking for invitations to the bash. The American Electronics Association woke up first. This trade group represented U.S. manufacturers of consumer electronics, and it began putting together a study to show how HDTV might revive the industry. When this study was complete, the organization planned to suggest the level of direct aid Washington should offer. Prodded by the AEA, seventeen of the nation's leading corporations, including IBM, AT&T, Apple, Hewlett-Packard, and Motorola, agreed to put together a joint "business plan" for an HDTV consortium that would be paid for, the industrialists said, "with government and private funds." These companies would put up 51 percent of the money, just enough to hold control. The government would hand over the rest. Not to be left behind, the AEA's larger competitor, the Electronic Industries Association, was working on its own proposal.

By summer, the revelry got so loud that many of the partiers couldn't hear anything else. So at first they didn't even notice that some of the guests were quietly slinking out of the room. Secretary Mosbacher, a willing witness at earlier HDTV hearings, began declining some congressional invitations. The Commerce Department's HDTV plan kept getting delayed, though everybody said nothing was amiss. At the National Association of Broadcasters' convention in early May, Mosbacher said his department was in the final stages of drafting the legislation. The president told him he wants HDTV "to move ahead," he said. But Mosbacher also cautioned that he had "grave concerns" about a larger federal role in funding and developing HDTV. Competition, not government planning, should drive the U.S. economy, he said. To anyone paying careful attention, that should have been a clear warning. But the revelers didn't seem to notice.

By mid-May the American Electronics Association had finished its study and was preparing to join the party, apparently unaware that the winds had shifted. Testifying before the Senate Commerce Committee, the AEA revealed that it was proposing a government-private partnership to develop, design, license, and possibly even produce HDTV equipment. What would the government's role be? Well, the AEA officers said with no discernible qualms, we are asking Washington to hand over $1.35 billion. Yes... *$1.35 billion.*

If any other lobbyist had come to Capitol Hill and asked the government to give him even half that amount, the laughter would have chased him out the back door, down the Capitol steps, through the parking lot, and all the way down to the Greyhound station. But this was the HDTV jamboree. Anything goes. Chairman Hollings said he thought the AEA plan was just great. It happened, though, that Secretary Mosbacher was testifying before another committee just down the hall that afternoon. Somebody asked him what he thought of the AEA plan, and he didn't offer his usual amiable answer.

"Don't depend on Uncle Sugar to fund it," Mosbacher snapped.

The party was over.

Earlier, Mosbacher had been quoted as saying something encouraging about HDTV, and Richard Darman, President Bush's budget director, hadn't liked it. Maybe the Bush administration was more moderate than Reagan's in some areas, but economic policy wasn't one of them. Bush and his advisers believed that the government had no business interfering in the economy (a policy they grew to regret at reelection time). But in 1989 nothing was more likely to get a dismissive slap than any proposal that smelled like a "national industrial policy"—or "picking winners and losers in the economic market," as Bush put it later. While Darman and other Bush aides watched with pursed lips, the Commerce Department had stepped uncomfortably close to doing just that by promising to promote the development of HDTV. As John Sununu, the White House chief of staff, explained, "We told [Mosbacher] it is not our policy to talk up a single slice of industry." The Bush administration was shutting down the party, sending everybody home.

The congressional revelers didn't face the morning after with grace. They were bitter, shrill. The Commerce Department never did put forward that HDTV plan, and in October a large group of senators and congressmen wrote a letter to the White House trying to rekindle interest. "HDTV has become a symbol of our national willingness to compete on the cutting edge in strategic industries and technologies of the 1990s," they said. "Even the threat of benign neglect is too great a risk."

"This is the Battle of the Bulge for electronics!" Senator Al Gore shouted up Pennsylvania Avenue.

In answer, the administration slashed Darpa's HDTV budget by one-third, from $30 million to $20 million. Congress had actually been planning to increase funding for Craig Fields's program to $50 million, maybe even $100 million. With this cut the mewling complaints from Capitol Hill turned to angry caterwauling. "The administration has decided to concede the field to the Japanese," Gore griped. "If this happens, the administration will have sanctioned the complete and irrevocable demise of the American consumer electronics industry!"

The White House wasn't finished. In March of 1990, Congress tried to add $10 million to the Commerce Department's budget as new money for HDTV research and development. Commerce said it didn't want the money. "This is the most idiotic thing I've ever heard from this government," snarled Senator Warren Rudman, a Republican. Then in April the Bush administration put out word that it wanted to take even more money out of Craig Fields's HDTV budget. That money, about $10 million, would be used for a new study on streamlining the

Pentagon, or for aid to Panama and Nicaragua, or . . . they weren't exactly sure. The screeching from Capitol Hill grew even louder.

Two weeks later came the final blow. The White House ordered Craig Fields demoted and transferred out of Darpa. A short time later he quit. Congress was furious. The House HDTV Caucus called Fields's transfer "a serious error, a slap in the face." Senators Gore and John Heinz said "the firing and silencing" of Fields was "at best shortsighted, at worst a major breach of our future economic security." Negative reverberations rippled through editorial pages across the country. Andrew Grove, the president of Intel, actually sent a violin to the White House along with a note saying, "You'll need this to fiddle with as the high-tech industry burns."

But through all this, no one seemed to notice what was going on over at the FCC.

Dick Wiley had worked out his Advisory Committee's bureaucratic problems and plunged forward, as was his wont. He set deadlines, then coaxed, prodded, complained, and threatened when they weren't met. He consulted and massaged the committee members, and before long these men (as usual for this industry, there were no women among the voting members) became his rubber stamp. Despite its cynical origins, Wiley was determined that his committee would accomplish something. He was going to make a difference.

Wiley was a Midwesterner, in his fifties, with perfectly parted graying hair and carefully enunciated speech, delivered with precise, emphatic punctuation. And he was a driven man; he rose before dawn weekday mornings and got to the office by seven o'clock. A brass plaque on his desk read "Thank God it's Monday," and that was no joke. After a decade out of government, he had hoped to get a job in the Reagan or Bush administrations, and his name had been mentioned for several high-level posts. Nothing had come of that. So when Dennis Patrick called him, Wiley had been hungry for "an opportunity to provide public service again," he says. The Advisory Committee was giving him that chance.

"He thought this was very important," said Lex Felker, who left the FCC to join Wiley's law firm. "Part of him is really a corny, old-fashioned kind of guy." Wiley and his wife were regular churchgoers, and he had always been quick to volunteer for nonpaying public service positions in his profession and in the Republican Party. He had a strong sense of duty. As if that weren't enough, he was Washington's leading attorney in the field of communications law. Heading an FCC committee working on the next generation of television certainly wouldn't be bad for business (though, as he frequently pointed out, he had to give up many billable hours for the unpaid committee work).

Through the congressional frenzy, Wiley "watched the histrionics and kept my head down," he said. He was canny about the low ways of politics; he knew perfectly well when it was best to keep quiet. Instead of adding to the din, he began collecting HDTV proposals from two dozen engineering labs. The FCC was barely paying attention, so Dick Wiley had to decide which direction American television ought to take. Though he was not allowed to say anything in public, Wiley was not about to turn his panel into a coronation committee for the Japanese. If in the end they had the best system, so be it. But others had to be given a chance, too.

America had not changed the broad technical standards for television since the National Television Standards Committee had set out the first ones almost fifty years before. That had certainly been a messy endeavor, and the FCC had performed no better when it set the rules for color TV a few years later. Wiley knew his work would be the stuff of legend in the broadcasting industry for decades to come, and unlike his predecessors he was going to *get it right.* But when he looked at the task ahead, at first he didn't know what to do. He had "all these people with different ideas." How to choose?

In Japan, government and industry had come together and proclaimed Muse to be the standard. And the governments of the European Community were pouring several hundred million dollars into the EC's own HDTV research program. Clearly, the Bush administration wasn't going to support anything like that. What was he to do?

The solution didn't come to him as an instant, blinding insight. It grew incrementally as he made one decision after another. We have all these applicants, he thought, and we have to choose among them. The only way is to test them, see which ideas work and which do not. But this has to be an open process. Otherwise whatever we decide will be subject to legal challenge. So we have to give everyone with worthy ideas the chance to come forward and be tested, too. Still, this can't stretch on forever; we've got to set some deadlines.

In 1988 and 1989, Wiley announced these precepts one by one. Then one day he looked up and realized that he'd started a race!

Nothing like this had ever happened before. Wiley's rules had set off a grand, international competition, sanctioned by the United States government! Anyone in the world could enter. The contestants would be tested and graded. Finally Wiley and his committee would choose a winner, who would hold licensing rights for the next generation of television. Everyone who built and sold HDTVs in America would pay this winner royalties, which would be worth millions—billions! Sniffing the scent of all that money, just about everyone in the world with an interest in television—major corporate conglomerates, people with a few tools in a backyard shed—started writing and calling for entry forms. Wiley's race was launched, and contestants were off and running.

Wiley's plan got an unexpected boost in the spring of 1989. Dennis Patrick announced that he would resign from the FCC to enter the business world. And as the Patrick reign ended, the FCC's view of high-definition television took a hundred-eighty-degree turn. The man who stepped forward to fill Patrick's chair was none other than the radioman from Missouri, Al Sikes. He had refused Bob Tuttle's offer to be on the FCC when he first came to Washington in 1985. Now, however, Sikes wanted the job. To some Bush aides, appointing Sikes amounted to letting the fox into the chicken coop. But Sikes had powerful friends in Congress, and he had been nominated several months before the White House and Congress broke into their open HDTV war.

Sikes's enthusiasm for HDTV could not have been better known, and he embraced Wiley's plan for an international race. That gave Wiley even more power; he could threaten and cajole. That was important, because it seemed as if a problem or delay lay behind every door. "The next call you get will be from the chairman of the FCC," Wiley liked to bellow into the phone every time someone threw a new obstruction in his path. Generally the threat worked.

On Capitol Hill, Congress was in high dudgeon over the administration's refusal to spend hundreds of millions of dollars to support HDTV. Out of their view, just a few blocks away, Dick Wiley was barreling forward, spending not a dime of government money. And in time, his race would accomplish far more than the congressmen could have anticipated even in their most self-indulgent dreams.

In Tokyo through all this, the engineers at NHK had been on a wild emotional ride. Their 1987 demonstration in Washington had been very well received, and when they put on the show in Ottawa a few months later, the Canadians had embraced Muse even more enthusiastically. "It was very successful," Sugimoto said with a smug smile. "We've proved our system in the U.S.," added Keiichi Kubota, one of NHK's engineers. He and others at NHK didn't see much of the news coverage that followed, didn't fully understand at first that their visit had set off a protectionist frenzy. To them, NHK's North American foray had seemed a grand success. The following fall, John Abel, Eddie Fritts, and a couple of their NAB colleagues came to Tokyo for a visit. First they thanked the Japanese; NHK had saved the broadcasters from Land Mobile. Then Abel and the others broke the bad news. We've started a process now, the NAB officers said, and your system is not acceptable for the United States. Muse is intended for delivery by satellite. It just won't work in America.

The NHK officers were shocked. "They asked us for the demonstration, and we were successful. They told us it was good," Masao Sugimoto would later complain. "Then they say, 'No, thank you.' "

Seeing how upset the Japanese were, Fritts and Abel offered them a window, suggesting that Muse be redesigned to suit the American market. Change the system so it could be broadcast from a TV tower, not a satellite, they said. The NHK officers weren't at all sure they wanted to do that; it would mean a huge investment of time and money. But if the American broadcasters thought that that was the way to break into the American market, then they would consider it.

After the NAB officers left, Joe Flaherty got a call from Tokyo. Flaherty was the senior vice president for technology at CBS and a longtime friend of NHK. He had been an important American supporter while NHK was fighting to have its system chosen as the world's new standard. Flaherty knew what Abel and the others at the NAB were really up to. He tried to explain, though he couldn't be sure the Japanese really understood. NHK "was raped," Flaherty concluded.

NHK had already poured two decades of effort into Muse, and they knew their system was the world's best. So after weeks of discussion they began work on a new version that could be transmitted from a TV tower, not a satellite, just as Abel and the others had suggested. They also started to redesign Muse so it could be broadcast over one TV channel, not a channel and one-half. They planned to call it "Narrow Muse."

Then the FCC appointed its HDTV Advisory Committee, and the

Japanese were confused again. So a couple of NHK officers flew to Washington and took Dick Wiley to lunch. After the pleasantries, Wiley said they asked him whether "Ronald Reagan had called me in to tell me to start this committee," presumably to stop Japan's advance. Wiley assured them that was not so. His committee would show no nationalist partiality. NHK was welcome to enter the race.

NHK went back to work, "very confident we had the best equipment and would win," Kubota said. At an industry conference in the fall of 1988, Masao Sugimoto exuberantly announced that Japanese companies expected to begin exporting HDTV equipment soon and might even manufacture the TV sets in the United States. That raised some eyebrows. Then in January 1989, *Business Week* put a story about HDTV on its cover with the headline "High Stakes in High Definition TV." The article described a panel discussion at the consumer electronics show in Las Vegas. Don Ritter, the radio-bashing congressman, asked plaintively, "Shouldn't the U.S. be a major player in one of the most important inventions of the twentieth century?"

"Our future competitiveness in high-tech electronics could be at stake here," another panel member said, "and we're not in the best position."

"At that point," *Business Week* wrote, "a couple of Japanese executives in the audience looked at each other and grinned."

Articles like that sparked a nasty backlash. William Verity, who had become commerce secretary after Malcolm Baldridge was killed in a freak rodeo accident in the summer of 1987, warned the Japanese not to try to "dominate the market for high-definition television. We have seen this happen in TV sets, VCRs, and other communications equipment. And it appears that Japan Inc. is trying to launch a similar effort to preempt our domestic market for high-definition TV. If this occurs, I can assure you there will be bitter friction."

In Tokyo, that set off an earnest debate: Should we keep quiet, lay low? Several important NHK officers thought the answer was yes. Why rile the Americans? It would only cause problems. Sugimoto disagreed. We're ahead of the world in this, he argued. We should stand up for ourselves, be proud of what we've done. Still, after that, the Japanese were noticeably less exuberant, though they had no doubt they would win Wiley's race. Hardly anyone in the United States disagreed.

2

The Players

d television, they would say. We
n home. We know best what the

ishments behind him, David Sarnoff
y toward enshrining his legacy. He'd
r II and had helped General Eisen-
nications system to serve the invading
r, he lingered in the army for many
eded, pushing and prodding the Penta-
el to general. As soon as the promotion
e, and from that day forward he insisted
al Sarnoff."
-five years of age in the 1960s, the General
short, round man who seemed incapable
No one was more aware of his own accom-
General began lobbying for awards, medals,
blic testimonials, and honorary degrees to
as a pivotal figure in modern world history.
build a monument to himself—a memorial
his papers and display all the parchments,
ary certificates, trophies, trinkets, testimonials,
ed photos he had carefully collected.
e, Sarnoff had consolidated his prized engineer-
campus, 350 choice acres just outside Princeton,
y rolling property had originally been farmland;
pers lovingly landscaped the ponds and ferried in
s the waters. They planted willow trees and tulip
ter of the property, they built a vast research lab.
d, would be his shrine. The library would be a new
d he plunged into planning the facility down to the

we doing today?" he'd ask his aides as he arrived at
taking off his jacket and rolling up his sleeves.
sent an RCA officer to visit several presidential li-
elt's, Truman's, Eisenhower's—to see how these great
their papers. When Sarnoff's library was complete, all
s, trophies, and trinkets were displayed in backlit glass

3 They just gave us away!

Few people were more certain of their place on the planet than the twelve hundred or so engineers and technocrats of the David Sarnoff Research Center, the research and development arm of the Radio Corporation of America. It's no wonder; RCA created the broadcast industry and most of the major inventions that propelled it through the twentieth century. If anyone was going to save America from another Japanese conquest, if anybody stood ready to win the HDTV race, surely it would be the engineering giants of RCA. Everything about them—their history, their traditions—convinced these men that they were best equipped to take on this problem, solve it, and offer their solution to the world. And as Dick Wiley's race took form, RCA's illustrious history and the company's determined, certain point of view proved to be a powerful, sometimes malign, force that helped shape the rules and direction of the contest.

By the 1980s, newcomers to the RCA labs counted themselves fortunate if they felt simply humbled. "Please meet the father of television, the father of color TV, the father of this, the father of that"—those were the introductions offered to one new engineer. It almost seemed as if hubris, 100-proof, was coursing through the engineers' arteries. They lived in the clouds.

Sarnoff engineers had been working on high-definition television even before Wiley's Advisory Committee was formed, and they did indeed become important players in his race. But they had to suffer through several years of humiliation first. Given their history, they were singularly unprepared for this.

David Sarnoff immigrated from Minsk in 1900, when he was nine years old, and scrabbled for a difficult existence with other newcomers in the swarming Jewish ghetto on Manhattan's Lower East Side. He was an intense, determined young man, and by the time he was fifteen he had learned English and enough about American business to talk himself into a job sweeping floors at the American Marconi Wireless Telegraph Company. That was the barest of steps out of the ghetto, but Sarnoff never looked back. He was gradually promoted within the company and its successor, RCA. And in 1915 Sarnoff conceived the modern concept of broadcasting. At that time the wireless, Guglielmo Marconi's invention, was seen only as a limited, point-to-point communications instrument—a more versatile telegraph. But as a midlevel manager with Marconi's company, Sarnoff wrote a memo to his superiors:

I have in mind a plan of development which would make radio a "household utility" in the same sense as the piano or the phonograph. The idea is to bring music into the house by wireless.... The receiver can be designed in the form of a simple "Radio Music Box" and arranged for several different wavelengths which would be changeable with the throwing of a single switch or the pressing of a single button. [That way] hundreds of thousands of families [could] simultaneously receive from a single transmitter.

His idea was revolutionary, and not surprisingly it wasn't adopted immediately. A few years later, though, Franklin D. Roosevelt, then assistant secretary of the navy, encouraged the General Electric Company to help create the Radio Corporation of America. Wireless communications had played an important role in World War I, and Washington was concerned that Marconi, an Italian, controlled the most important wireless company in the United States. As soon as RCA was born, Sarnoff became the young company's most important creative power.

A Westinghouse radio station is actually credited with the first corporate radio broadcast, on November 2, 1920: Station KDKA in Pittsburgh broadcast the Harding-Cox presidential election returns. A few years later, though, Sarnoff persuaded RCA to found the first national radio network, the National Broadcasting Company (NBC). Then in 1930, Sarnoff was named president of RCA, a position he held for almost forty years. Addressing a stockholders' meeting five years later, he laid out his plans for a new medium that he had been pushing his engineers to pioneer: television. Experimental research had been under way in Europe and the United States for decades, but RCA, Sarnoff

said, woul
United Sta
velop an ex
here.

With fulso
service at the 1
in RCA's grand,
From that beginni
ish. It didn't turn

With Sarnoff's
RCA's engineers cre
was designing a color
acceptance, even thoug
a color wheel that spun
was, the wheel had to b
cabinet for today's large-
Buick. Still, the FCC appr
pressed his engineers to u
tem that wasn't really ready
ment. Viewers saw women
bananas blue. The next day
Lays a Colored Egg."

But Sarnoff still pushed. T
averred. Just wait. We invented
color system will prevail. In fa
mind and adopt RCA's system.
the market in 1954, and within a
color. "The All-Color Network" b

The fight wasn't over, however.
standard, but CBS and ABC refuse
ufacturers continued making only bla
lost money making color sets that alr
color shows that hardly anybody saw.
believed he could bulldoze past the con
brace color TV. He was right. By the
begun to collapse. ABC and CBS started
again Sarnoff had prevailed. And in the
Sarnoff's fight for color TV grew to be a
pany. Chests puffed, chins jerked high, R

cases. Every one of his speeches, memos, and pronouncements was carefully cataloged in handsome leather-bound volumes. But Sarnoff's collection didn't end with his official papers. The leather-bound books also memorialized every tribute he'd ever received. *Forrestal Award Dinner* was the gold-lettered title on one two-inch-thick volume. Inside was every scrap of paper generated during the planning of this event—memos, invitations and responses, menus, remarks, and thank-you notes. The Veterans of Foreign Wars Certificate of Merit generated an inch-thick, leather-bound volume. All these "books" and other volumes of papers and detritus were meticulously arranged inside glass-doored cabinets. Upstairs, Sarnoff built carrels for the researchers and writers he was sure would come—five booths with sufficient seating so that up to twenty writers and scholars could pore over the General's works at any given time.

At the back of the memorial library a wide stairway led to the Presidential Suite. Upstairs, on both sides of the center hallway, hung signed photos of other pioneers of radio, television, and science: Marconi, Edison, Einstein, Angstrom, and others. To the left was a formal dining room that seated eighteen around a long walnut Queen Anne table. Behind that was a fully equipped restaurant kitchen. Across the hall were a small bedroom, dressing room, and bath adjacent to a magnificent wood-paneled office notably larger than the Oval Office. More trophies and tributes hung on every wall.

A staff of four was hired to manage the Sarnoff Shrine, and they kept it open six days a week. Outside, twenty-five parking spaces were set aside for visitors. Seldom were they filled.

The David Sarnoff Research Center now occupied 660,000 square feet, and the wing holding the Sarnoff Shrine sat right at its heart. The research laboratory complex wrapped around the General's new library. There Sarnoff's engineers, encouraged perhaps by their close proximity to his monument, were to create for him the industry's next great generational change—and the next, and the next. Once every week or so, one of the General's limousines ferried him down from Manhattan. He held court in his office, lecturing his engineers from behind the presidential desk, occasionally reaching into the bottom right-hand drawer to pull out a gold-monogrammed plastic holder for his rich-smelling cigars. Sometimes the General would stay the night, rise early from the president's bed and take a light breakfast served by his stewards at the long Queen Anne banquet table, occasionally reaching for the telephone bolted into place on the table's underside, just beside his knee. When breakfast was done, he would lay down his white linen napkin and wander through his labs, stroking and pampering his pets, the engineers who brought the inventions that would continue his greatness.

The General grew ill not long after his Shrine was finished. Soon he was confined to bed. In 1971 he died, and in the following years his company was passed around among far less able men. Without the General, RCA began a slow but inexorable decline. At the Shrine, however, the General's engineers worked on, though the company's new leaders in Manhattan were paying scant attention now. As a result the lab's work grew to be less focused. Meantime, RCA's profits started to shrink. Nonetheless, its employees carried themselves as if they lived on Olympus still.

In the 1970s, Sarnoff's engineers created the VideoDisc—movies on a platter that looked like a phonograph record. People were to buy the discs and slip them into relatively inexpensive RCA players that would plug into the backs of their TVs. This was a few years before the VCR had become a consumer product, and RCA's marketeers marveled at the possibilities. Just imagine: People could watch movies whenever they wanted to. What a great idea!

"These things are just like records," they loved to boast. "And people buy *lots* of records." Some within the industry pointed out that people listened to records over and over again, while they weren't likely to watch a movie more than once or twice. Besides, they cautioned, the Japanese were working on videocassette recorders. When those came to market, people would be able to record movies for free, right off the air. Who'd want to buy video records then? RCA's marketeers paid no attention. We're the ones who *invented* television, they'd say with that Olympian air. *We* know what will sell. But this time they were wrong. VideoDiscs were a flop; the company lost $575 million. By 1980 RCA was spiraling toward trouble.

Down at the Sarnoff Shrine, in Princeton, the groundskeepers still fed the geese and cleaned the ponds. Curators continued to dust the books and polish the plaques in the memorial library. Maids fluffed the pillows in the presidential bedroom (used most often now for one RCA executive's secret trysts). Managers of the labs, meanwhile, continued to recruit promising young engineers and send them off for graduate degrees to Princeton, Rutgers, or other local universities at company expense, as they always had. And the veterans still believed they could walk on water. So it was little wonder that in 1985 they began looking in their own leisurely way at high-definition television, even as other American manufacturers slumbered.

The General probably would have pushed them hard. Here was a generational change in broadcasting. Let's get it to market—*now*! RCA will make the high-definition sets and send thousands of them to the stores. NBC will begin broadcasting high-definition programming. We'll *create* a revolution, just as we did with color TV. Once again we will prevail.

But none of the men now sitting in Sarnoff's estuary atop the RCA Building at Rockefeller Center had his vision or his drive. So in Princeton the engineers worked along at their own casual pace. Seldom did anyone from New York show any interest in what they were doing, and left to their own devices engineers will *never* quite finish a project. There's always a new approach to explore, another refinement to test. Left alone, they'll tinker forever.

In 1985, they had an idea for a different way to transmit the TV signal that would give them a somewhat cleaner, clearer picture. They had run a few tests, pushed some theories. It was a concept, really, not a product. They were in no rush. Sure, Sarnoff's men knew that NHK's production specifications—1125 lines, 60 fields per second—were to be proposed as a world standard, and it was clear that the State Department was supporting this proposal. But they weren't worried. How could some *Japanese broadcasters,* of all people, hope to compete with the engineering gods of RCA? At the National Association of Broadcasters' convention in Las Vegas in the spring of 1985, some of RCA's leaders began talking up their vague, barely formed ideas for HDTV. RCA was climbing into the ring now; the others would simply have to step aside. Then, just days later, a letter from Washington arrived at RCA headquarters.

Diana Dougan, the director of the State Department's Bureau of International Standards and Communications Policy, hardly seemed awed by the RCA of 1985. She wrote to Robert Frederick, the president of the company, noting that the Consultive Committee of International Radio was due to consider setting a world television standard in just a year's time. The months leading up to the vote in Dubrovnik were known as the study period, and Ambassador Dougan wrote:

We have concluded, together with interested industry representatives, that failure to attain a worldwide HDTV standard during this study period will probably result in a failure to attain such a standard at all, with significant adverse consequences to U.S. trade and information interests.

I have to share with you, however, concern about news which has surfaced in my office in recent days. Our staff has been advised that at least one member of the U.S. Congress visited the RCA exhibit stand at NAB in Las Vegas [and] concluded that RCA does not share the U.S. determination to attain a worldwide standard this year.

The message was clear: Back off. Get with the program. Stop talking up your own research. Support the Japanese.

Sarnoff's men were stunned. Was the State Department actually saying "that if we're opposed to the Japanese system, we're *un-American*?" as one engineer put it. Roy Pollack was RCA's executive vice president in charge of consumer electronics, and to him the whole dispute was "just plain silly," but he couldn't ignore it. RCA owned a television network and several TV stations around the country. They needed licenses from the government to operate, and "when a government agency talks to a person with a license, it's a veiled threat," Pollack knew. He took the Dougan letter up to Frederick and told him, "We had better show some restraint." Frederick nodded agreement, and with that RCA backed away from HDTV. Some of the engineers in Princeton wanted to write their congressmen, make some noise. But the corporation's leaders in New York put a stop to that.

With that, a chill settled over the Princeton Shrine. The HDTV work, such as it was, essentially stopped. The Olympians felt as if they'd taken a blow to the belly. But far greater indignities lay ahead.

Thorton Bradshaw, RCA's chairman, was a rumpled, professorial sort of fellow. And in December 1985, as every year, he called the company's senior executives to a meeting the morning after the annual Christmas party. When he stood up, everyone expected another rambling recap of the year and a look ahead. Instead he dropped a bomb. RCA had been sold, Bradshaw announced. By mid-1986 the once proud company would be a subsidiary of General Electric.

Around the table, the stunned silence was so intense that it seemed as if the very air might shatter. Even with rich stock options in their hip pockets, each one of the executives was stabbed with mortal fear. GE was a no-nonsense, bottom-line kind of company, and John F. Welch Jr., GE's president, wasn't about to do any of them favors just because RCA had a grand history. Welch was known as "Neutron Jack," after the bomb that wiped out whole populations but left the buildings standing. He'd reduced GE's payroll by almost 20 percent—132,000 jobs—in the last five years. That had won him *Forbes* magazine's title: Toughest Boss in America. All of them also knew that GE used to make televisions. But Welch had decided the field wasn't profitable enough. In fact he'd been moving GE out of consumer electronics altogether, dumping one division after another. The executives immediately began to worry: What's to become of us? And that concern soon spread throughout the company. It was particularly acute at the Sarnoff Shrine. After all, GE already had its own research lab, in Schenectady, New York. Even given the Shrine's glorious history, why would GE need a second research center?

In the labs little got done through the following months. Work on HDTV had already stopped and now most of the engineers drifted through the day, updating résumés, calling up payroll to see if they were vested in the pension plan. A cartoon on the bulletin board captured the mood. Two goldfish are swimming around in a Waring blender full of water, and one of them says to the other, "The tension is unbearable."

Early in 1987, Welch turned the blender on. With more than twelve hundred well-paid employees, the Shrine was just a drag on his bottom line. Given the sad state of the industry, no one was interested in buying a consumer-electronics research lab. So in February GE announced that the Sarnoff Research Center was being *donated* to SRI International, a diversified consulting and research firm based in Menlo Park, California. The Shrine would go with a generous "endowment"; GE would give Sarnoff $250,000 in research contracts every year for five years. Though the payroll would certainly be reduced, at least the labs would remain open. But inside the Shrine, Sarnoff's engineers—direct descendants of the men who had created the industry—walked around with their mouths hanging open, their minds numbed by the one stunning thought:

They just gave us away!

Jim Carnes was a son of the Sarnoff labs. He grew up in Hagerstown, a pleasant little out-of-the-way town in northwestern Maryland, and joined the labs as a junior engineer in 1969, while General Sarnoff was still alive. He had a Ph.D. in electron-device physics from Princeton, and he quickly showed his talents, winning nine patents for research in camera technology. Then in 1977, RCA moved him to Indianapolis, home of RCA's largest TV factory. There he moved his way up through the managerial ranks in the Consumer Electronics Division. A big bear of a man with a bushy mustache and a meaty, Germanic face, Carnes was surprisingly soft-spoken and amiable. Unlike many of RCA's exuberantly boastful engineers, Carnes could assume an understated, aw-shucks manner. Asked once to lecture a broadcasting convention on the company's current work, he played down what his men were doing, and later he explained, while looking at his shoes, "You know, if you do too hard a sell, they won't want you to come around anymore." Behind that, though, Carnes was fully vested in the Sarnoff state of mind. He considered himself one of the Olympians, no doubt about it, and his true view never lay far behind that self-effacing front. Trying to arrange a joint venture with another manufacturer once, he was shocked that the other company didn't leap at the offer. Nostrils flaring, eyes wide, he snorted, "I sort of thought, We're the lab that invented color television. I kind of thought maybe they might want to talk to us."

Carnes was in his mid-forties in 1987 when he was asked to go back to Princeton and help manage the Sarnoff Shrine after Welch gave it away. The Shrine had to make the transition from a research facility that had been the favored pet to a consulting lab that had to win contracts to survive. He was the new vice president in charge of consumer electronics, and he realized this was going to be no easy task, especially given that the staff had to be reduced right away—from 1,221 to about 850. The staff was so demoralized that it was nearly dysfunctional. They needed a new project—something important they could start working on right now.

All the theoretical work on high-definition had stopped soon after the State Department letter-bomb arrived in 1985. Later some of the engineers had started working on what they called "improved-definition" TV. They tinkered with a regular TV, put in some better circuits and filters so that the TV offered slight improvements in picture quality. Then, a few months before GE bought the company, Carnes flew to Princeton from Indianapolis with Jack Sauter, one of RCA's senior marketing officers, to have a look at this improved-definition TV. Sauter took a seat, and Carnes asked one of the lab techs to roll the tape. Sauter watched for a moment and then shook his head.

"Doesn't look any better to me," he growled.

"Well, you gotta get a little closer," Carnes told him, stepping toward the TV. "Come up here and have a look."

"Damn it, Jim," Sauter snapped back, "they aren't gonna stand that close in the store."

That ended the discussion of improved-definition TV. As long as he was there, however, Sauter looked around the lab a bit. In the back he spotted an odd-shaped TV monitor. It was wide, like a movie screen. Let's have a look at that, he said.

Carnes and the others didn't want to show it to him. They hadn't been able to get the damned thing to work right. You don't want to see this, they told him. "The picture's terrible." But Sauter insisted, so they turned it on.

"Look," Carnes said, pointing at the distortion. "See how lousy it is?"

Sauter stood back and pondered this strange-looking box. The wide screen intrigued him. "If you can get that working," he said at last, "I think I can sell some of these wide-screen TVs." Right away, Carnes got the Sarnoff engineers working on that. He didn't know it then, but he had begun work on the lab's entry in the HDTV race. Advanced Compatible Television was born. Sarnoff had found its cause.

From the beginning, David Sarnoff had espoused a straightforward philosophy about new products. With each new invention, RCA took care to protect the devices still in use in people's homes. In the 1920s, anyone who chose not to buy one of Sarnoff's Radio Music Boxes could still listen to the phonograph. Americans who didn't want to buy a television in the 1940s still had their radios. And when NBC started broadcasting in color, black-and-white sets could still pick up the signal.

So it was to be with Advanced Compatible Television. The way Sarnoff's engineers conceived it, anyone who bought an advanced compatible television set would get a wide picture, just like the one that had caught Sauter's eye, along with slightly improved resolution. "It will immediately obsolete every TV in America," Carnes noted with enthusiasm. Just like the good old days, RCA would sell lots of new sets. At the same time, though, the *C* in ACTV stood for *compatible*. Conventional sets would be able to pick up the ACTV signals and display a normal TV picture, just as black-and-white sets did with color. It seemed the perfect solution.

Even after Carnes got the engineers back on their feet again, they had to endure one more kick in the belly. In July 1987, just five months after Welch had cast them off, General Electric sold the rest of RCA's Consumer Electronics Division to Thomson Consumer Electronics, a company wholly owned by the French government. Faceless bureaucrats in Paris, of all places, now held rights to the proud RCA name. In fact, however, very little had changed. Joseph Donahue, a longtime RCA executive, had been one of the RCA-GE officers who managed the Sarnoff Research Center. Thomson quickly hired Donahue, and Donahue hired the Sarnoff Center as Thomson's research labs in America. The names and faces were the same. But General Sarnoff's men had become employees, in essence, of a foreign government.

They pressed ahead with the ACTV project anyway, and by September they had a "computer simulation" ready. A computer program was able to produce a simulated ACTV image, complete with a wide screen and a slightly improved picture. They wanted to take it to a trade show in Ottawa, make a public announcement: The Olympians are back, and we have the answer—ACTV! But now they had to ask the French for permission.

The word from Paris was, We want to see this new system before you do *anything* with it. Carnes didn't see this as particularly encouraging news. Thomson would send a few executives to Princeton, but they couldn't get there until the day the equipment truck had to leave for the trade show. Carnes told his men to load the truck. They should be ready to leave the instant the Thomson men gave the nod.

The French arrived late, of course. It was close to noon. As soon as Carnes could do so without being impolite, he showed them an ACTV demonstration. When it ended, he paced back and forth, waiting for their reaction. They chattered among themselves in French, almost as if he weren't even there. While the truck idled outside, Carnes took them to lunch around the eighteen-seat Queen Anne table in the General's dining room. As they ate, Carnes nervously waited for the Frenchmen to tell him something, anything. But they just kept talking among themselves, and Carnes couldn't understand even a word. Outside, the drivers had climbed out of the cab and begun pacing back and forth. Still the

Frenchmen said nothing to Carnes, who sat there lost in his own dark ruminations: They must have hated it.

Finally, over dessert, Carnes burst into their conversation: "Can the truck leave?"

They looked up, startled. "Oh, yes, of course," one of them said with a wave of the hand, then resumed prattling in French. Carnes didn't know what to make of his new employers. But at last, ACTV was a go.

Over the previous sixty years, RCA, NBC, and the Sarnoff labs had build up a mighty public relations dynamo, capable of powering elaborate circuslike pageants to promote real and imagined triumphs. In the General's later years, he'd used it to paint himself as an American folk hero. Now the machine was turned on again for ACTV, and it roared back to life with a loud, guttural rumble. Releases poured forth, industry leaders gushed lavish praise. At trade shows, satin-tongued announcers stood on carpeted rostrums and spread arms wide to the passing crowds as their radio voices crooned, "Come see the TV of tomorrow. Come see ACTV!"

But there was one big problem. As the Sarnoff labs proposed it, ACTV would fit neatly into a single television channel. It was compatible with existing TVs. A simple and clean engineering solution. In fact, with ACTV, the broadcasters wouldn't need a second channel. So the NAB and other broadcasting organizations immediately asked, What in the hell are these people thinking? The whole idea is, we *need a second channel* for HDTV!

Wally Jorgenson, chairman of the NAB board, called the announcement of ACTV "good news," but immediately added that it "should not lead the FCC to make a premature decision on what the bandwidth needs will be for the next generation of American television." Tom Paro, president of the Association for Maximum Service Television, was even more direct. ACTV "is not the ultimate HDTV system," he insisted. "A new generation of television technology providing even higher-quality service will be developed sometime in the future, and that new technology will almost certainly require additional spectrum for local station implementation."

At the Shrine, Carnes and his men heard the call. They got the message loud and clear. They'd been arguing about ACTV, trying to decide whether a one-channel system was good enough or whether they should produce an enhanced version that required part of a second channel, too. Then the broadcasters started screaming that they would not support anything that didn't save them from Land Mobile. This, Carnes decided, "makes the technical arguments moot," and within a few weeks the RCA public relations machine was pumping out a new message: This is ACTV-1, just the first step in a gradual evolution toward *true* HDTV. Soon we will offer the next step, ACTV-2. It will offer a better picture. And ACTV-2 will require *more than one channel!*

All of a sudden Sarnoff's men had another hit. The nation's broadcasters started stumbling all over themselves in support of ACTV, showering money and praise on Carnes and his men. Capital Cities/ABC donated $500,000 for ACTV development. NBC's affiliate stations across the country threw in $3.3 million. Even some of the Cable Mongols—HBO and another major cable company—announced that they would provide "substantial funding." And when Sarnoff's men set up a computer simulation of their system at the NAB convention the following spring, lines for the viewing room stretched around the back of the booth. As they left the demonstration, people weren't talking about how great the picture was; they hadn't even seemed to notice the wide screen. "It's so inexpensive," they said. "It solves a lot of problems."

With all the new money and support, Sarnoff's men focused on creating a real system, not just a computer simulation, and they had a deadline in mind: April 20, 1989. When that day came, they were ready for the first over-the-air demonstration using real hardware. They were going to beam an Advanced Compatible Television program from the WNBC tower atop the World Trade Center to wide-screen ACTV receivers in an auditorium at the Princeton Shrine. Two hundred invited guests gathered for the show.

The date had not been chosen by accident. April 20 was the fiftieth anniversary of the first TV broadcast, General Sarnoff's historic demonstration at the 1939 World's Fair. Now that same heady feeling filled their heads again: Sarnoff's soldiers were making history once more. Certainly the world around them had changed, but they were certain that when the advanced compatible televisions clicked on, the old magic would return. For any of the reporters, broadcasters, or other invited guests who remained unconvinced, four wide doors had been flung open at the back of the auditorium. Beyond them, down a short flight of steps, was the General's Shrine, and for the occasion all the plaques had been dusted, the crystal bowls polished, the backlighting turned all the way up. The message was clear: Wander through here if you have any doubts about ACTV, and surely when you leave you'll fall into a swoon. For visitors so dense or obdurate that they left the General's Shrine unbowed, the RCA public relations engine rumbled away in another room. "Today's broadcast is to television what Kitty Hawk is to flight!" intoned Michael Sherlock, a longtime NBC executive.

"I've seen progress; I've seen ACTV!" said buttons pinned to almost every employee's lapel. The General's men had wandered in the wilderness long enough. Now the Olympians were back!

The clock struck 8:00, and a hush settled over the crowd. The lights dimmed, and at the World Trade Center veteran Sarnoff engineer Liston Abbott felt the heavy mantle of history settle on his shoulders as he pushed the button inaugurating the first ACTV broadcast. In the auditorium at the shrine, a tape of the 1989 Rose Bowl Parade filled the wide screen. After that came tape from a Sea World show and other vignettes.

Some people in the audience were startled. For all the hype and hoopla, this TV picture wasn't really that much better than the one on the conventional television sitting beside it. Many of these people had seen the Japanese demonstrate their HDTV system. This wasn't *nearly* as good. And this wide screen—well, it looked as though panels had been stitched to each side. Sometimes you could even see the dividing lines.

In the following weeks, questions grew sharper, though many of them came from rivals. Sarnoff is working for the French government, they'd say. ACTV is just a European plot to stall the United States while Europe works on *true* HDTV. But Sarnoff's men were unconcerned. After all, the first experimental black-and-white broadcast in 1939 didn't have such a great picture, either. And don't forget the blue bananas in the first demonstration of color. Those pictures got better. So will this one. The General showed everyone back then, and we'll show you now.

"We're the ones who *invented* television," they'd say with that lordly air once again, leaning back in their chairs, hands clasped over their bellies. "*We* know what will sell."

Dick Wiley understood the game better than anybody, and to him ACTV looked like a scam. As head of the FCC's Advisory Committee, he couldn't say anything. But he had seen the demonstrations of ACTV, and "frankly," he said, "it just didn't look that good." Joe Donahue, the former RCA executive now with Thomson, was promoting ACTV at every opportunity. It was "a convenient two-step evolution for broadcasters," he'd say with silky self-assurance, thumbs hooked through his suspenders. The final step to *true* HDTV, he would add, "will be in 2004." Around the office, Wiley started calling him "2004 Donahue." Actually, Wiley found the whole American effort discouraging. By now, two dozen labs had come forward with HDTV proposals, but none of them looked very good. It was so depressing. He'd go to the trade shows and look over "these pathetic jury-rigged systems.

"Over here some guy's got a box with dowels going 'round and 'round." Wiley shook his head. "And then you go look at the Japanese system, and it looks just great!"

ACTV didn't improve his mood. It was "the perfect deal" for the manufacturers and the broadcasters, Wiley would say. "The perfect deal!" The manufacturers could sell everyone an expensive new ACTV set and then another one in ten years or so when the next generation came along. What could be better? The broadcasters, meanwhile, could stave off Land Mobile while pretending to give America HDTV without spending any real money or improving picture quality very much. The Sarnoff Center was promising that new ACTV equipment would cost each station only about $100,000—nothing, really. To the broadcasters that sounded awfully good because by 1989 they were in a fevered search for a low-cost escape from the HDTV trap they had set for themselves.

John Abel could rightly be called the father of HDTV in America. His lobbying scheme to defeat Land Mobile had given birth to all that happened later. But by 1989 Abel and his brethren were beginning to fear that their child might eat them. High-definition television really seemed to be coming, and Abel was humming a new tune. HDTV will mean "huge expenses for broadcasters," he complained in a speech that April. Every one of them will need a new transmitter, maybe even a new TV tower, if the old one is full. Given the zoning requirements and regulatory problems, Abel went on, "it sometimes seems easier to invade Austria than to build a new tower." The NAB had ginned up a cost estimate for conversion to HDTV. By piling on every conceivable expense and then some, they had managed to find that TV stations would have to spend somewhere between $4 million and $18 million each. As the months went by, the estimate seemed to bloat. Broadcasters started talking about $30 million in expenses, maybe even $35 million. That, Abel and others pointed out with relish, was more than many stations would be worth if they were put up for sale. Even if the NAB estimate was exaggerated, nobody wanted to spend *any* money on high-definition television. HDTV had already served its purpose. Land Mobile was vanquished and still quiet after more than two years. So the father of high-definition television took on a new mission: Kill the baby.

In a contemplative mood, Abel would sometimes reflect on what he was doing. In 1986, "I really didn't realize what I was starting," he said later. Soon after, though, "I began to see the problem. We'd started this, but it was turning into something we might not be able to stop"—though he certainly did try.

Abel and his confederates had lobbied hard to be sure that Dennis Patrick's Advisory Committee on Advanced Television Service was stacked with friendly members, and they had largely succeeded. But Abel hadn't bargained on how quickly Dick Wiley would take over. Wiley *was* the Advisory Committee. All the NAB's handpicked members became ciphers, nothing more.

The NAB lobbyists needed a new forum, some other means to bring this brushfire under control. It arose from an unlikely source—a speech, another one of those hopeful, futurist talks that their president, Eddie Fritts, liked to give. This one had come in September 1987, when the NAB was still actually *promoting* HDTV. Broadcasters, Fritts had said at an industry meeting in Washington, should create a sophisticated broadcast technology laboratory, a special facility for evaluating different approaches to high-definition television. The goal, Fritts offered, would be to bring "HDTV to reality for terrestrial broadcasters." Now here was a lofty idea, born of the noblest intentions. Or so it had seemed. Over time, though, Fritts's proposal began taking form with an entirely different motivation.

If Dick Wiley was going to run a race, everybody realized, he would need some means to evaluate the contestants—test the new machines to see how well they worked, what kind of pictures they could produce. The United States had no independent lab capable of making the sophisticated measurements that would be needed. Broadcasters didn't want somebody else deciding what kind of system they would get. So they began planning to create their own test center, just as Eddie Fritts had suggested. John Abel saw it this way: "This is going to be a test center for *broadcasters.* We control this test center. And we will do this for the government. It will give us even more control. And the principal reason for all this is to pick the best system *for broadcasters!*"

Abel's detractors thought they knew just what that meant. Joe Flaherty, the CBS executive, was arguably Wiley's most important lieutenant on the Advisory Committee, and he also sat on the board of directors for the proposed new lab, to be built in Alexandria, Virginia, and called the Advanced Television Test Center. As he watched plans for the center move haltingly forward, Flaherty believed he knew exactly what was going on. The test center was to be governed by broadcasting officials and nobody else. Most of these people were businessmen, not engineers. So the lab being created to pick the best high-definition television system for America would be directed by many of the very same people who were trying to *kill* HDTV. In Flaherty's opinion, the test center would have only one goal: "To put ACTV over on America."

Plans proceeded anyway, and in the summer of 1988 Peter Fannon got a call from John Abel. Fannon was a broadcast industry consultant, and Abel wanted to know if he would like to be the executive director of the Advanced Television Test Center. Fannon's first reaction was "You've got to be kidding; I don't know anything about this." But that was not entirely true. Fannon had been the president of the National Association of Public Television Stations until recently, and he had been involved in the broadcasters' world for many years. A careful, not to say punctilious man, he had also worked for the General Services Administration and the Office of Management and Budget. He was not an engineer, but Fannon was a reasonably astute politician, and, as he would soon learn, that was a more important talent for this job. Fannon didn't know much about HDTV, but he figured he could learn, and this could be an exciting assignment. The FCC was running this grand, international race, and the new lab would be the finish line. He accepted the job and moved into a temporary office at PBS headquarters in Alexandria, a Washington suburb, while he searched for the test center's permanent home.

When he arrived, the rough working estimate floating among board members had been that the test center would need about $3.5 million over its several-year life. The member companies making up the board—the TV networks; broadcast industry groups, including the NAB; and the Electronic Industries Association—were to split the costs among themselves, and $3.5 million seemed a bearable expense. Fannon thought this estimate was absurd; he figured the center would need at least twice that much, and when he suggested that to some board members, they didn't seem to want to listen. Soon, though, a bigger worry started growing in him.

The test center board met once a month, and board members were careful to stick to test center business. But when they'd worked through the last item on the official agenda, often they would declare the meeting formally over and then start talking about their larger concerns. Sometimes Fannon was allowed to stay and listen, and the discussions were eye-openers. What can we do to slow the FCC down? they would ask themselves, leaning forward at the conference table, hands clasped in front of them. How are we going to get out of having to pay for all this new HDTV equipment?

These people don't care anything about HDTV, Fannon realized. He would leave one of these meetings, think over what he had just heard, and conclude that "I didn't hear anything positive, only negative. All the talk was defensive. They'll be perfectly happy if the whole thing falls apart of its own weight. These people are having a hard time figuring out why exactly they are even *doing* this."

There was another problem: NBC. The network was the prime backer of ACTV, and as Fannon watched, the NBC representatives "continually pushed anything that gave a leg up for ACTV." They tried to stack some of the test center's committees with their partisans. They kept insisting that ACTV be placed on an equal level with true HDTV. They're trying to manipulate the process, Fannon concluded. When he looked around the table at the other board members, he realized that several of them had actually donated money for ACTV development.

Fannon grew depressed. What had he gotten himself into? Months passed, and the test center's board continued to bicker and complain. Nothing seemed to be coming together. John Abel had hired him, but Abel's attitude at every meeting seemed to be: Hold back. Go slow. Spend little. Meanwhile, NBC kept pushing for special advantage.

We're getting nowhere fast, Fannon thought. He'd given up a good job to take this new position, but this test center was beginning to look like one big joke.

Then in January 1989, Fannon was finally ready to present the test center's formal budget proposal. The board members were gathered around a conference table at PBS once again, and they called on Fannon. In his methodical monotone, he laid out the budget categories: rent, equipment, salaries, insurance, utilities, and the rest. To pay for all that over three years, he said, the total came to $10.5 million.

Around the table jaws dropped. As Fannon looked at each member one by one, almost every one of them was openly appalled. Why do you need fifteen or sixteen employees? Abel was demanding to know. Why can't you do it with five or six?

"My company's going to choke on this," said the representative from ABC. Now Fannon saw everything clearly, and he was angry: this whole thing was a scam! "Nobody," he said later, "wants to be anybody's fall guy!"

4 *Burning the furniture*

Hardly anyone was surprised at the Zenith Corporation's dismissive reaction to Advanced Compatible Television. "ACTV's just a ploy to sell more picture tubes," sniffed Wayne Luplow, vice president for research and development. That was hardly a surprising remark for a Zenith executive. From its earliest days, the company had been belittling RCA, its largest competitor, while feeding off it at the same time. Zenith had never been known as a wellspring of major innovation. Through most of the last seven decades, RCA had come up with the significant inventions; Zenith would add a few tweaks, manufacture the sets, and then make lots of money.

But this time was different. In 1989 Zenith actually had an HDTV proposal of its own, and Luplow honestly believed that his company had a *better idea*—an unaccustomed position for the engineers at Zenith's headquarters, in Glenview, Illinois, just north of Chicago. It felt pretty good, but more than a little bit scary, too. They'd entered Dick Wiley's race with great fanfare and promise. Then just as soon as Luplow and his men had jumped forward, heads swollen with the idea that they could actually be *leaders* for a change, they'd found that the rest of the company was depending on them for Zenith's very survival. The corporation was foundering. Only high-definition television could save them!

Two young engineers just out of the navy started the company in 1918; they made radios on a kitchen table, heating their soldering iron on the stove. Within a few months, they moved their "factory" into a shed on Chicago's Northside and started an amateur radio station, too. Its call letters were 9ZN, and from that they coined a name for their enterprise: Z-Nith.

On New Year's Eve in 1920, Eugene McDonald, a Chicago auto dealer, happened to hear a radio playing in the garage where he parked his car. After he'd had a look, McDonald wanted to buy one but was startled to learn that he would have to wait several months. Production of radios couldn't keep up with the growing demand. That got him thinking. McDonald had been a naval intelligence officer during the war, and upon his discharge he'd been promoted one rank, to lieutenant commander. From then on he preferred to be known as Commander McDonald, and in 1920 his largest business achievement had been to introduce the first successful concept of auto loans.

McDonald found his way to the Z-Nith shed. He bought a radio, and a short time later he bought the tiny company and changed its name to Zenith. As demand for radios mushroomed—thanks largely to David Sarnoff's efforts at RCA—Zenith grew quickly and in the following decades introduced a number of novelties and innovations, including one of the first portable radios. It was the size of a suitcase. Later came the first radio with push-button tuning, and the company was among the first to embrace FM. When television came along, Zenith offered the first workable remote control.

Still later, a Zenith engineer created the standard for broadcasting FM stereo that is still in use today. Nonetheless, by then the company had already fallen far, far behind. Over the previous half century Zenith's leaders had created a corporate culture that proved a liability in the new era. Others set the agenda; they were followers. "It doesn't make sense to be first" came to be a slogan of sorts. While Commander McDonald was pursuing new approaches for marketing radios in the 1920s and 1930s, over at RCA Sarnoff was already aggressively pushing the company into television. As Sarnoff staged his TV demonstration at the 1939 World's Fair, Commander McDonald watched, and Zenith followed lazily along. What was the rush? They were making lots of money selling radios. After General Sarnoff introduced color TV in the early 1950s and pushed the nation to adopt it, Commander McDonald remarked with fatuous indignation, "We will never experiment on our customers!" Zenith built exactly one color set, just to prove the company could do it. What was the rush? Let RCA take the risks while Zenith made lots of money in black and white. Zenith finally put its first color TV on the market in February 1961, after every other American manufacturer had already begun marketing color TVs.

In the 1960s, the electronics industry embraced the invention of the transistor, and television manufacturers started making solid-state TVs. Without vacuum tubes, they were far more versatile and reliable. But not Zenith. Wayne Luplow was a young Zenith engineer by then, and as he saw it, solid-state technology "scared them; they didn't understand it. Why rock the boat? We were making a lot of money selling tube sets." Zenith even tried to make a virtue of its aged designs. Through the late 1960s, its advertising touted Zenith's "handcrafted chassis," an implicit slap at those newfangled circuit boards. Why not? Making TVs the old way, the company remained fat, comfortable, and secure. That's exactly why Luplow and others like him had chosen to come on board.

Luplow's home was just up the road in Milwaukee; his father was the credit manager at a dairy. But he'd started his engineering career working for RCA in New Jersey. He hadn't much liked the East Coast and quickly realized that RCA occasionally laid off employees as markets changed. With little seniority, that scared him. Besides, he wanted to go back home, and so he called an old college buddy who was working at Zenith. He told Luplow that Zenith was a secure, low-pressure kind of place. Most of the engineers were homeboys, recruited from local universities. Summer evenings, everybody gathered on baseball diamonds laid out on Zenith's huge campus in Glenview, a quiet suburban community, and competed in company-sponsored baseball leagues. Best of all, Luplow's friend said, "There's never been a layoff here." To Luplow, that sounded just great. A job with Zenith was a comfortable home for life.

Or so it had seemed. Almost as soon as Luplow took his desk, Zenith's fortunes began to change. In the 1970s, Japanese televisions began entering the market, and profit margins for everyone began to slip. Other TV manufacturers started diversifying; some left the business altogether. Motorola, for example, moved out of televisions and into the semiconductor business. But Zenith declined. Their archrivals over at RCA had bought Coronet Industries (a carpet manufacturer) and Banquet Foods, among other new, diversified holdings. Zenith executives used those acquisitions to disparage the whole concept of diversification. We don't want to sell rugs and frozen dinners! they growled. Why should we? Though the profits were not as great as before, Zenith was still making lots of money selling TVs. In fact, through the '70s, Zenith and RCA were alternately America's largest and second largest manufacturer of televisions. After half a century of animosity, Zenith remained bitterly obsessed with its chief rival. Occasionally a Zenith engineer would quit to take a job with another company, and if he was going to GE, say, he would be allowed to give a month's notice, pack his things, put his house on the market, and say good-bye to all his friends. But if an engineer announced that he was going to work for RCA, he had to leave the building that very day. He barely had time to clean out his desk.

Then, in 1986, GE bought RCA and sold the television business to Thomson Consumer Electronics. The people at Zenith were as surprised as everyone else. Suddenly their archrival was gone. Before Luplow and the others could catch their breath, they looked around and realized they were the last remaining American manufacturer of televisions! And not so long ago the field had been crowded with proud names: GE, RCA, Magnavox, and Motorola. Philco, Sylvania, Emerson, and Admiral. In the '50s, more than ninety American companies had made TVs. Most of those brand names had disappeared. Others were, like RCA, the property of foreign firms. Now Zenith was all alone, and a look at the state of its business helped explain why the others had pulled out of the field.

By now Zenith had laid off thousands and thousands of employees. The first big cut came on September 27, 1977. A banner headline in the *Chicago Sun-Times* that day had screamed, "Big Zenith Layoff: 2,100 Here, 3,500 Elsewhere. Imports Cited." It was the beginning of a slide that would not stop, and by the early 1980s Zenith was losing money. As the end of the decade neared, Zenith was assembling most of its TVs in Mexico, but the company *still* wasn't making money and seemed to have no prospects. The losses grew deeper every year: $7.7 million in 1985, $10 million in 1986. About the time Wayne Luplow was promoted, the company released its 1987 results: Zenith lost $19.1 million. In 1979 Zenith had 20 percent of the American TV market; nine years later, it was 15 percent and sliding south.

These were dark times in Glenview; employees dragged through the days. A few engineers saw no future at Zenith and resigned. For the others, the workdays grew longer, as each of them had to take on the duties of the guy who used to sit at the next desk.

Driving to work, the engineers passed big white signs beside the highway just south of the plant—billboard-size announcements of the company's deepening troubles. "Lots for Sale," they said. The company was selling off its land. Pulling into the parking lot, nobody had to fight for a space anymore. Inside, long hallways that once bustled were dusty and dark. Office furniture was so worn and battered that Goodwill probably would have declined to take it. One day employees got a new directive: It's time to start reusing paper clips. Even the walls gave away the company's sorry state; they were mustard-colored and brown—tints that evoked their era, the mid-1970s, when the company was still fat. They hadn't been repainted since. Summer weekends, as employees gathered for the baseball games, a twilight feeling hung in the air. Everybody knew this season might be the last. The company continued selling off lots, and Zenith's property line seemed to creep closer and closer to right field.

Zenith was alone against the world. As its troubles deepened, people began treating the company as if it were already dead. "They've been burning their furniture for years," snipped Joe Donahue, the former RCA executive now working for Thomson. "They seem to be in a real bunker mentality over there," said Greg DePriest, now an executive at Toshiba. That was the environment as Luplow accepted his promotion.

As he crept toward middle age, the new vice president was balding, wizened. Some might say scarred. His nasal voice sometimes lapsed into a whine. Like so many at Zenith who had once approached their careers with relaxed confidence, Luplow had grown to be defensive, even combative at times. Zenith headquarters truly did feel like a bunker now. As other American industries—steel, automotive, computer—had learned the hard way, there was no place in the market any longer for followers. Companies had to innovate to survive. Just look at some of Japan's recent new products: the VCR, the Walkman, the portable CD player. None of those marketing successes had bothered Zenith much—not even NHK's introduction of HDTV. Luplow and the other veteran engineers had seen the Japanese demonstrate their system at trade shows, but they had thought little of it. Sometimes they talked about Muse, but finally they convinced themselves that they could dismiss NHK's work as irrelevant. Muse had been designed to be transmitted by satellite, after all. It wouldn't work in America.

Then early in 1988 NHK announced that it was redesigning Muse for the American television market. Now the Japanese were marching into Zenith's own backyard, and Luplow clearly heard the cry from everyone around him: "The TVs and VCRs are on the boat!" That finally got even Zenith's attention.

Luplow's division was working on several new projects, though the R&D budget was pitiably small. Still, Zenith had begun making cable

TV converter boxes and was number three in that market, behind a New York firm, General Instrument, and Scientific Atlanta. The converter-box business wasn't big, but Zenith believed that cable TV might be important for the company's future, now that half of American homes were hooked up. Luplow's engineers had been working on a new, more efficient transmission system. It would give cable operators a scrambling system that pirates couldn't easily break. The Zenith system worked by "digitizing" part of the signal.

By 1988 much of the communications world was going digital, meaning that information was moved as computer code, a series of digits—zeros or ones. Each group of digits was reordered to represent each new piece of information. Computers carried information digitally. So did compact discs. But TV and radio broadcasters still sent their signal in conventional, "analog" form.

Analog television signals are sent from the TV station tower in electromagnetic waves whose size, shape, and intensity are modified as the original images and sound patterns change. These analog signals are roughly like waves rolling across the ocean. Though the general shape of the waves can vary, each one rolls along as a preformed unit. Little can be done to manipulate a wave before it crashes onto the beach. Digital signals, on the other hand, are more like armies on the march—large groups of individuals who make up a whole. Before the soldiers set out, their commanders can arrange them into squads of whatever size or configuration they want and direct them to carry out an array of missions once they reach their objective.

The advantages of digital television were manifold. If a TV signal could be translated into computer code, it couldn't be distorted during transmission. Interference might knock a few "soldiers" out of the battalion, but that would have no effect on the rest. Every viewer would receive a near perfect picture—or no picture at all. More important, a digital signal could be manipulated in hundreds of ways—rearranged in infinite orders so that different homes received different signals. The digital "army" could be dispatched so that certain squads arrived at some homes but not at others. That meant viewers could order whatever programming they wanted, when they wanted it. They could page through video catalogs, play along with game shows, participate in video town halls; they could communicate with TV stations just the way computer users interact with on-line services. Digital televisions wouldn't be dumb boxes any longer. They would become interactive computers, potentially more powerful and versatile than any desktop PC. Digital television would represent a true technological revolution—as important as the transition from radio to television in the 1940s. Television would come alive!

That was the promise. But as Luplow and his men worked out their plans in 1988, there was one large problem. So many digits, or "soldiers," were required to create a complete TV picture that it seemed utterly impossible to send all of them over a single television channel—or two channels, or even three. Far less than half of the signal would fit through that figurative 6-inch pipe. Not even the channel-and-a-half schemes that NHK and the Sarnoff labs were offering could accommodate a digital signal. Nobody in the TV business knew how to get around the problem. So *fully* digital television remained just a dream. "People talked about going all digital," Luplow said. "That would be the best." But nobody knew how to do it. Without exception, broadcast-television engineers believed that digital TV was decades away.

Still, Zenith's engineers had managed to convert a small portion of the TV signal to digital form while leaving the rest in the conventional, analog format. That proved useful immediately, because digital information was much easier to scramble. They also discovered that even this incremental step made the signal smaller and cleaner, less likely to interfere with the signals on adjacent channels. Using this innovation, Luplow's engineers also found they could transmit a larger number of channels down the cable lines. Cable operators would be able to offer forty-two channels, say, instead of thirty-five, without running new cables. Now here was an innovation they could market!

Rich Citta, a manager in the R&D division, led the team that devised the new cable transmission scheme. Citta was another local boy; he grew up in Chicago, and he'd been with Zenith for twenty years. Because of his wild hair, thick salt-and-pepper beard, and dark sunken eyes, some called him "Wolfman." But he was a bright, voluble, and enthusiastic employee, possessed of an ingenuous warmth. One afternoon in the spring of 1988, Citta was perched near the top of a ladder repainting his house, when suddenly the ladder's feet started to slide away from under him. He scrabbled to grab at the wall but then plummeted to the ground, his legs tangled in the ladder. His left ankle was crushed. A few hours later he found himself flat on his back in a hospital bed, with the knowledge that he would be stuck there for several weeks.

Citta quickly grew bored; he wanted to be back in the lab. Staring at the ceiling, he began thinking about the cable project. Right there it hit him. It was so obvious. Why hadn't this occurred to anybody before? By digitizing part of the cable signal, we're sending more information down the line. Why can't we use the same principle for HDTV? Why can't we digitize part of the regular broadcast television signal, send more information over the air with less interference, and thereby increase the number of lines on the screen? We'd improve the quality of the picture. Wouldn't that be HDTV? Couldn't we enter the FCC's race?

Lying in bed, he jotted down some preliminary diagrams and equations. Back at work a few weeks later, he told his superiors about his idea, and they told Jerry Pearlman, Zenith's president. Everybody was enthusiastic. Best of all, it was cheap. They had already done the research. Entering the HDTV race would cost very little. So Pearlman decided, "We have some technology to take to the party."

All they had then was a theory—"vaporware," Luplow called it, using a common bit of industry slang. But the idea seemed sound enough. If it worked for cable, it sure ought to work for broadcast TV. They figured they could offer a TV picture with 787.5 lines, a 33 percent improvement. With that, in September 1988 Zenith formally entered Dick Wiley's race.

To Luplow it was scary. Zenith, he said, was "fighting NHK, Japan Inc., with all its money, and RCA with all its tentacles"—their bitter rival for half a century, still competing now, even if under different ownership. And with ACTV, the Sarnoff labs seemed to have the whole broadcast industry behind them. In addition, North American Philips, the Massachusetts Institute of Technology, Bill Glenn, and more than a dozen others were talking about entering the race. Still, most of the serious contestants planned to get a higher-resolution picture by broadcasting over more than one TV channel, just as the Japanese had. That caused Luplow to wonder, "Where are they really going to get that extra channel? And even if they get it, that's going to be messy."

Rich Citta was calling his proposal a "simulcast" scheme. In other words, during whatever transition period the FCC set, HDTV would be broadcast all by itself on its own separate TV channel—a neat, clean solution—while the conventional television signal would continue to be broadcast simultaneously in its regular place. The broadcasters liked that idea; they would get a second channel, continue to stave off Land Mobile.

In the broadcast engineering community of the late 1980s, notable innovations were rare. The profession wasn't attracting the best and the brightest any longer. Broadcast engineering conventions—hotel conference rooms full of middle-aged white men—had grown to have a certain sclerotic air. In that environment, Zenith's simulcast idea was greeted as a revolution—"a major, major step," Charlie Rhodes opined. He was a veteran engineer over at North American Philips, and to him the two-channel systems were "evolutionary" while simulcast was "revolutionary." As soon as they heard about Zenith's proposal, Rhodes and many others looked to Zenith "as the leaders." After all the dark times, the bad news, inside the Glenview Bunker that was sweet music indeed.

Soon enough, though, financial realities broke in on their reverie. Several years of losses had left Zenith with a mammoth debt, and by 1988 the company was reaching the point when it could no longer make the payments. Something had to give. Somebody had to go.

Almost despite itself, Zenith had diversified a bit by now. In 1980, the company had bought Heathkit, a little company that made mail-order TV, stereo, and computer kits. From that, Zenith had spun out a thriving little computer business. In 1988, Zenith Supersport portable computers were a hot product. If not for profits from that, Zenith's losses would have been far, far larger. The company would have gone out of business for sure.

Financial analysts began urging Jerry Pearlman to sell off Zenith's consumer electronics division, presumably including Luplow's team. It was nothing but a leaky boat, while Zenith's computer business was "the crown jewel," as the *New York Times* put it. Pearlman was tempted. Sure, Zenith held a certain revered position as the last American manufacturer of televisions, but romantic attachment wasn't paying the bills. The consumer electronics division was dragging the company toward bankruptcy, possibly even dissolution. Sell it! Sell it! Wall Street was crying. At a stockholders' annual meeting in the spring of 1988, that cry grew to be an angry demand. "We are bound and determined to explore all options to improve our corporate profitability," Pearlman promised the shareholders. "We plan to boost our earnings by reducing significantly the drag that consumer electronics has had on our earnings."

Then came Zenith's entry in the HDTV race. Suddenly the pressures began to reverse. Washington was reeling from its HDTV sickness. High-definition television was going to pay off the budget deficit, solve the trade deficit, end unemployment, bring forth the Messiah. Soon the infection spread to Wall Street. The analysts began thinking, Maybe HDTV can save Zenith, too.

"Now show us some vision, Jerry Pearlman," *Forbes* wrote. Sell the computer division. Compete in the HDTV race.

In the end, however, it wasn't "vision" that directed Pearlman's decision; he didn't have any choice. He hired an investment bank to put the consumer electronics business up for sale. Several foreign firms said they would bid. One European company—Pearlman wouldn't tell anyone who it was—called just two weeks before the bids were due. We want to be preemptive, they told Pearlman. Send a division vice president over here to sit down with us for two or three days and help us prepare our bid. Tell us how we should cover employee benefits, what other inducements we ought to offer to be sure our bid is best.

Pearlman sent one of his senior officers. But when that officer got off the plane and met with the company's leaders, he found to his dismay that the bidder's mood had suddenly changed. They weren't interested in submitting a bid any longer. Neither was anyone else. Pearlman was furious. He investigated and found enough information to convince him that the U.S. government had put out the word: Don't bid for Zenith; if you do, there will be significant trade ramifications. The United States cannot allow its last remaining television manufacturer to slip into foreign hands. Pearlman complained to his congressman, raged around the office. But there was nothing he could do. He was stymied. Even the U.S. government was against them. *Everybody* was an enemy. Truly, Zenith lived in a bunker now.

Pearlman made the only decision he could. Zenith sold its computer division to Groupe Bull of France for about $635 million, and Pearlman fumed about the hypocrisy. The way it was turning out, Washington was allowing him to sell his *profitable* division to a foreign company, while insisting that he keep the money loser. He announced that Zenith would use the money to pay down its debt and pursue new research—particularly the work on HDTV.

As Wayne Luplow walked around the building, employees would stop him, look up into his face, and ask with wide eyes and imploring tones: How's the work going? Please, tell us about the race. Please, tell us about HDTV. The consumer electronics division continued to lose money, and soon the cash from the Groupe Bull sale would run out. Competitors were disparaging Zenith's chances for survival. "We just want to know who's going to own what when Zenith sells off the rest of its parts," Joe Donahue observed. Up and down the halls of the Bunker, everybody was saying: We've just got to win the race. Only HDTV can save the company now.

In his heart of hearts, Pearlman knew that wasn't really true. Even if Zenith won, HDTV sales probably wouldn't bring significant profits for years. If Zenith wasn't turning a profit again long before then—well, the HDTV team would probably be working someplace else. Still, the enthusiasm was infectious; even Luplow's engineers got swept up by the mood. "I like Zenith; I've been here since 1975. I hope it can continue," said Wayne Bretl, a team member. "If we don't have high-definition, obviously we aren't going anywhere."

In the labs, Bretl and the others labored to turn Rich Citta's idea from vaporware into a working machine. The pace was frantic; they were in a race for survival. So Luplow stopped work on all his division's other projects, even those that were bringing profits. Every engineer in research and development turned to HDTV. Zenith was going for broke.

Reporters started coming by. It was a good story: America's last remaining manufacturer of televisions was fighting to set the standard for the next generation of television even as it flirted with bankruptcy and dissolution. John Taylor, the company's young public relations director, suggested to the engineers that they take the plastic protectors out of their shirt pockets before they sat down for an interview. All the attention was like a warm bath: Zenith was being recognized for its technology!

Soon after, layoffs rippled through the Glenview Bunker once again. Hundreds of employees were let go, some from almost every division. Only Luplow's group was spared. High-definition television had, as Luplow put it, become "the Great White Hope."

5 See if you can do better

Larry Dunham was no engineer, but he knew a hot business opportunity when he saw it. So when Stephen Petrucci, a vice president with Hughes Communications, asked him in 1988 if his company could do high-definition television, Dunham answered without missing a beat. "Of course," he said. "Absolutely"—all the while thinking to himself, What on earth is high-definition television?

Dunham was the president of a small company in San Diego that led the market in satellite television equipment for the home, particularly the boxes that decoded the signals from backyard satellite dishes. Hughes was planning to launch a couple of new satellites soon and begin offering a national television service in a few years called DirecTV. Subscribers who bought a small dish would get movies, sports, and other programming beamed to them from orbit, and Hughes was hoping Dunham's company—the VideoCipher Division, a subsidiary of a high-tech conglomerate called M/A-Com—would make the decoder boxes. But with all the talk about HDTV coming out of Washington, Hughes officials had started worrying: What if we put decoder boxes in millions of homes only to find that all of them are obsolete a year or two later because they can't receive HDTV? So they had called Dunham.

As soon as Dunham got back to his office in San Diego, he asked Jerry Heller, one of the division's founding engineers, what he knew about high-definition television. To Heller, HDTV was Muse, the Japanese system. He'd read about it but not given high-definition TV a great deal of thought. It didn't seem to have much relationship to his company's business. Still, as he focused on it now, the realization crept over him like a cold wave: HDTV could *destroy* us!

The problem was clear: The Muse signal was wide; it filled up more than one television channel. If HDTV were to be transmitted by satellite, the signal beamed to the ground would include so much additional data that a normal backyard satellite dish wouldn't be able to receive it. Any customer hoping to pick up high-definition television would need a dish the size of his living room!

Heller's little company was already loaded down with crippling problems. But if HDTV came along, he knew, the market potential for backyard satellites "goes down to about zero." The company had to do something. So he called one of his best engineers, Woo Paik, and asked him to go out and have a look at some of the American ventures into HDTV. Maybe we can work with one of them, Heller said. Maybe we can collaborate, come up with a joint HDTV proposal that's satellite-friendly. Paik readily agreed. He didn't know the first thing about television, but this looked like it could be an interesting new project. Still, to Heller at this moment, HDTV was just one more headache, and he despaired. This was the *last* thing the company needed! Already the VideoCipher Division was trying to crawl out from an avalanche of troubles. What a change this had been! Not so long ago he'd been gliding down a greased slide toward glittering success.

Everyone who knew Jerry Heller thought him exceedingly bright. No doubt about it. He was also a private man, soft-spoken and largely unknowable to his colleagues. "Not a relationship kind of guy," one of them said. Born and raised in Queens, he was of average height. Quiet, precise, and determined, he spoke with the barest hint of inflection, looking out with little expression from behind round wire-rimmed glasses. Heller seldom shared all of his thinking with others. His thick beard served as a partial mask. As a result, fair or not, some of his employees saw him as a puppet master, though one they generally admired and respected. After all, half a dozen successful companies had been spun off from Heller's earlier work. And until recently the VideoCipher Division had seemed to be at the very top of the world.

Heller got his undergraduate degree at MIT and stayed there for his graduate work. In the late 1960s, he was working on his doctoral thesis; it was a complex and ambitious bit of research on error correction and coding for communications with satellites and deep space probes—an interest of his academic mentor, Professor Irwin Jacobs. In 1969, midway through Heller's work, Jacobs decided to start his own company.

The professor moved to California, home of Jet Propulsion Labs, the company that was building many of NASA's satellites. Heller went along. He continued research on his thesis and worked for Jacobs's new company, named Linkabit. The company's mission fit naturally with Heller's own research: Linkabit would provide sophisticated digital communications systems for satellites and deep space probes. This was a cutting-edge technology in high demand.

A digital system offered distinct advantages, since the signal left earth as a series of coded digits, not the pulsating waves of conventional radio communications. That meant it couldn't be distorted as easily. Conventional radio waves could be bent and deformed as they encountered obstructions, and when that happened the signal would be distorted when it was received. That couldn't happen with digital signals. They could be destroyed, of course, but they couldn't easily be changed. Each bit of the signal represented either a zero or a one, nothing else. If a few of those digits were knocked out along the way, the others sailed on unaffected.

In 1969, Heller was Linkabit's only full-time employee. He'd leave the office late in the evening and, walking to his car, push his way through the long line outside the movie theater next door. The film was *M*A*S*H.*

The new company started slowly. But once Heller, Jacobs, and the others began marketing their ideas they found that NASA and the military eagerly paid them to improve the accuracy of earth-to-space communications. That wasn't surprising; it was far cheaper to improve signal quality than to build the huge satellite dishes needed to receive less robust transmissions. So Linkabit won important contracts to provide communications equipment for the Pioneer and Voyager missions to the planets. Later, Heller finished his thesis and received his Ph.D. as Linkabit began taking contract work from the Department of Defense, too. The company moved from Los Angeles to San Diego and began designing secure digital systems for communicating with submarines and strategic bombers, including the B-52s that circled the earth twenty-four hours a day, in case nuclear war broke out. As Linkabit won more and more big contracts, Heller, Jacobs, and the others had more work than they could handle. So in the winter of 1978 Heller returned to MIT to recruit some young engineers.

Woo Paik had come to MIT from Korea on a student visa, but by the time he was close to winning his doctorate, he decided to stay in the United States. America offered more opportunities. The problem was that he didn't have a green card, just a student visa, and when corporate recruiters started coming by, that proved to be a disqualifying handicap. Most of the big companies did contract work for the military, so their engineers needed a government security clearance, which was seldom granted to foreigners.

One morning Paik spotted a notice on a bulletin board saying that a representative from Linkabit was coming by for interviews. Maybe a former MIT professor would give him a break. Paik put his name on the interview list, but the woman in the recruiting office told him Linkabit wouldn't be interested. They had sent word ahead that they wanted only U.S. citizens, so she refused to give him an appointment. Well, put me on the standby list, Paik told her.

The day before the Linkabit recruiter was scheduled to arrive, snow started to fall, and by midafternoon a blizzard had paralyzed Boston. The airport was closed, and the recruiter's visit was postponed. A few days later, MIT mailed Linkabit a list of the students who had signed up for interviews, and Paik's name was sent out to San Diego with all the others. A few weeks later, Paik got a letter from Jerry Heller. I hope I can see you when I come, it said. Paik smiled; apparently the list hadn't said anything about his residency status. He took the letter over to the placement office. See, they're asking for me, he said. You have to schedule an appointment now. Paik got in to see Heller a few days later, and Heller was impressed.

Heller saw a big man. Paik had a square jaw and straight black hair. He also had a friendly, easygoing, confident manner. Paik's English was good enough, and his training was in exactly the right area, digital signal processing. He was clearly bright and obviously motivated. Heller offered him a job. Only after that did Heller think to ask about Paik's residency status. Paik already knew Heller wanted to hire him, so with a bit of cheek he said, "I assume you can help me with that." Heller nodded.

In 1980, M/A-Com, a high-flying company that had been buying up all sorts of small but promising high-tech firms, offered to acquire Linkabit. Heller, Jacobs, and the company's other principals agreed; being part of a larger company would bring more opportunity, and each of them stood to make quite a lot of money in the deal. After a couple of years, though, Jacobs left and eventually started a new digital communications company of his own. Meanwhile, Heller and the other employees continued winning contracts for satellite communications systems. Paik stumbled along at first, feeling he was in over his head. Heller asked him to do the engineering work for a new military communications modem, the device that translates the signal so that it can be transmitted. Paik didn't say anything, but he had no idea what a modem was; he looked it up and soon enough was deep into the work.

After a few years, Heller decided it was time to change his company's focus. Sure, they were winning lots of contracts, and they routinely wowed their clients. They were *stars*! The problem was that they would win a contract and do the job—great work, more often than not. But once they created a product, the invention belonged to someone else, no matter how brilliant their work had been. When the contract ended, they had nothing left. Really, Heller was just renting out engineers. What was the long-term profit from that? It was time they created their *own* product—something they could manufacture and sell. Heller didn't know what they ought to make, only that it should be something that took advantage of their expertise in digital communications. Beyond that he had few clear ideas. But it didn't really matter. By now the men of M/A-Com were seldom bothered by doubt. They were MIT graduates, the smartest guys around. Certainly they would dominate whatever field they chose.

Just before Christmas in 1982, Heller called Paik into his office and got quickly to the point. I've got a project for you, he said. Home Box Office has put out a request for proposals for a satellite scrambling system. This could be just the opportunity for us.

HBO sent its movies and other programming to cable systems across the country by satellite. A cable company would pick up HBO's signal with a big dish, then scramble the signal and send it out over the cable. Customers who paid for the service got a cable box set up to unscramble HBO's signal. The problem was that the cable companies weren't the only ones with satellite dishes. Two to three million homes across the country had dishes in their backyards, and all those people could pick up HBO for free. What was worse, hotels and other institutions in

Caribbean and Latin American countries were also picking up the signal—stealing it, HBO said—and distributing it as widely as they wanted. The company's paying customers grew to resent this. Why should we pay $8 a month for HBO, they asked, when everybody in Nassau is getting it for free? The complaints grew so loud that HBO decided to scramble the signal sent out from New York. Cable companies would be given an unscrambling device, while all the backyard satellite people, and all those foreign hotels, would pick up a scrambled mess. So HBO put out a request for proposals for a scrambling system.

Heller heard about the HBO competition late. Already fifteen companies had entered, almost every one of them mainstream firms with long histories in the broadcasting and cable TV industries: Scientific Atlanta, RCA, and Zenith, among others. Heller's company was only now emerging from the cloistered world of government contracting. To HBO, they were unknowns—underdogs, to say the least. And Heller had no experience in this arena. Neither he nor anyone else in his division knew much about cable TV. But they *did* know about sending coded signals to satellites. Wasn't that what this competition was really about? The men from MIT were full of confidence, not to say arrogance. They didn't hesitate.

The other competitors have already submitted proposals, Heller told Paik. We are way, way behind. Why don't you see if you can come up with something different. Sure, Paik said. I'll get started on it right after Christmas. In late December Paik came back and said he was ready to have a look at the HBO problem. Good, Heller said, because HBO is coming in to see our proposal on January 23rd. Paik struggled to keep his jaw from dropping. *Three weeks?* he thought, panic surging up from his belly. And New Year's falls in between! Paik didn't say anything. He had been with the company long enough to be accustomed to this. They always seemed to be underdogs with impossible goals facing crushing deadlines. But, the thinking always went, we'll win anyway because we're the brightest of them all.

In the end Paik didn't have to consider very long about how to approach the HBO project. He decided to use a coded digital signal, just like all those the company had created for NASA and the military. It was the natural thing to do. No big deal. And for the purposes of a proposal, Paik didn't actually have to make anything, just be ready to describe what he planned to do.

On the 23rd, the HBO officers gathered around a conference table.

Paik and Heller *wowed* them. Nobody else had suggested a digital solution. A short time later, they got the call: You've won! Now all Paik had to do was make the scrambler and unscrambler boxes, just as they had done so many times before, on contract for the government. It seemed so simple. The men from MIT would be stars once again. Little did they know. . . .

The people who bought backyard satellite dishes were a special breed. Most of them lived in isolated areas not served by cable television, and generally they were well off; a complete satellite system could cost thousands of dollars. Many of these people were western ranchers, gentlemen farmers—fiercely independent individualists accustomed to being left alone. To them, the satellite signals falling on their property from the sky were *theirs!* Nobody had any right to take them away.

HBO quickly discovered that these people were a powerful lobby. They had money. They voted. They made noise. Some of them even sat in the United States Senate. Barry Goldwater had a satellite dish, and so did a few other senators. An industry had formed around them: Dish manufacturers, dealers, and servicemen. Satellite TV magazines and guides, satellite TV talk shows—a thriving, diversified business serving people with money and power. A close-knit fraternity had grown up.

HBO began scrambling its signal in January 1986, and suddenly across the country dish owners started to howl. Right away, sales of backyard satellite dishes took a dive, so dealers began screaming, too. Nobody had ever riled these people before, but they were clenched-fist angry now.

Home Box Office decided to give them back their HBO—for a price. So the company gave Heller a second contract: Create a consumer box, a decoder for people with backyard satellite dishes. They'll buy the box, pay a monthly fee, and start receiving HBO once again. No problem, Heller and the others said. In fact, this seemed a wonderful opportunity. For the first HBO contract, they had been asked to make unscramblers for cable systems across the country. They'd called this device VideoCipher I, but it had a rather limited market—cable operators—that was unlikely to grow. Now they were being asked to make boxes for thousands of customers, and the market could grow to millions.

Finally, they had their *own* product with vast potential. At last they were breaking free.

They named their new product VideoCipher II, and now their offices came to be known as the VideoCipher Division of M/A-Com. Given how quickly the division would be growing, the parent company wanted to give them stronger business management. So M/A-Com hired Larry Dunham, who'd managed another company that had entered the HBO competition. Dunham was a big, enthusiastic man with a linebacker's physique. He grew up just outside Goodwater, Alabama, and had never lost his southern drawl. He was only in his mid-thirties, but his baritone voice belied his youth. He was smart and eager, and after he spent a few months at the company headquarters in New York, M/A-Com made him the president of the VideoCipher Division, even though some of his employees thought he was too young for that. But the division's engineers had larger problems to worry about. They had never made a consumer product before.

There had been few financial constraints in creating coded signals for the military. The Pentagon just wanted a signal that was secure, no matter how expensive or complex. The coders and decoders they had made for HBO were professional devices, so they could be relatively complex, too. Now they were being asked to make a box for home owners. That meant it couldn't cost more than a few hundred dollars, yet it had to be even more secure than the professional devices, because somebody somewhere would certainly try to bust it. Some compromise had to be found. In the end, they decided to send out sophisticated scrambling just for the audio signal, not the video. Anybody with a little technical knowledge could unscramble the picture by swapping a couple of wires inside the box. But Paik created a sophisticated digital coding system for the sound, and it seemed unbreakable. The coding was embedded in a single computer chip. Paik and Heller were confident that nobody could break into it. Certainly some people would figure out how to restore the picture, but who would watch TV with no sound? Nobody except maybe a few die-hard fans of sports, or pornography.

In April 1986, VideoCipher II boxes came onto the market, but that only made the satellite dish owners angrier. Other cable companies were now scrambling their signals, too, and as the dish owners saw it, a constitutional right was being taken away. Angry dish owners railed. Satellite magazines ran heated editorials; satellite TV talk-show hosts ranted and raved. And as soon as the VideoCipher II boxes began appearing in the stores, all that anger and contempt grew to be concentrated on just one target: the VideoCipher Division. Sometimes Heller, Paik, and the others would watch the satellite TV talk shows, and they were shocked. Nothing in their experience had prepared them for this. These people were *crazy!* One weekly show was nicknamed *Piss on VideoCipher,* and it opened with a sequence whose crude production was matched only by its vulgarity. A cameraman walked down a long hall toward the men's room, his camera at eye level pointing straight ahead—giving viewers the impression they were accompanying a man on his way to the bathroom. A hand came into the frame to push open the door, and then the camera moved toward the urinals along the back wall. Finally, you were looking down into one urinal, and in it lay a VideoCipher II box. A moment later, a yellow stream splashed over the faceplate.

As Heller and Paik watched with some distress, the focus of the discussion began to evolve from irate sputtering to practical, determined talk of revenge. The satellite dish community was intent on cracking VideoCipher. Ads for new chips began to appear in the magazines. Just pull out the VideoCipher chip, they said. Plug in the new pirate chips, and you'll get all of your shows again. These chips turned out to be fakes, and Heller didn't really think anybody could beat their box. Were some backyard hackers really going to outengineer a company full of Ph.D.s from MIT? Nonetheless, every time a report came in that a box somewhere had been compromised, the company had the unit flown to San Diego so it could be checked out. By autumn 1986, they'd sold tens of thousands of VideoCipher II boxes, and as far as they could tell, none had been beaten. It looked as though they had another hit product.

In the fall, the VideoCipher Division won an Emmy for technical excellence for its HBO work. The division also picked up an important suitor: it had all of a sudden become an important player in the cable TV world, and so General Instrument, a New York firm, wanted to buy it. GI, as the company was known, owned a diverse group of businesses, but the most important was Jerrold, the nation's largest maker of cable TV converter boxes. GI was offering a handsome price for the VideoCipher Division, and the division's principals—Dunham, Heller, Paik, and a few others—stood to become multimillionaires.

In September, they flew to New York to accept their Emmy. On the trip back, they posed for a picture in the first-class section of the airplane, standing behind the front row of seats on a 747. Heller held the award, Paik stood beside him, and the others clustered around, all of them smiling with cozy contentment. Everything seemed right with the world. Once again the men from MIT had proved they were smarter, better than everybody else.

On a Thursday afternoon in late November, a call came in to the VideoCipher Division from a dealer in Phoenix. A customer had just walked in, the dealer said, and told him about a guy who was selling pirate VideoCipher boxes. Go buy one, the VideoCipher worker told him. Send it to us by overnight express. Saturday morning, the package arrived, and a technician took the "pirate" VideoCipher box over to the lab and put it up on the bench. He removed the cover and peered inside. Look at this; there's wires hanging out all over the place in here. Looks like another scam. The technician screwed a coaxial cable into the back of the box, feeding in the scrambled satellite signal, and then connected another cable to a lab TV. The technician clicked on the TV, expecting to hear the same fuzzy scramble that always came up.

Not this time. Even before the tube had warmed up, a clear audio signal blared out of the speaker. Uh-oh. Something's wrong. And when he looked up, there was an HBO movie coming in sharp and clear—with perfect sound. He groaned as he reached up and switched through the other channels. *All* of them were clear. Somebody, finally, had found a way to bust the box. Time to send out the alarm.

Heller was at home when the call came in. So were Paik and most of the other VideoCipher executives and engineers. Heller was still wearing shorts when he showed up at the office a short time later. He went straight down to the lab and had a look. By now the workers had found the problem. Somebody had pulled out the special coded chip and substituted one of their own. It sure looked like a cheap knockoff. But the damned thing worked. Heller couldn't believe it. How could these basement hackers beat *us*? We're the best. The brightest. But they've outsmarted us with this dumb little chip! Around him, people were standing in tight clusters. They babbled—confused, nervous chatter. Everyone seemed lost, and in Heller's head the ivory tower suddenly started teetering. Then it fell to the ground and shattered. "Oh my god," he thought, panic racing through him like a fever. "They've broken it. They've really broken it, and we're *never* going to fix it!"

After Heller, Paik, and the others had calmed down enough to give the pirate chip a serious examination, they had to admit: Maybe it's crude, but this thing's actually rather clever. Under the VideoCipher system, when a viewer paid to subscribe to HBO, an authorizing signal would go out from the VideoCipher Division, up to HBO's satellite, then down to the subscriber's box. Inside the box, this signal would flip an electronic "switch" to turn off the scrambling for HBO. The creators of this pirate chip had designed it so that subscribers who ordered any pay service would get all of them. One authorizing signal would flip all of the switches. It came to be known as the Three Musketeers chip—one for all and all for one. And it spread like a virus.

The cable companies regularly sent copies of sales orders to the VideoCipher Division, and if these sales orders were to be believed, at the end of 1986 Americans were suddenly seized by an extraordinary interest in the news. In state after state, satellite dish owners were signing up for the Cable News Network and canceling all their other pay services. CNN, it turned out, was the least expensive service; it cost about $2 a month. But with the Three Musketeers chip, a CNN subscription would turn on all the other pay services too.

At the VideoCipher Division, crisis meetings started early in the day and stretched on for many hours. Out of that came a plan of attack. At the factory in Puerto Rico, the circuit boards going into new VideoCipher II boxes were coated with epoxy so nobody could pull out the chip. That certainly seemed to be a simple answer, the engineers believed, but they were wrong. Within just a few weeks, the satellite TV talk-show hosts began loudly broadcasting the solution: Just put your circuit board in the freezer. When the epoxy is frozen, it cracks, and you can pick it right off.

They tried countless strategies. On the air, the satellite talk-show hosts mocked and sneered. VideoCipher was a *joke!* And even as the division figured out how to deal with the Three Musketeers chips, the pirates perfected others. They began advertising videotapes that showed how *anybody* could break a VideoCipher box. The problem broadened daily, and soon HBO began losing patience. We hired you to fix the problem, they said, and you've made it *worse!*

Just as he had planned, Heller had moved his young company into its own new business. But now they were in the soup! And then, as if things weren't bad enough, the dealers began complaining about an entirely different problem: quality control. Untold numbers of VideoCipher II boxes were coming off the assembly line with defects. Even the *pirates* were bringing the boxes back to the stores because they didn't work.

"We went from a production level of zero in the first month to more than one million in the twelfth month," Dunham explained. "It was amazing the wheels didn't fall off." Now HBO and the other cable companies were howling mad. The situation got so bad, the satellite TV lobby's complaints grew so loud, that Congress held hearings and called Dunham to testify. "As many as half the decoders on the market have been illegally modified," he told the congressmen. Everything was falling apart.

During all this, the deal with General Instrument went through; it was already far along when the real troubles began. Now, however, the leaders of the VideoCipher Division had new employers, who worried, What are we to make of our new acquisition now that it's the bête noire of the industry?

The men from MIT had lost their confidence. When Dunham went to bed each night, he lay there in the dark "wondering if the business would even be there in the morning." Sometimes Heller, too, found himself seized by the panic. "My God," he would think. "This business could just go away!" And then, when their mood could hardly have been worse, along came a new problem, an even greater threat: *high-definition television.*

The division's clients were angry now, but by and large they remained customers. But that might not be so after the advent of high-definition television. With its wide signal, HDTV could destroy the whole satellite-dish industry in a stroke—"kill a market of three to four million people," Heller observed.

What on earth could be next? The bubonic plague?

Woo Paik spent about six months looking over the HDTV proposals others were putting together, visiting labs, studying schematics. Then early in 1988 he came back and told Heller, "I don't like any of them." Every one of the proposals was so complex that the signal could not be received on a satellite dish smaller than twenty feet in diameter, and Paik could see no way to improve them. If any of these systems was chosen, sooner or later the VideoCipher Division's core business, such as it was, would dry up and blow away.

Heller thought about that for a minute. Even if somehow, someday, they managed to win the pirate wars, high-definition television lay in wait. There was only one possible solution. They'd have to create their own HDTV system for satellite. What choice was there?

Heller told Paik, "Woo, why don't you see if you can do better."

Paik got up and trudged back down to the basement.

6 It must be digital, it must be digital!

From his corner office perch in the headquarters building I. M. Pei had built for him, Nicholas Negroponte looked down on the high-definition television race and found it wanting.

Negroponte was a college professor, but he saw a grander place for himself in the world. He wanted to be known as a seer, the nation's oracle for high technology, and he'd made quite a life for himself at that. In the early 1980s, Negroponte and his colleague Professor Jerome Wiesner had coaxed the Massachusetts Institute of Technology to let them build a fancy $45 million postmodern headquarters, even though it couldn't have seemed more alien to its environment, at the very heart of the MIT campus. By promising cutting-edge research in the broad field of communications and the media—electronic newspapers, holographic movies, and the like—Negroponte and Wiesner had convinced an assortment of companies to bankroll their new home: the MIT Media Lab. Annual budget: $6 million.

Out of all this, particularly after Wiesner retired, Negroponte lived the life of an academic grand vizier. He traveled to Paris, Tokyo, Los Angeles, almost anywhere he pleased, to meet with his corporate sponsors and speak of his visions. At each stop, a chauffeur greeted Professor Negroponte as he stepped off the airplane, and whisked him in a limousine to the city's best hotel. All this was his due, Negroponte believed, and he wasn't shy about any of it. One of his aides, Suzanne Neil, said the professor liked to tell them, "I have enough arrogance for several people."

Negroponte's secret: He forecast the future. Or, more accurately, he placed demands upon it. "Nick always has to be five or ten years out," as Russ Neuman, one of his acolytes, liked to put it. In other words, Negroponte told everyone who would listen what he thought the shape of things *ought* to be a decade hence. That way he could reasonably ask his corporate sponsors to pay for the research that might help move the nation toward his selected goals. If reality ever caught up with his predictions, well, Negroponte would drop interest in that area and move on to something else. The present was always lacking. It had to be. Was a futurist of any value if he found today's state of affairs agreeable? Who would give the Media Lab money if Negroponte professed satisfaction with things the way they were?

From that point of view, the Grand Vizier turned his withering stare toward HDTV. The Media Lab was home to MIT's Center for Advanced Television Studies. Its director was William Schreiber, the professor who had tangled with Masao Sugimoto, the father of Muse, at the first congressional hearing on HDTV in 1987. High-definition television was a natural area for Negroponte. Here, too, he came up with a futurist demand.

"HDTV is a lot of hot air," he liked to say, and the quote was printed in magazines and newspapers over and over again. "There's something better—digital TV." The FCC, he would add, should not choose a winner until somebody had invented digital television, which could lead to "the complete integration of computers into television sets."

None of his rhetoric seemed to conflict with the work going on in his own lab. Luckily for Negroponte, Schreiber was pursuing pure research. Through the mid-1980s, he had directed his students to stay away from actual system designs, so the Grand Vizier was free to promote his futurist vision for HDTV without seeming to criticize Schreiber. And for Negroponte this formula was perfect, really *perfect*, because creating a digital television was impossible. Everybody knew that. "We'll have digital television the same day we have an antigravity machine," Joe Flaherty, CBS's senior vice president for technology, told an industry conference in 1989. So there seemed to be little chance that the present would ever catch up with Negroponte's vision.

In Washington, meanwhile, Al Sikes had promised to put HDTV "on the front burner" as he took over leadership of the FCC in August 1989. But like Dick Wiley, he wasn't entirely satisfied with the contestants in the race. Most of the proposals seemed like compromises; none was clearly better than, or even as good as, Muse, the Japanese system that got the whole process started. The engineers on his staff and those who had entered Wiley's race were telling him this was the best that could be done. But Sikes wasn't so sure, for he began hearing the siren's song from the north: *It must be digital, it must be digital!*

Sikes had never met Nicholas Negroponte, didn't know the first thing about the man. "But you come in here from the Midwest," he said, "and MIT—well, that's a big deal." If Professor Negroponte said digital TV was the thing, maybe Sikes had better take a look.

Up on Capitol Hill, Representative Ed Markey, the man who had held the first HDTV hearing, wasn't sure what to believe, either. MIT faculty lived in his district, so Markey had a strong relationship with the school's technologists. Even after other members of Congress, stung by the Bush administration, had dropped their interest in HDTV, Markey stayed with the issue. So one day early in 1990 MIT invited him to Cambridge for a conference on HDTV, and he found himself seated at a table in front of an auditorium packed with students. Negroponte was beside him, along with several of the professor's protégés. All afternoon Markey, too, heard the digital call.

Markey chose to be "agnostic on this," he recalled. Unlike Sikes, the congressman *did* know a bit about Negroponte and the others. But Markey also thought: As we sit here, there aren't any companies working on digital. Are we going to go forward with this HDTV race and have digital come in at the last minute? As a "TV junkie," he compared this eventuality to a late-night Perry Mason special: "Is Della Street going to come in during the very last scene with the critical clue when it's almost too late? We don't want that." When Markey got back to his office, he told Larry Irving, his aide, "I want the FCC to do an inquiry into digital technology." A few days later, a letter went out to Sikes:

In recent months, there has been increasing discussion about the prospects for digital video. According to some analysts, over the next several years digital video technologies will improve, thereby accelerating the convergence of the television and computer industries. These analysts further suggest that the "digitalization" of video, or the movement from analog to digital in video technologies, will have profound repercussions, not just for broadcasting applications, but for our entire telecommunications infrastructure.

So, Markey's letter went on, "I believe it is important that the Commission initiate an Inquiry to consider the implications of this new technology."

Sikes wasn't interested in starting some sort of special inquiry, but he was curious about this digital business. He asked Dick Wiley to find out if digital television was possible, and whether the FCC should wait for it. Wiley, in turn, summoned several of the major HDTV contestants to his office for a talk. By now the field had been narrowed to seven finalists. Earlier, one of Wiley's subcommittees had interviewed everyone who wanted to compete, fourteen groups with twenty-three different system proposals. The inquisitors had carefully weeded out the proposals that seemed unlikely or impractical. When "Hell Week," as the contestants called it, was over, Zenith, Sarnoff labs, NHK, William Schreiber of MIT, the Dutch company Philips, and two others remained as finalists.

Wiley sat some of these people around the conference table across the hall from his office and asked them whether there was any chance that a fully digital proposal would come along soon. Around the table, one by one, each of them shook their heads as they told him:

"Impossible."

"Never."

"Not likely."

"No way."

"Sorry, Dick."

By now, in early 1990, Wiley had been chairman of the Advisory Committee for more than two years, and the start of testing for the HDTV prototypes was still many months away. He certainly didn't want to delay the race for an indefinite period, waiting for "a vague development," as he put it, that might never come. Wiley adopted a saying that summed up his view: "Analog in the '90s; digital in the next century."

For him, that seemed to settle the issue. But softly in the background, Al Sikes still heard the siren's call: *It must be digital, it must be digital!* He wasn't altogether sure that the Grand Vizier was wrong.

All their protests to the contrary, several of the major American contestants were debating digital. In the Glenview Bunker, Zenith was already working on an HDTV system that digitized a small part of the TV signal. The company was so shriveled and weakened, however—the latest annual report showed a $52 million loss—that Zenith just didn't have the manpower to build it. Wayne Luplow's team scrambled day after day, seeming to fall further and further behind, trying to do the work with phantom bodies.

Jerry Pearlman, Zenith's president, saw the problem clearly and realized that there was only one possible solution. Pearlman started looking for a partner. After a short search, he settled on Bell Labs, of Murray Hill, New Jersey—the fabled research arm of AT&T. Zenith already had an agreement with AT&T; the company was going to make some of the semiconductors for Zenith's new TV. Hiring Bell Labs to build a small but complex part of their proposed system seemed a natural step. For their part, AT&T's leaders loved the idea. They had flirted with HDTV back in 1988 and 1989 but found no easy way to enter the race. Though AT&T had vast experience in several of the technologies that go into television, the company didn't know much about the TV business. So AT&T welcomed the partnership with Zenith, the last American manufacturer of TVs. All in all, the leaders of both companies considered the partnership ideal.

As they would soon find out, however, they were wrong. This was a marriage arranged in Hell—a Vanderbilt suddenly engaged to a butcher. Down in the labs of both companies, resentment began to seethe from almost the first instant, and in time it would molt into loathing.

Pearlman had worked out the partnership with Saul Buchsbaum, executive vice president of Bell Labs. Buchsbaum, in turn, passed the assignment on. One afternoon late in 1989 he bumped into Arun Netravali, one of his most valued engineers, in a cafeteria food line. Netravali, a first-generation Indian American, was a warm, pleasant fellow with a sharp mind. He'd been with Bell Labs since 1972 and was appointed director of the Visual Communications Research Department in 1978. More recently, he had been working as an adjunct professor at MIT's Media Lab, so he was quite familiar with the HDTV race. Netravali also had little trouble in theory with Negroponte's mantra: *It must be digital!* He didn't know much about the TV business, but Bell Labs had done a lot of work on digital coding for videophones and other projects.

Standing there in the cafeteria line with plastic food tray in hand, Buchsbaum asked Netravali if he'd like to run the Zenith project. Netravali agreed without hesitation and handed primary supervision over to Eric Petajan, promoting him to the position of engineering supervisor. But Petajan didn't greet the assignment with the same equanimity as his boss. The new supervisor wasn't sure he liked the idea of serving as Zenith's *subcontractor.*

Petajan was a determined young man with long but thinning ash blond hair drawn into a ponytail, a blond beard, and cold blue eyes. Blessed with a quick mind but possessed of a sharp tongue and an acid manner, he was certain of his place in the world and not at all shy about letting others know what it was. True, he was just one of about twenty thousand engineers employed in Bell Labs facilities around the world. He worked in a small, unremarkable office along one of the mile-long hallways that coursed through the massive Bell Labs headquarters, punctuated only by the emergency sprinkler heads that dropped from the ceiling every hundred yards or so. But this was *Bell Labs.* Petajan knew that engineers and scientists in these very same cookie-cutter offices had invented the transistor, the laser, the communications satellite, solar cells, light-emitting diodes, cellular radios—more seminal inventions than anyone could remember. Bell Labs scientists had won seven Nobel Prizes and five National Medals of Science. Petajan knew he worked for the premier research institution in all of world history. Now he was going to be a subcontractor of sorts, an employee of a bankrupt little television company that in its entire existence hadn't invented a single notable thing he could think of? This wasn't the way it was supposed to go. But Petajan was a team player. He would make sure his group did the best job it could. They'd show Zenith what kind of work *premier* engineers could do.

Zenith's work order arrived at Bell Labs early in 1989. It was a schematic diagram of the proposed HDTV system, showing which part AT&T was supposed to build. Looking at the drawings, Petajan saw "just this little box." In truth, though, that little rectangle represented quite a complex part—a small computer, in essence, that provided digital motion estimation, a function that lay at the heart of any sophisticated video signal.

TV pictures can include a great deal of motion—cars racing, people walking, starships flying through space. Usually, however, the motion is predictable. A car tearing through city streets usually travels in a relatively straight line; a space vessel flying across the galaxy seldom makes a sharp right turn. If a computerized TV system can make logical estimations of how the picture is likely to change from one frame to the next, it can operate more efficiently. A computer might look over a few consecutive frames of a car chase and guess that in the next one the car was likely to move a little to the right. So it would send a single message—"move existing car to the right"—instead of a far more complex message re-creating the entire car from scratch. If the computer guessed wrong, it could send a correction in the following frame, and it would all happen so fast that nobody would notice.

This is digital motion estimation. It improves efficiency so that more room on the airwaves is available for data that improves picture quality. AT&T's charge was to build a motion estimator for Zenith's system, adding another digital element to Zenith's partly digital TV. The way Petajan saw it, "If Zenith could have built this, they would have. But they came to us for a reason. They realize we have decades of experience in video coding." So Petajan assembled a team and plunged into the work. Maybe Bell Labs didn't know much about the TV business, but Petajan's team was going to build the best digital motion estimator a television manufacturer had ever seen.

At first the Zenith engineers found their new partners baffling. When people in Zenith's lab came up with an idea, they would often write it down on a scrap of paper snatched from the floor, then pass it around. They were only slightly less informal with their new partners. But every time AT&T wanted to share a thought, polished formal papers would arrive at the Glenview Bunker—bound, laser-printed volumes with scaled fonts, "ready for the Institute of Electrical and Electronics Engineers journal," Tim Laud, one of Luplow's engineers, liked to say.

When the AT&T engineers graced Glenview with a visit, they could not have seemed more foreign if they had arrived in a space vessel. Petajan, the team leader, wore a *ponytail,* and boy, did he have an attitude! Besides, these people didn't seem to have any idea what they were talking about. Every time a theoretical problem came up, Petajan's men would confidently promise to solve it with a diverse assortment of sophisticated electronic parts. Luplow's men would listen wide-eyed, mouths dropping open. As Laud put it: "Look at that architecture; it's got six thousand components on it!" Ray Hauge, a twenty-year Zenith veteran, was thinking: That's great. It'll work if you're building a space probe. But this is a consumer electronics product.

When engineers design a new electronic device, they build the prototype with dozens and dozens of inexpensive, off-the-shelf chips and integrated circuits. When the work is finished and the final product is designed, one or two highly sophisticated integrated circuits, or "ICs," are designed to carry out the functions of all those cheap prototype chips. For Hauge and many of the others, the bottom line was: "All this looks good on paper, but in the end they've got to get all of this down to a $7 IC. The way this is looking, it'll be great if it's a $700 IC."

Watching the Bell Labs boys lard their design with ever more expensive, complex components, the Zenith engineers grew to believe that Zenith didn't really need digital motion estimation at all. Sure, Zenith had digitized a small part of the proposed system, but at an industry conference in the fall of 1989 Pearlman, their president, had made clear his view of fully digital television. "By the time such a system is implemented," he said, "an analog HDTV system may have become entrenched and upgraded to the point" that a fully digital system would not be considered worth the price.

Luplow and his men started thinking: AT&T's digital estimator is just too expensive. We don't need more digitization than we already have. Those AT&T prima donnas don't know anything. And besides, we're getting really sick of their condescending attitude. "They have the perception that the consumer electronics industry is the bottom of the barrel," Dennis Mutzabaugh, one of Luplow's engineers, said bitterly. "And they're the top of the heap."

At Bell Labs, Arun Netravali and Eric Petajan had like misgivings about Zenith. The company simply came from a different world, and not one they particularly respected. Up and down the horizonless hallways, Bell Labs engineers saw their job as challenging assumptions, trying the impossible. But the Zenith people, Netravali complained, "are in the business of counting pennies."

No matter. At the Glenview Bunker, Luplow's team decided to build a whole system of their own. Who needed digital motion estimation? Their device would do without it and be every bit as good. At one of those poisonous partnership meetings, the Zenith team let it slip. "We're looking at other ways of doing things," one of them said. Petajan and the others weren't entirely sure what that meant. Zenith and Bell Labs had installed parallel facilities in Glenview and Murray Hill, so that Zenith could examine AT&T's work, and Petajan's team had already begun sending Zenith videotapes that showed how they were progressing. If Luplow's men wanted to try their own approach, Petajan knew, they could study everything AT&T had done until now.

Then one morning early in 1990 a package arrived in the mail at Bell Labs—return address: Zenith Corp., Glenview, Ill. Inside was videotape and a note: Why don't you have a look at this? Down the hall a short time later, Petajan's team assembled around a TV monitor. One of them slipped Zenith's tape into a player and punched "play." There on the screen was a Chicago street scene showing tulips in a park, traffic, pedestrians, office towers. Here was a demonstration of Zenith's approach, with no digital motion estimation. And without hesitation every member of the AT&T team went on the attack: Look at that—it's terrible. It's fuzzy. See the lines? Where's the resolution?

This is a *dog!*

Right away Petajan and Netravali prepared a videotape of their own—not just a casual research update of the sort they'd been giving Zenith before, but a clear demonstration that their approach was *better.* They sent it off to Glenview thinking: We'll show them. There's no way their analog ideas can compete with our digital approach. Have a look at this, and *nobody* will be able to argue that digital isn't best.

With that, the tape war was launched. Luplow's men offered their own merciless appraisal of the AT&T tape. Then the Zenith team set out to make a *better* tape. Soon both teams flung themselves full force into "trying to break the other's system, trying to kill each other," as Laud put it. Both companies were "doing biased tests to make each other look bad," Pearlman realized. He and Saul Buchsbaum spoke on the phone several times a week, trying to keep the tape wars under control. The engineers were throwing nearly all their energies into this rivalry, and work on the rest of the system was falling behind. But for Pearlman, the bottom line was: "We really believe our analog approach looks just as good. And it's cheaper."

Netravali and Petajan were just as certain that digital was inherently superior. Still, as the videotape packs jet-expressed through the air, Petajan started getting nervous. We're not in a secure position, he was thinking. We are basically a subcontractor to Zenith. Even if Bell Labs won the tape wars, Zenith could still decide to use its own approach.

At the Princeton Shrine, work on ACTV was moving rapidly forward, and the old RCA publicity engine continued to promote it, rumbling on at 6,000 rpm, spewing clouds of noxious fumes in its wake. Meanwhile, NBC executives were still twisting arms wherever they could. Like everyone involved with the project, Glenn Reitmeier, an engineering team leader at Sarnoff, believed that "ACTV makes a lot of sense," even knowing that it offered only a bare technical advance. He understood that ACTV meshed nicely with the business and political needs of the broadcasters and TV manufacturers.

A native of New Jersey, Reitmeier had been with Sarnoff for more than a decade and had shown his talent early with several important television patents. He was a loud, even pugnacious fellow who pursued arguments with predatory tenacity. He wore the RCA arrogance as if it were a custom-fit suit.

In 1989, Reitmeier believed digital was "inevitable, it's going to happen." He quarreled with the engineers who said digital TV was so far in the future that it wasn't even worth contemplating now. Reitmeier's view was that "it's maybe ten years out." Though Reitmeier's division was working exclusively on ACTV, elsewhere at the Shrine engineers were using digital video images for other projects—video teleconferencing systems, computer-vision designs, and related research. Why couldn't the television team learn from these other people? Casually at first, Reitmeier asked those engineers about their work. Then he began some exploratory experimentation—computer simulations and the like. Really, he was just dabbling, but over time Reitmeier came to believe that digital television might not be quite the futurist fantasy everyone assumed.

"Maybe this can happen sooner," he concluded. "Maybe this can be a deployable technology by the middle of the nineteen-nineties," six or seven years hence. He had no conclusive proof, just some mathematical calculations and preliminary theories, but toward the end of 1989 he took his ideas to Jim Carnes. Let's pursue this, Reitmeier said. We can go forward with ACTV now, while also making a major push to create a digital system to be introduced later.

Though Carnes was certainly not one to stifle innovation, he was torn. The most important inventions in the history of broadcasting had come out of the Sarnoff Shrine, but now the labs had a different role. To survive, they had to recruit clients and produce salable products and concepts. The days when the Shrine's engineers could devote most of their energies to uncertain, long-range research were gone.

More important, at the end of 1989 the Sarnoff Center was on the verge of signing a very important new client—Philips, the huge Dutch corporation that had invented the compact disc and was a major European power in consumer electronics. Philips was a contestant in the HDTV race, too. The company's American labs, just outside New York City, were working on a system similar in some ways to ACTV. Actually, it was closer in design to HD-MAC, the high-definition system that a consortium of companies including Philips was putting together in Europe. Started three years earlier and swimming in government money, the HD-MAC research consortium was Europe's great hope for beating Tokyo.

Though Philips had a large presence in the United States, the company decided they'd have "more chance of success" by joining with Sarnoff, said Peter Bingham, a former RCA executive who headed Philips's HDTV program in America. He and Carnes reached a tentative partnership agreement, and Carnes didn't want to jeopardize that deal by starting a competing line of research. Philips was fat and rich. With Philips as a partner, *nobody* could match this consortium's strength, and Sarnoff's future as an independent lab would seem far more secure.

Still, Reitmeier's idea was tempting. Maybe the General's labs could show the way once again. . . .

A final strategy meeting with all the partners—Sarnoff, Thomson, NBC, and now Philips—was scheduled for early January 1990, so Carnes invited Reitmeier to attend and make his case. They met over lunch around the Queen Anne table in the General's private dining room: Carnes; Reitmeier; Michael Sherlock, of NBC; Joe Donahue, now a senior manager for Thomson; Peter Bingham and Larry French, of Philips. Everyone at the General's table was a former RCA employee. Carnes used to work for Bingham when they were at RCA. Bingham used to work for Donahue. And at one time a few years before, French had been an RCA vice president and senior to all the others. So everyone here knew everybody else's game—Larry French not least of all, and he was skeptical of the whole enterprise. ACTV "is a real loser," French believed. "They're always demonstrating ACTV as part of the HDTV race. But ACTV never was HDTV. This is truly trying to sell hamburger as steak."

Coming into this meeting, French held the deciding vote. Everyone at the table wanted Philips in the consortium, and he was the company's senior officer. Still, when he thought about the tortured politics holding all these companies together, he realized that he didn't really have many options. He knew full well that NBC and all the other broadcasters "really don't give two hoots about HDTV. All they care about is getting a second TV channel" to stave off Land Mobile. Now here he was at the General's dinner table, and French knew he would be urged to abandon his company's own system and accept ACTV. He'd already come to terms with that idea.

Carnes gave the floor to Reitmeier. The young engineer talked about his digital research, told the others he thought it held real promise, and then gave the pitch, saying, "OK, folks, let's stick our necks out here. It's time for a bold, revolutionary step. Let's do it. Let's start a highly focused, fast-running program to do an all-digital HDTV system. Let's go for it. Maybe it's five years out, maybe more. But I'm convinced it can be done." ACTV today, digital tomorrow—but only if we pour real resources into digital research now.

French couldn't believe it. Digital—this is just concepts, pieces of paper, he was thinking. They don't really have anything, just fuzzy work. Besides, the real game here is more spectrum. I don't know if any of us is ever really going to make any money on HDTV. Now Sarnoff is proposing to spend more money on something even *more* speculative? No way!

French got up from the table and called Carnes, Donahue, and Bingham out to the hallway. They stood in a semicircle. On the walls around them, the General's signed portraits gazed at them: Marconi, Edison, Einstein, and the others in whose ranks Sarnoff had believed he belonged.

Look, French said, we're not going to get into this digital business. If you want us as partners, we're going to have to go with what we have, and that's ACTV. The others nodded. They went back into the dining room and announced the decision. Reitmeier was so disappointed that Carnes was afraid he might quit. But the deal was done, and it was time to announce the new partnership—the Advanced Television Research Consortium, as the group chose to call itself. A few days later, trumpets could have blared as the announcement went forth. "We are making a very large first step in forming a major consolidation of all the HDTV proponents," crooned Sherlock, the NBC executive. If the other contestants will line up with us behind ACTV, "then it'll be just us against the Japanese."

Others began to see a problem with that line of nationalist reasoning. Jim McKinney, who'd recently left the FCC, noted that with Philips and Thomson as the senior partners, "this new consortium certainly seems to have a heavy European orientation."

7 *Let's go for the gold*

No one was surprised when William Schreiber entered Dick Wiley's race and eventually became a finalist. As he neared sixty-five years of age, Schreiber was a genial little man just over five feet tall, with a full head of white hair—a respected college professor with a coterie of former students scattered throughout the broadcast engineering world. Since 1983, he'd been the head of MIT's Center for Advanced Television Studies, known as CATS. And as Schreiber knew only too well, hardly anyone in the United States had been working on high-definition television longer or with more intensity. In fact, to many people Schreiber was the dean of academic broadcast engineers.

In 1982, Schreiber had already established his reputation as an expert in image processing when the television networks and several TV manufacturers began shopping for someone to lead a broadcast engineering research consortium. The industry was already withering. Often when a network executive encountered a technical problem, he had to drum his fingers on the desk until evening, when he could place a call to Tokyo, where the workday was just beginning for the engineers who held the answer. As a result, in 1983 NBC, CBS, ABC, PBS, and several major equipment manufacturers each started contributing $100,000 a year to create the Center for Advanced Television Studies. Schreiber often complained that $800,000 or $900,000—it varied a bit from year to year—was "less than the networks spend on one hour of prime-time TV." But as an academic he was accustomed to working with limited resources, and the mission of CATS was quite similar to his own: Get top students interested in broadcast engineering and explore the potential for new, advanced forms of TV broadcasting.

But the broadcasters and other sponsors had another, overriding credo: this is to be an *American* effort, they insisted. We want to breed *American* engineers, create *American* innovations. We're tired of calling Tokyo every time we have a problem. When other American companies—Zenith, General Instrument, and HBO—said they might be interested in joining, they were welcomed into the CATS fold. But inquiries from Japanese companies were routinely turned away. Schreiber didn't have any particular problem with this, if it was what his sponsors wanted, though he was hardly a xenophobe. He had happily worked with Sony on an earlier project. Now, however, he went along with his sponsors' wishes. CATS seemed an exciting opportunity.

Schreiber spent the first year or so slowly building up his equipment, but once that task was complete, he gave his students a clear instruction: Nobody here is to try to design a television set. "For now," he said, "let's look at what's out there and see what the problems are." Part of the reason was that he hadn't yet settled a disagreement with his sponsors. Led by NBC, they were insisting that when Schreiber did design an actual system, it had to be compatible with existing TVs. In other words, conventional TVs had to be able to receive the signal, though new sets would be able to receive the enhanced picture. Even then, before Dick Wiley's race had started, before ACTV had begun to take shape at the Sarnoff Shrine, an NBC vice president had written a letter to Schreiber insisting that he had better not spend "one nickel" of the sponsors' money on any research that would lead to a noncompatible TV system. Most of the other CATS sponsors agreed; they weren't interested in creating a system that would render their equipment obsolete, cost them a lot of money. But Schreiber saw this as "a really bad idea"—one that would cripple any new system, because the next-generation TV would have to work within the narrow technical limitations of today's televisions.

Schreiber was a master of the memo, and he wasn't hesitant about sending them. He sent so many of them to so many places—long, typed documents with footnotes detailing his views on a host of technical and political issues—that some people wondered how he had time to do anything else. Over the years, he sent more than a hundred of these Schreibergrams to his sponsors, and in many of them he argued that making a compatible HDTV system was not only stupid but largely impossible. How could it be high-definition *and* compatible with standard-definition television? Can't be done.

A few buildings away, meanwhile, Nicholas Negroponte was busily building his own empire, the Media Lab. And in time the Grand Vizier decided he wanted to bring the CATS program into his domain. He began courting Schreiber with blandishments and bribes: a large lab, a powerful computer, copious equipment. Schreiber's research had already grown to the point where it strained the resources of his department. So when "Nicholas made me an offer I couldn't refuse," Schreiber agreed to move. Later, Schreiber would add, with a rueful smile, "Maybe I just didn't know him well enough."

By 1986, Schreiber had moved most of his equipment and students over to the Media Lab. Negroponte paid him little notice, though on occasion he did help Schreiber obtain equipment. Once, Schreiber needed a dozen expensive computer disks. Negroponte said he'd see what he could do, and the next thing Schreiber knew, Fujitsu of Japan had offered to donate several of them, worth about $500,000. The way Negroponte tells the story, Schreiber would not accept the disks because they were Japanese, saying that his sponsors would be furious. But Schreiber says he refused the disks only because Fujitsu wasn't willing to provide as many as he needed, not because they were Japanese. Whatever the reason, Negroponte was angry and embarrassed. Fujitsu wasn't going to understand why the Media Lab was turning down this gift, and in Negroponte's view the very future of the Media Lab depended on an ever closer relationship with the Japanese. Already Japanese companies were important financial sponsors of the lab's work. Several Japanese scientists were actually in residence at the Media Lab. They wanted to replicate it back in Tokyo.

Schreiber wasn't paying much attention to any of this. He was lost in his own world—pushing his research, writing papers, dashing off memos. By the late 1980s, he had finished the exploratory phase and had begun designing an HDTV system. The steady stream of Schreibergrams had left the sponsors weary, and when Zenith introduced its noncompatible system Schreiber finally convinced his board that he ought to be allowed to design one. So the professor began putting together a system that was similar to Zenith's in some ways. It relied on "simulcasting." The HDTV programming would be broadcast on one channel while conventional TV broadcasts continued on a different channel. And like Zenith's, Schreiber's proposed system was partly digital. He and Zenith argued about who really introduced these ideas first.

"Zenith stole my idea but took out all the good stuff," Schreiber liked to say, while at Zenith Wayne Luplow had his own view: "I have my doubts about how much technology or ideas they really have. And a bunch of grad students building something this complex?" He gave a derisive laugh.

In any case, with his new proposal Schreiber entered Dick Wiley's race and managed to survive Hell Week, the Advisory Committee inquiry in 1988 that narrowed the number of entries from more than twenty to seven. With that, Negroponte began taking note of this embryonic idea growing up in his own lab. From his pulpit the prophet was preaching the gospel: "HDTV is a lot of hot air," because *it must be digital. It must be digital!* Yet an engineer in his own lab was working on an HDTV system, and it was partly analog.

Meanwhile, Negroponte's Japanese sponsors were complaining about another piece of work under way at the Media Lab. Russ Neuman, a political scientist who had a particular interest in communications technology, had carried out an audience reaction survey in a local shopping mall. Neuman's group had set up an HDTV demonstration on the mall's concourse using NHK's Muse system and asked random shoppers to stop by, have a look, and say what they thought. Most people hadn't seemed particularly interested, so Neuman began saying HDTV was a flop. This infuriated the Japanese.

High-definition television boosters complained that Neuman used small-screen TVs so that the difference in picture quality was not readily apparent. Lighting and viewing conditions were not particularly good, they added. They also pointed to other studies that had produced markedly different results. So, they said, Neuman's study was invalid. Regardless, working out of the Media Lab, Neuman had created a little cottage industry for himself as the HDTV naysayer. He was invited to offer the contrary view at countless HDTV hearings and conferences nationwide.

Everyone could see that the Grand Vizier didn't like it. After all, noted Suzanne Neil, who helped with the shopping mall study, "Sony was one of his great sponsors." Sony was also one of the leading manufacturers of Muse equipment, and the conflict was clear. Sony officials wanted to have a talk with Negroponte. While he was on a trip to Tokyo, a senior Sony officer stopped by his room at the Imperial Hotel and told him, as Negroponte recounted the conversation later, "You have to get rid of Bill Schreiber and his colleagues, because they are dabbling in things they don't know anything about."

Actually, Negroponte explained, the Sony officer was more concerned about Russ Neuman than about Schreiber. "But I told him we welcome a plurality of views. That's academic freedom." Still, this HDTV business was growing awfully messy, and Neuman says Negroponte was "a real control freak." Finally, the Grand Vizier concluded, we must invite the Japanese into the tent.

The corporate sponsors of the CATS program renewed their contract with MIT every three years, and the last renewal had been in 1986, while Schreiber was moving to the Media Lab. Now, in 1989, the CATS contract still listed Schreiber's old department as the formal home for the program. As the renewal date approached, Negroponte began urging Schreiber—"beating on me," the professor says—to move the administrative home over to the Media Lab. Negroponte nagged and badgered, and finally Schreiber agreed.

Formally moving CATS to the Media Lab would give Negroponte the leading role in deciding terms for the new CATS contract. That didn't bother Schreiber; the two men had maintained a reasonably cordial relationship. They flew to New York the night before the meeting with the sponsors. Over breakfast the next day, Negroponte told Schreiber what he intended to tell the sponsors: You've got to let foreign companies into the program if you want to stay at the Media Lab. You've got to bring in the Japanese.

Schreiber didn't hesitate to tell Negroponte what he thought of that. CATS was created to be an *American* consortium training *American* engineers who would produce *American* inventions and innovations. That was the consortium's raison d'être. "If you do that," he told Negroponte, "then you're going to lose the whole project." Negroponte was unmoved.

When he and Schreiber walked into the boardroom at 30 Rock, NBC's headquarters, the sponsors—presidents or executive vice presidents of the nation's television networks and leading equipment manufacturers—were already seated around the table. As these powerful business leaders saw it, they were here to listen to a college professor ask them for a grant. They didn't seem to realize that a grand vizier had graced them with a visit. Negroponte stood before the long table and opened, as he usually did, by citing the magnificent accomplishments of the Media Lab, both real and imagined, that had come about under his wise direction. All of you, he suggested to the executives, can share in my vision. Listening with bemused tolerance, some of them may have expected a papal gesture, arms arching forward, palms open to the sky. Then, in a slightly scolding tone, Negroponte told them, "You cannot continue at the Media Lab if you do not allow foreign companies in."

With that, jaws dropped, and around the table the corporate officers exchanged stares of disbelief. The Grand Vizier continued. Learn from me, he was saying. Learn from my foreign friends. "Some foreign companies, such as Sony, have a great deal more experience in these fields than you do. You could benefit from cooperation with them." And then he proceeded to offer a lecture on Japanese business practices and industrial policy.

Sponsors began to erupt. CATS was set up to be an *American* program, they barked back. The very reason we exist is to *reduce* our dependence on Japan. Now you want to let the Japanese in, and give them full access to all of our research? That's absurd!

Howard Miller, a senior vice president with PBS, had just returned from a four-year assignment in Japan. He couldn't believe that this *college professor* had the gall to stand up there and "pontificate on corporate relationships and things in Japan because he just made a *two-week trip*! I spent *four years* there dealing with these issues. Negroponte's perspective of the way things work in international trade is incredibly naive, incredibly naive." While the PBS officer's lips pursed and his jaw tightened, the other officers, unaccustomed to being lectured by *anyone,* were growing angrier and angrier. Just who did this guy think he was? But it got worse.

The Grand Vizier could see that his pupils were restive. He'd known from the start that he would be "swimming upstream at best," as he put it later. You don't understand who I am, he seemed to be telling the executives. I'm explaining the way things are. This is truth, and you are simply calcified, out-of-step leaders of an archaic industry heading quickly toward its grave. I can fix your problems. Come to me. I can save you.

Those weren't Negroponte's precise words, but that was the message the executives heard, and now some of them were actually pounding their fists on the table as they cried out: This is outrageous! Just who do you think you are? This is unbelievable!

Schreiber watched all this quietly, shrinking back into his chair. He began to realize that he was going to lose his program. Sure enough, as Negroponte droned on, one after another the sponsors decided with a firm shake of the head: We're going to take the program away from you, you arrogant son of a bitch! By the time the meeting broke up, they'd made it clear: When the contract expired at the end of 1989, CATS would be moved to another university.

Negroponte appeared to take the decision with a shrug. The Media Lab doesn't need these people, he started saying. We'll create our *own* program that fits *my* rules. But Schreiber was crushed. At the same time, though, he felt relief. Being a contestant in Dick Wiley's race was wearying. He was just a college professor, up against some of America's largest corporations, and he had already decided that "even if I am right about every technical issue and they are wrong about every technical issue, there is no way I can win." Wiley had been insisting that everyone submit an actual prototype for testing, not just a computer simulation. Schreiber had been urging him to reconsider that decision. The professor couldn't afford to build such a complex and sophisticated machine. He had buried Wiley in Schreibergrams, but Wiley hadn't budged. Schreiber knew he'd have a hard time putting together the money to build a prototype even *with* his corporate sponsors. Now, thanks to Negroponte, he was going to lose them. The professor started to feel that "this is just too painful a road." Soon he would turn sixty-five. A friend who lived in Switzerland had been urging him to come for a six-month visit. Suddenly that sounded awfully appealing. Maybe it was time to retire.

More than a dozen universities expressed interest in taking over the CATS program, and after that appalling meeting in the NBC boardroom several of the sponsors were convinced they needed to make a clean break from MIT. But even as Schreiber announced his retirement, he grew concerned about his students. They'd be unable to finish their doctoral research if the program was moved away. Besides, Schreiber didn't want to let go completely. He'd put several years into this HDTV project. Maybe he could find another MIT faculty member to take it over, and Schreiber could serve as "something of an elder statesman." The sponsors were willing to consider that and, perhaps, stay at MIT. You've just got to find the right person, they said. In answer to the call, along came Jae Lim.

Lim was born in Korea and moved to the United States when he was seventeen. And from the moment he landed on U.S. shores, he wore his Asian values like a garish tattoo that refused to fade. Quite frankly, he found America wanting. And if anyone wanted to know what should be done to bring the nation back, well, Jae Lim was only too happy to show the way. After he entered an American high school for his senior year, he quickly noticed that students seemed far more interested in fun than in achievement. Sports were more important than math. Lim was shocked; it certainly hadn't been that way in Korea. Even at this young age, he decided America needed to change.

The next year Lim enrolled at MIT, where the students seemed more serious. One of his classmates was Woo Paik, who would later work for General Instrument. They were the only two Koreans in their class and grew to be friends. They both learned poker and spent many evenings playing cards. After graduation, both Paik and Lim stayed at MIT to pursue doctorates in engineering. After they completed their doctoral work, Paik moved to San Diego and started working for Jerry Heller's young company, Linkabit. Lim stayed at MIT and became an assistant professor in the Research Lab of Electronics. Schreiber and Lim knew each other—but not well, because Lim's expertise was in digital image and sound processing.

By 1989 Lim was a respected, tenured professor—a resolute man with an earnest, preachy manner and an unshakable certainty in his own rectitude. He was a permanent U.S. resident now, but he still didn't like what he saw around him. If Americans would listen to him, he seemed to be saying, then the nation might rise above its troubles.

"Americans take their wealth and standard of living for granted," he would tell anyone willing to listen. "But we're going to lose it if we don't change." He would point out that back in Korea, "nobody ever bought foreign goods. You just didn't do that. It wasn't patriotic. But in America, people buy whatever they want, and it doesn't matter where it comes from. In fact, it's *better* if it's foreign. Well," Lim would continue, pursing his lips, "that's just not right." And if other Americans were stupid or heedless, Jae Lim was going to set an example. "I don't buy any foreign goods," he would say, with a lift of the chin. "I tell my secretary, 'Buy only American things—pens, pencils, paper, and supplies. We won't have foreign things here.' When I buy a car, the first thing I do is eliminate all the foreign models. When I fly, I don't go on foreign airlines."

His own field, digital audio processing, inevitably led him to the consumer electronics industry, and he was astounded to watch American companies drop out of the business as Japanese corporations took over. Few Koreans have much love for the Japanese to begin with, but Lim just couldn't understand why the United States so willingly allowed Japanese hegemony over this important industry. Americans actually *preferred* Japanese televisions and stereo receivers. The whole sordid picture made him ill.

When Schreiber first talked to him about the CATS program and HDTV, Lim was utterly captivated by the consortium's credo: This is an *American* effort. We want to breed *American* engineers, create *American* innovations. "I liked that," he recalled with a rare, warm smile. Even better, the CATS sponsors had given Negroponte a swift kick when he had tried to bring in the Japanese. Lim realized that there could hardly be an opportunity better suited to him. In 1989 everybody was saying that HDTV could save the consumer electronics industry, bring America back. Maybe he could lead the way.

Lim looked at the front-runners: NHK was still ahead of everybody else. Meanwhile, all the American entries seemed flawed. Sarnoff's ACTV system wasn't even true HDTV. "It's the wrong direction," Lim decided. Besides, Sarnoff was being bankrolled by *foreigners*—Philips and Thomson. At Zenith, Wayne Luplow and his men had a good idea, but Zenith was weak, virtually bankrupt. The company might not survive long enough to bring this product to market. If Zenith went under, it was even possible that the company would be bought by the *Japanese*!

There was nobody else—"no other Americans who can provide a solution for this country," Lim concluded. He had no choice. "Our nation needs this." Maybe, just maybe, Jae Lim could *save America!*

The sponsors turned CATS over to Jae Lim in June 1989, and at first running the program seemed like a warm dream. But reality caught up quickly. Lim found he had to spend more time raising funds than doing research. As Peter Fannon's Advanced Television Test Center was moving toward completion outside Washington, Dick Wiley was talking about collecting fees from the contestants—$200,000 per system—to help defray the costs of testing, which was to begin in a year. Where on earth was Jae Lim going to get $200,000? MIT was unwilling to put up large sums of money for a speculative enterprise like this. The CATS sponsors were already donating $100,000 a year to the program. What's more, several of them were also contributing money to Sarnoff's system, ACTV. They weren't about to give Lim any more money.

Lim hit the road trying to raise funds from former students. He scrabbled together some money, though not nearly enough, and he soon saw that he couldn't spend all his time at that. He had to build the HDTV system, too, and quite a lot remained to be done. Schreiber had outlined a system that would broadcast half its signal in digital form and half in conventional, analog form. Most of the proposal was still on paper, Lim found, "with only a few small pieces built here and there." He puzzled over Schreiber's design month after month. He tried every approach he could think of, but he just couldn't get the damned thing to work. As he saw it, analog and digital systems each offered certain distinct advantages and disadvantages. When he tried building Schreiber's system, he concluded that it offered "the disadvantages of both worlds and the advantages of neither."

Schreiber was in Switzerland now; and, besides, Lim was not the sort of man quick to ask someone else for help. He didn't know what to do. He was unable to raise enough money, and he couldn't get the HDTV system to work. Lim had nowhere to turn. He still kept in touch with his old college friend Woo Paik now and then. They would reminisce about their days in graduate school, the evenings they had spent together playing poker. Paik had often won those games, and now he liked to tell others, "I can read Jae's mind." As the months slipped by, Lim gossiped with Paik about his work and his many frustrations. As 1989 ended and the time for filing the $200,000 testing fee approached, Paik listened but never said a word about the project taking shape in his own lab.

Lim struggled on, planning, dreaming, begging for money, trying to solve his problems so that maybe he still could save America—no hint in his head that his old college friend was about to shatter his dream.

Dick Wiley was having trouble with Al Sikes.

At the beginning of 1990, Wiley was in his third year as chairman of the Advisory Committee, and finally his race was about to enter its decisive stage. Testing of the contestants would start in a year, and the trade press was beginning to take note of his work. But over at the FCC, Chairman Sikes was growing restive. He had been pushing HDTV since 1986, back when hardly anyone else, including Wiley, was paying attention. Now that Sikes was chairman of the Federal Communications Commission, Wiley's Advisory Committee reported to him. So Sikes was actually in charge of the HDTV race. But Wiley was getting all the attention; anytime the press wrote about HDTV, they called Wiley.

This is supposed to be an FCC process, he complained to Wiley as the two of them stood in a hallway at the commission. But all I read about is the Advisory Committee, the Advisory Committee, the Advisory Committee. Wiley thought fast. As much as he liked attention in the press, he knew that the whole process would fall apart if he lost Sikes's support. "Look," Wiley said, "you're the chairman. I don't want to be chairman. I've already been chairman. The trade press is going to call me every week, but if you want I will defer to you." That seemed to satisfy Sikes—for the moment. But when Wiley got back to his office, he started thinking about how to give Sikes a larger role in the HDTV race. As he considered the situation, he began wondering if he might kill two birds with one stone.

For more than a year now, Wiley had been unhappy with the contestants in his race, particularly Sarnoff. The fact was, hardly any of the contestants was offering true high-definition television. HDTV was generally defined as television offering resolution at least twice as good as conventional TV, and none of the contestants but the Japanese offered that. ACTV and most of the other entries were more properly defined as *enhanced*-definition television; true HDTV would come later, the contestants were saying. This was extraordinarily convenient for manufacturers, Wiley knew. They could sell everyone a new TV set now and then another one in a few years. But he feared that the nation would *never* get true HDTV under this scheme. We have all the resources now, he was thinking—the Advisory Committee, the test center, a committed FCC chairman. If we use all of this to set an *enhanced*-television standard and then fold up, we'll never get the momentum to do this again in a few years' time.

There, he realized, lay the seeds of a strategy. Sikes had been fully supportive of Wiley's race. At the same time, though, the chairman had been asking lots of questions about digital television—suggesting that he, too, might be unhappy with the existing contestants. The truth was, the FCC had not made a formal policy statement on HDTV since September 1988, when Dennis Patrick's commission had said the winning system had to be compatible with conventional television. That had given energy to ACTV and some of the other "enhanced" TV offerings.

Why don't I suggest to Sikes that he put out a new policy, announce that the FCC will accept only *true* HDTV, Wiley decided. That will give him a lot of attention. He'll seem like a visionary.

"Let's go for the gold," Wiley told the chairman. "Let's go for true HDTV." As he outlined his strategy, Sikes listened with interest and nodded enthusiastically. Wiley could see that "he got it right away." A few weeks later, toward the end of March 1990, the FCC announced its new policy. It would not select an enhanced system, nor an "augmentation" scheme using all of one TV channel and part of another. The commission would accept only true HDTV, one that offered twice the resolution of conventional TV. The other, lesser ideas would be considered only if HDTV failed.

Most important, the commission directive said that the system had to use a "simulcasting" system—that is, it had to utilize one TV channel for the HDTV programming and allow conventional TV to be broadcast on a separate channel during the transition period. This would make the new system compatible with today's television sets; people unwilling to buy HDTVs would still be able to get conventional television broadcasts on their old sets. Finally, the commission directive said, after June 1, 1990—the Advisory Committee's application deadline—no new entries would be accepted unless they were based on "new technologies."

Sikes got all the attention he might have wanted. Almost every newspaper in the nation ran a story on the new FCC policy, and most of the articles spoke of Sikes's initiative, with no mention of Wiley. Everyone agreed that Sikes was trying to encourage the creation of a digital system. When the commission said it would leave the competition open for "new technologies," what else could it be talking about but digital TV?

The new policy struck the Sarnoff Shrine like a thunderbolt. The consortium's leaders were slow to see it and even slower to admit it, but with a single stroke Al Sikes had pulled the plug on the old RCA machine. To evade the obvious, the partners were trying to slip through a loophole. The FCC says it will consider adopting an *enhanced* standard if HDTV fails, noted Michael Sherlock of NBC. Well, we think HDTV may very well fail. When it does, we'll be here with ACTV. Behind that public stance some of the old RCA hands were still thinking, We can *power* ACTV into the stores. We've done it before. Don't forget—nobody believed the General when he said he could move America to color TV. Remember that first demonstration of color. Remember the blue bananas!

Most of the other contestants offered no such evasions. Bill Glenn, the early pioneer down in Florida, felt he was really out in the cold. He barely had the money to build his "augmentation" system, and now Sikes was saying that the FCC wouldn't accept it. Glenn dropped out. Over the next several months, so did the other small contestants. One element or another of the Sikes decision disqualified them.

NHK was still a contestant and remained a strong entry. The new version of Muse—Narrow Muse, still under development—was intended to fit in one TV channel, so it was a simulcast system. As planned, it would offer twice the resolution of conventional TV. Though Narrow Muse wasn't even partly digital, the Japanese realized that they still had the world's only working HDTV system, and NHK was due to begin regular high-definition programming in Japan in just over a year. The Japanese remained convinced that they still led the race.

Jae Lim had a simulcast system, and it was partly digital. In theory the decision was good news for him. But he couldn't get it to work and didn't have enough money to build it in any case. So what did it really matter?

Only Zenith greeted the news with untainted glee. "We're in fat city now!" Luplow declared. Zenith had a partly digital, simulcast system, and they had every confidence that they could make it work. "This is terrific," Jerry Pearlman said with a broad smile, and every other member of the team came to the same view: We have no real competition now. Press accounts called the FCC ruling a big boost for Zenith. Their hometown paper, the *Chicago Tribune,* practically crowned them the winner. Now that it looked like they'd won, Zenith's leaders believed they could begin selling the new sets in just a couple of years. The company was losing more money than ever now. If they could hang on a little while longer, HDTV really was going to save them. And with that intoxicating conclusion, another, even more satisfying realization dawned: If we have no competition—if we're really going to win—why on earth do we need those arrogant bastards from AT&T?

A few weeks later, Arun Netravali and Eric Petajan were summoned to Glenview for a meeting. The tape wars had continued all these months, and by the spring of 1990, both sides saw that they had battled to a virtual draw. Now, Zenith was saying, it's time to settle this fight. Let's put both subsystems—AT&T's digital approach and Zenith's analog ideas—up on a screen at Zenith headquarters, with engineers from both sides in the room. Right there we'll decide which is best. A shoot-out, this was called in the trade, and ahead of time Zenith and AT&T each sent the other "our nastiest sequences to code," as Petajan put it. At the shoot-out, the challenge was to run both tapes through the system and see how they looked.

When the day came, in mid-May 1990, everyone was there in the screening room: Pearlman, Luplow, and others from Zenith, along with Buchsbaum, Netravali, Petajan, and some of the other men from AT&T. For Zenith's tape sequence, Luplow had sent a team out with a camera, and they'd come back with footage of a large, leafy tree swaying in the wind—difficult to handle because of all the random motion. For another shot, Zenith superimposed fast-scrolling text over a Chicago street scene. AT&T had put together a tape of computer-generated geometric figures. Another segment showed a rich-looking Persian rug rapidly panned from many angles.

Each side ran its "killer tape" through the systems. And the verdict was clear: AT&T's digital approach looked better. The difference wasn't dramatic, but it was apparent to most people in the room. Even the Zenith officials who argued that there was no appreciable difference had to acknowledge that because AT&T's system was digital it could be improved, while Zenith's analog approach was about as good as it would ever be. What was worse, Zenith's system "still couldn't even account for all the bits that were being transmitted," Petajan averred. In other words, part of the signal that went into Zenith's computer just got lost; the system couldn't handle all of it. For months now, Petajan and his men had been sneeringly calling this the "fiber to the home" approach: "Viewers would get some of the signal by antenna," Petajan quipped, "and we figured Zenith must be planning to run fiber-optic cable to each home so they could send the rest."

AT&T had won the shoot-out. That was clear. But to the Bell Labs boys, Zenith's reaction was unnerving. Standing around in the hallway afterward, Luplow's men just didn't seem to care. They were complaining about secondary matters: Our sub-band coding is better than yours.

As for AT&T's larger showing, Zenith seemed strangely unconcerned. A few minutes later, the leaders of both teams retired to the wood-paneled conference room adjacent to Jerry Pearlman's office. Pearlman sat at the head of the table and led the meeting.

For weeks Zenith had been complaining that AT&T's approach was too expensive, no matter the technical merits. To offer digital motion estimation, each TV set would need a small extra part called a frame buffer—a little chip that could store part of the signal for a fraction of a second. Both Zenith and AT&T had been estimating how much this part would add to the cost of each set. AT&T's figure was $20 or $30; Zenith said it might cost more. In the conference room, Pearlman cut right through that argument with a wave of the hand. It doesn't matter, he said. Given the state of things today, we just don't need digital motion estimation. It doesn't improve the picture enough to be worth the extra cost.

Petajan could see that Zenith was "planning to come out with these new TVs in just a couple of years, and they think it's going to save them." So Zenith didn't really care whether the picture quality would be as high as it could be. Pearlman wasn't saying exactly that, but his bottom line was: We don't need you anymore. Thank you very much. Good-bye.

The Bell Labs boys couldn't believe it. We're being terminated, they realized as they slunk out of the room. These Zenith cretins have actually *fired* us!

Standing in the doorway, Pearlman seemed to be savoring the moment—watching, as he put it later, while the Bell Labs boys left "with their tails between their legs."

8 Can it be? Can it really be?

In Chicago, New Jersey, New York, Tokyo, and Washington, Wiley's race was thrumming right along. Corporate chieftains were pouring millions of dollars into the quest. Dozens of men were putting in thousands of man-hours a month. Everyone was running in high gear. But at the midcourse point, most of the energy was being expended on gamesmanship, scheming, and political maneuvering. The idea of looking for technological innovation seemed far from most people's minds.

For Jerry Heller and Woo Paik in San Diego, all that foment might just as well have been taking place on the planet Saturn. Out there in the nation's southwest corner, the East Coast broadcast television world was the farthest thing from their minds. The engineers of the VideoCipher Division lived on a different world. They were caught up in their own deep problems; they still worried about ordinary *satellite* television. In fact, after Paik had trudged down to the basement to begin work on that satellite HDTV project, the VideoCipher Division's piracy wars had grown only hotter. One day Jerry Heller realized that his entire engineering department, including Paik, was caught up in the fight. Over and over, Heller would pull groups of engineers off the front line and tell them to concentrate on one special project or another. Then the Federal Express man would deliver a new pirate VideoCipher box, and inside they would find an ever more clever approach to defeating their satellite scrambling system. Suddenly every engineer in the division, even those on the special-project team, was pulled from what he was doing and plunged back into the war. Paik was being yanked away just as often as everyone else.

Finally Heller threw up his hands and decided he needed to create a wholly separate department within the company: Advanced Development. Larry Dunham, the division's president, heartily agreed. They named Paik the director of Advanced Development. He was given a handful of engineers and a lab of his own, along with strict instructions: You report to Heller and work *only* on special projects, especially satellite HDTV. No matter what happens, keep out of the VideoCipher wars. Paik readily agreed; in fact, he couldn't have been happier.

Ensconced in his own lab, the head of his own department, Paik approached the HDTV problem as he had many others. In his first days at Linkabit, Heller had asked him to work on a digital modem, and he had said, "Yes sir," even though he hadn't known what a modem was. He had looked up the word in the dictionary. Paik didn't know the first thing about broadcast engineering, either, so he ordered several textbooks and technical manuals and read them over a period of a few weeks, taking copious notes along the way. Seemed simple enough. Then he and Heller began talking about a plan of attack: Just what kind of high-definition television should they build? As they considered that question, they both thought about what had worked for them before.

When Heller arrived in California in the late 1960s, he and his mentor, Irwin Jacobs, built digital modems for NASA, and the two men grew to be stars. Later, the military was impressed with Linkabit's digital systems for communicating with B-52 bombers circling the globe. More recently, when Heller and Paik entered the civilian market, they had won the HBO competition against all the odds because they had offered a *digital* system, while the other competitors had stuck with analog designs. By now, Heller and Paik regarded themselves as genuine authorities on digital communications. So when they sat down to figure out what sort of HDTV system they ought to design, Heller told Paik: Let's try a digital approach. It seemed only natural.

Back on the East Coast, all the broadcast engineers were telling Dick Wiley that digital television was impossible. But Heller and Paik weren't broadcasters. They had no connection with that community; they'd never even *met* Dick Wiley. Nobody had told them digital TV was impossible, so Paik plunged blithely ahead.

One basic challenge lay behind every one of the plans to create high-definition television: each of the designers had to find a way to fit more information into a single 6-megahertz television channel—that figurative 6-inch water pipe. Some designers, including Bill Glenn and the Sarnoff labs, had avoided the bulk of that challenge by using all of one TV channel and part of another as well. Others, notably Zenith and MIT, had decided to digitize part of their signal; they could compress the digital part, and that allowed them to fit more information into a single pipe. Digital information could be compressed in ways that a conventional, analog signal could not. Digital compression of video was a relatively new field; only recently had computers grown sophisticated enough to manipulate complex video signals. But Heller and Paik had been among the pioneers, and to them compression was the key to their project.

The fundamental principle behind digital compression is quite simple: most forms of communication include a great deal of repetition, and by taking redundant material out of the transmitted signal, its size can be greatly reduced. Look at any TV show close up, and this seems obvious.

Think of lifting a 35 mm film out of the can and holding the film up to the light. Suppose it's a *Star Trek* movie and you're holding part of the film that shows the starship *Enterprise* soaring through space. Examine the sequence frame by frame and you'll see that the image generally changes only slightly from one frame to the next. In each successive frame, the starship remains near the center of the picture and most of the frame around it remains black—though a few dots of light, representations of distant stars, slide slowly past.

Other television images are similar. Most TV images change little from one frame to the next. To take advantage of this, Paik planned to divide the picture into hundreds of tiny blocks. For each new frame, new information would be transmitted for only those parts of the picture, those blocks, that actually changed. For all the blocks that remained unchanged—the black of space in the *Star Trek* movie—a digitally compressed signal would send just one notation: Repeat what was shown in the last frame.

The problem was that when a full change of scene came along—from a view of the *Enterprise* in space, say, to the captain on the bridge—there was no redundancy, nothing to compress. Occasionally any digital compression system would be overwhelmed. With no redundancy, no possible compression, not all of the information would fit in the pipe. Some of it would have to be discarded. But human vision has well-established limits, and Paik realized that if the compressed signal could catch up quickly, viewers wouldn't notice when a frame or two was transmitted at a lower resolution.

Those digital compression principles had worked for transmissions from space, for music and other audio signals. But transmissions from space probes were easy to compress because they included almost no motion, while music and voice transmissions were far less complex than video. Compared to Paik's challenge now, those tasks were positively simpleminded. No one had ever done what Paik was setting out to do. Zenith had managed to digitize and compress a tiny portion of the TV signal. MIT proposed to digitize even more, but Jae Lim hadn't been able to figure out how to do it. Nobody had succeeded in converting a complete television signal—full-color pictures and stereo sound—to digital form and then squeezing it into a 6-megahertz channel. The TV signal was so complex that the digital information representing it would fill twenty TV channels, maybe twenty-five. In fact, digital compression would have to be so extreme that 90 percent of the signal, maybe more, would have to be squeezed out. It was no wonder that TV engineers threw up their hands and said, You want us to throw away 90 percent of the picture *and* double the resolution for HDTV? That's impossible!

Paik began with an important advantage—and a head start. He was creating a system for satellite television, and a satellite TV signal was almost twice as wide as the signal for broadcast TV. That meant he had to compress the signal only about half as much. Paik also found that some basic work had already been done. A short time earlier, General Instrument had hired a young engineer named Ed Krauss, a bright, soft-spoken young man who had just earned his doctorate from MIT. Krauss had worked in the CATS program in its early years; he had been one of Bill Schreiber's star students. And when he moved to San Diego he brought with him some lessons from MIT's work in digital compression of video. When Paik got started, Krauss had already begun experimenting with digitizing and compressing high-resolution still pictures that the company had acquired from MIT. Of course, compressing a still image was easy, in theory, because no part of the picture changed. But Krauss's work had established certain fundamental principles.

From that base, Paik began establishing a research plan. The company had bought a "digitizer"—a complex and expensive piece of equipment that could convert a signal to digital form and store a short burst of it in random-access memory. Television signals were so complex that they consumed vast areas of computer memory, and even this sophisticated device was capable of storing only ten seconds of conventional television—or three seconds of HDTV. (Just a few years earlier, random-access memory chips had not been advanced enough to do even that.) Then Paik acquired some high-resolution sequences from recent movies. Hollywood films were already produced in a resolution that would be considered high-definition by television standards.

With all his equipment, material, and assistants in place, Paik set to work on a bench at the far end of the Advanced Development lab. The goal was straightforward: compress a complete television signal to a tiny fraction of its ordinary size, then display the compressed image without severe flaws *(artifacts,* TV engineers called them) on a high-definition monitor. Heller and the other General Instrument executives may not have known that broadcast engineers back East would have regarded Paik's assignment as hopeless, but they did know full well that their star engineer was attempting no easy feat. They had no idea whether he could do it, and frankly that's just the way Paik wanted it.

"I like it better when people don't think I am going to be able to come through," he said. "I feel less pressure than when people are certain I can do it and are expecting it."

Down in the lab, he and Ed Krauss worked side by side. They both knew the strategy: trial and error. It was going to be tedious—months, maybe years, of writing long strings of computer code, trying each one out, and then rewriting the program again and again until finally they settled on a version that worked. They started with an image from the movie *Top Gun.* Fighter jets in combat roared through the sky. Sitting at the table, Paik wrote out the first strategy—a long string of computer commands that would make up the first proposed compression algorithm—the coding that told the computer which parts of each frame to transmit and which to hold back because they were redundant. When Paik finished, he gave his notes to Krauss, who sat down at a computer terminal to write the complex microcode that would carry out Paik's plan—thousands of lines of arcane symbols that had to be painstakingly entered and checked.

All of that took more than a day, and as Krauss worked, Paik returned to his office to think about new strategies, "trying to figure out any way I can to make the signal better." When Krauss finished, he called Paik. They both sat at the bench and anxiously played the fighter-jet scene held in the memory of the frame storer. The three-second image coursed through the computer holding Paik's compression algorithm, and then the compressed image played on the small high-definition monitor sitting atop the workbench. The frame storer played the scene over and over in an endless loop, and at first both Paik and Krauss grimaced. It looked *dreadful.* The picture was dotted with small, odd-colored blocks—computer distortions. The airplanes seemed to break apart in the air. Well, Paik thought, no one said this would be easy. This is just the beginning. We've got a long way to go.

He studied the image as it played over and over, trying to figure out where he'd gone wrong. Then he headed back to his office to write a new algorithm, hoping the next one would be at least a little bit better—and the one after that a little better still. After several long months they finally settled on an algorithm that produced an acceptable picture. Paik felt a quiet satisfaction as the fighter jets zipped through the sky without significant artifacts or distortion. But there was no larger celebration. This had been an *easy* sequence. Far more difficult work lay ahead.

Finally, early in 1989, Heller and Paik told Larry Dunham that he ought to come down for a look. By now Paik and Krauss had acquired the ability to play somewhat longer sequences, and they were using a larger high-definition monitor, made by Sony. They didn't have a fancy demonstration, but they were able to put on a short show, polished enough so that nonengineers could appreciate it. When they played the *Top Gun* tape, another from an *Indiana Jones* movie, and a third showing Japanese models in traditional dress smiling at the camera—just about sixty seconds of material in all—Dunham exclaimed, "Eye-popping! Good work, Woo!" As he turned to leave, he urged Paik to keep at it.

This was the outline, at least, of a digital high-definition television system for satellite TV subscribers. True, the compression was only half as extreme as would be necessary for broadcast television. Still, Paik had gone more than halfway toward that impossible goal—digital TV! He was pleased with himself, and the company's leaders liked what they saw. Beyond that, however, none of them really understood what they had, and they weren't entirely sure what to do with it, either. Their thinking was a muddled mix of conflicting fears and flawed marketing concepts. As Dunham put it, "All I really wanted was an endorsement of GI as the source for the satellite HDTV system." Satellite TV was the VideoCipher Division's arena—their own small, familiar domain. With satellite TV, Heller said, the division could "have a lot of freedom to do things the way we wanted to do them." The last thing Dunham and Heller wanted was to be sucked into the wider world of conventional broadcast television. That was subject to FCC regulation and, worse still, to brutal, unfathomable East Coast politics.

Still, Heller began to realize that their game plan had one large, gaping flaw. There weren't any high-definition television sets on the market to display Paik's creation. They'd begun this project when it looked as if the Japanese sets were about to go on sale. Now, in theory, they had a hot new product. But it had no market and nobody knew when one might develop—if ever.

By now, the VideoCipher Division had been a part of General Instrument for just over two years—though, with offices in New York, GI remained a remote owner. The company's larger corporate interests seldom entered the minds of most people in San Diego. "It took a while to get a GI-wide perspective," Heller said. One of GI's subsidiaries, however, was the nation's largest manufacturer of cable TV converter boxes. And as Heller tried to figure out what his division ought to do with Paik's new invention, he seldom considered conventional cable television. But soon after Paik's satellite TV demonstration for Dunham, an idea began to dawn: A satellite signal was twice as wide as the signal for broadcast or cable television. Why not take Paik's system and "cut it in half?" as Heller put it. In other words, double the level of compression and create a high-definition system for cable TV. That would be good for General Instrument when high-definition televisions finally did come onto the market. The company could serve not only the small satellite market that was the VideoCipher Division's traditional base but also the far larger cable television audience, now more than half the nation's homes.

That potential lay somewhere in the indefinite future. In the meantime, though, Heller had an even better idea: If Paik could compress a high-definition TV signal, what was to stop him from compressing a *conventional* television signal? The digital compression technologies they were creating for high-definition television could be used for any TV signal, high-definition or not. If Paik could compress an ordinary TV signal, then cable operators would be able to transmit several different programs over a single cable channel—tripling, maybe even quadrupling a cable system's channel capacity with a relatively modest investment. Now *there* was an invention GI could market right now.

To Heller, none of this seemed revolutionary or bold. It was simply the natural thing to do. "Going digital is going back to our roots," he said. So when Heller gave Paik his new directions, he passed them on with something of a shrug. Let's try to double your compression, reduce the size of the high-definition signal once again so it can fit in a 6-megahertz cable channel. After that, we can get to work on compressing a *conventional* cable TV signal.

Sure, Paik said with a smile. He'd already had the same idea. He may have been unaware of the digital dogma so well accepted on the other coast: *We'll have digital TV the same day we have an antigravity machine.* He may not have known that he was being asked to take his present invention (one the broadcast engineering community would have classified as unlikely) and turn it into the impossible: fully digital television. After all, the compression ratio for a cable channel was no less severe than for regular broadcast TV. Though Paik didn't know he was taking on a preposterous, chimerical challenge, he was at least vaguely aware of the HDTV race raging in the East. He'd gotten into this HDTV business back in 1988, when Heller had asked him to have a look at the American systems under development then, to see if any of them could be made to work with satellites. Paik had met some of the players, read about the work of others, and followed the occasional small articles on the race in the professional journals. A couple of times a year he also chatted with his old college friend, Jae Lim, back at MIT. So Paik knew a race was under way for what he came to call "the terrestrial standard"—earthbound television transmissions—and he very much wanted to enter. "All the time I was interested in getting into the race for the terrestrial HDTV standard," he said later. "I'd always been interested in terrestrial television."

But Paik told this to no one. It was his own little secret. After all, General Instrument wasn't even in the terrestrial television business. Still, as the days began growing longer in the late winter of 1989, he returned to the lab table and tried to alter his compression algorithm so that it could serve cable TV. First he asked Krauss to double the compression using the same algorithm. Krauss said it wouldn't work, but Paik told him to do it anyway. Krauss prepared the new program, and they ran a signal through the system. When the picture—a car chase through city streets—appeared on the lab monitor, by Paik's own assessment "it looked terrible." The cars were disintegrating before his eyes.

And it's no wonder. For the satellite system, Paik had compressed the signal to less than half of its original size. To reduce the signal enough so that it could fit in the 6-inch pipe, the compression scheme had to be so severe that less than 10 percent of the original information remained. This was really going to be hard. He wasn't sure he could do it.

Paik and Krauss went back to work: Write the program, put it in the computer, look at the tape. Go back to the desk. Trial and error, month after month. By spring 1989 the new formulations were yielding tiny improvements here and there. The fully compressed picture still looked rotten, but it was getting better.

As Paik plodded methodically along, in Tokyo NHK appointed a new chairman—Keiji Shima, a former political reporter who had risen through NHK's ranks to the top of his network. In 1989, just a month after he took office, Shima was in Washington letting everyone know exactly what was on his mind. He met with Commerce Secretary Robert Mosbacher and the heads of the major TV networks, among several others. And he told all of them that in his view the United States was moving too slowly on HDTV. One reason was that Americans were more interested in international politics than in actually producing goods: "Japanese manufacturers of these home appliances work harder, and they are more efficient than their American counterparts," he liked to say. The reason for that, he would add, "is attributable to your laziness."

Somebody asked Shima if Japan might accept an HDTV standard developed in America. Sure, he said, but it wouldn't ever be as advanced as the one NHK already had. Nor did Shima appear to think much about digital TV. NHK was already transmitting experimental Muse broadcasts into a few homes, and full operation was due to begin soon. Nobody wanted to upset that boat.

Everyone who met Shima found him surprising; for a Japanese businessman, he was extraordinary. Though correct in appearance—he was in his early sixties, had gray, thinning hair, and wore modest, dark-framed eyeglasses—Shima was blunt, gruff. None of those gentle bows and insincere smiles. Shima had little trouble saying no. Within NHK, his friends considered him "outspoken," as one put it, and that in itself was an unusual trait in Japan. Shima's enemies were many, and the Japanese press called him autocratic, dictatorial, megalomaniacal. Shima gloried in his reputation, boasted about how close he had come to starting a fistfight with Peter Ueberroth, who was the chairman of the Los Angeles Olympics Committee, and another with a senior official at General Electric, which was working on one of NHK's satellites. After Shima made the rounds in Washington, official after official said they had never met a Japanese quite like him.

One of Shima's most important visits on this trip was to the National Association of Broadcasters. Masao Sugimoto, the "father of Muse," says Shima believed that the NAB *owed* his network. Hadn't NHK spent millions and millions of dollars modifying the Muse system for America "at the request of the NAB?" Sugimoto said, adding, "We got no thanks from the U.S. for spending all that money." Shima puts it this way: "I was so willing to make any improvement or modification in order to enhance the system we developed, so that they would take up our system." But after two long years of effort, Japan was *still* no closer to having its system accepted in America, even though no one doubted that Muse remained the most advanced HDTV system in the world. Now, with the FCC giving every engineer on earth a chance to catch up with NHK, Shima was angry. The Americans "got this tremendous impression that this kind of monstrous thing has come from Japan and will spread through the U.S. market, and the American industry could be damaged to a great degree." That was not his intention at all, he argued.

"America," Sugimoto said, "is like a big baby."

John Abel looked out the window when he got word that Shima was down in the lobby with his entourage—half a dozen aides and sycophants. Outside Abel saw two stretch limos idling at the curb. Upstairs in Eddie Fritts's office, Shima walked right up and leaned over Fritts's desk. We've changed our system so it can be broadcast like any other TV signal, he said, speaking through a translator. We've done just what you asked. We want to make a deal. Why don't you start using Muse as soon as possible? If the Muse system is chosen for America, he added, "I'm not going to talk about licensing, royalties, and things like that. I'm not going to charge even a cent."

They were stunned by his manner. This is the most direct Japanese I have ever seen, Abel was thinking.

Still, Fritts and Abel told him: That's very generous, but we don't control this competition. The FCC's running this. It's a government process now.

Shima didn't seem to believe them. "He thinks we're looking for a better deal," Abel concluded. Whatever Shima thought, he strode off with his coterie. They climbed back into their limos and drove away.

Washington was deep into autumn when Shima returned with a new strategy in mind. The FCC controls this, Fritts and Abel had told him. If the FCC controlled things, he would go to the FCC. So Shima's aides made an appointment to see Patricia Diaz-Dennis, one of the FCC commissioners. In her office, Shima got right to the point. NHK has the most advanced HDTV system in the world, he said, and "if you don't deal with us, we will have to deal with governments not friendly to the United States."

Mrs. Diaz-Dennis was an easy, cheerful lady with an ever ready smile. But at that her jaw dropped. Wide-eyed, she just stared at Shima for a moment. He stared straight back. I'm not hearing this guy right, she was thinking. A Japanese executive is telling a United States government official that if we don't buy their product, they'll sell it to unfriendly nations—presumably the Soviet Union? This has to be a mistranslation. She turned to her aide and whispered, "Isn't this the most bizarre meeting there ever has been in government? I can't believe this guy is actually threatening a United States government official!" The Japanese translator either didn't catch that or chose not to repeat it, and Mrs. Diaz-Dennis decided to overlook Shima's remark; she couldn't believe he was serious. A few minutes later, the meeting ended.

Shima returned to his suite at the Willard Hotel and summoned another visitor: Larry Irving, the senior counsel to Representative Edward Markey's telecommunications subcommittee. On Capitol Hill, Irving was known as Mr. HDTV; he'd arranged all the committee's hearings and reports on HDTV. As Irving walked into the suite, Shima and his aides were seated, waiting for him with cookies and tea. The first thing that caught Irving's eye was one of Shima's aides, who was wearing "a really bad suit—baby blue polyester." Irving thought, I'm allergic to that suit. He smiled to himself, but Irving was given little time for private bemusement. As with Diaz-Dennis before, Shima got down to business fast. If the United States doesn't use our system, he said, NHK will have to deal with governments not friendly to the United States. Now it was Irving's turn to be stunned. What did I do to provoke this guy? he thought. Markey is sometimes regarded as a Japan-basher.

Maybe that's it. But this guy's really obnoxious. A few minute later, as Irving pushed the button for the elevator in the hotel hallway, he was thinking: This guy didn't really say that to me. He just couldn't have.

But he had, and to others as well. After Shima made a similar remark to the leaders of the Association for Maximum Service Television, Greg DePriest thought, Whoa! This is really getting out of hand.

As Shima explained his thinking later: "If Americans don't want to use our system, I have to look for somebody else who will adopt it. I thought it was my job to do that at that time. I really don't know whether I meant it to be a threat."

Shima didn't end his visit with that. No trip to Washington was complete, it seemed, without a stop to see Fritts and Abel at the NAB. If, as Abel suspected, Shima had left the last time thinking the NAB was looking for a better deal, this time he brought one.

The stretch limos were idling outside once again. Fritts and Abel had been in a meeting with the NAB board, but they got up as soon as they learned that the NHK chairman was waiting to see them. The two of them walked down the hall to another conference room and greeted Shima. And by their account (Shima has disputed parts of it) the chairman got right to the point and made a startling offer.

We've done what you asked (they say Shima told them). If you get our system selected, we'll give you 50 percent of the worldwide licensing rights.

Nobody said anything at first; cold shock was washing over both Fritts and Abel. Finally, Abel spoke. "Licensing rights?"

The chairman nodded.

"Fifty percent?" Abel went on, speaking very deliberately now. Shima nodded again. Fritts listened, dumbfounded at first. After a moment, though, the potential of this proposal began to become clear, and he was thinking, What's that worth? To find out, Fritts asked Shima a straightforward question, one he had been curious about for quite a long time: "How much has NHK spent on this effort so far?"

Five hundred million dollars, Shima told him.

Fritts repeated the number slowly: Five hundred million dollars?

They're certainly going to want to recoup that from licensing rights, he thought. At a minimum, this has got to be worth $250 million to us.

I could do this, Fritts thought to himself. Right away he began figuring out how he could explain taking all that money. "We're not the only players in this country who have been co-opted by Japan," he figured he could say. "There are a number of people in the U.S. strongly pushing the Muse system, for their own reasons." Standing there looking at Shima and his entourage, Fritts also realized that "with the horsepower this organization has, this process in its infancy, we can operate clandestinely and undo some things that have already been done." In other words, Fritts believed his organization had the lobbying power to stop the Advisory Committee race and install NHK as the winner. "We can manipulate the process," Fritts thought. "We can do this!" He looked over at Abel, who by now was so giddy he could barely speak.

"Eddie," Abel whispered, "we've gotta caucus, we've gotta caucus," and the two of them scurried over into a corner of the room.

"Eddie," Abel hissed, "we're in the soup. We're in the soup!"

Shima had offered them all this money, and Abel could see the wheels turning in Fritts's head. He also thought about how deeply he had pulled the NAB into Wiley's race. The NAB had helped found the Advanced Television Test Center. An NAB officer sat on Wiley's Advisory Committee. The broadcasters got the whole HDTV race started. So how on earth could this organization steer the competition toward NHK?

"Eddie, you gotta say no," Abel said. As happily as Fritts had been flirting with the idea at first, he reluctantly concluded that Abel was right.

"I know," he said at last. "But what are we gonna say?"

Abel considered for a minute. He thought about a demonstration of Narrow Muse he had seen at the last NAB convention. He had left the booth thinking, "the picture's not for shit." So he suggested to Fritts, "Let's just tell him his system isn't good enough to be the U.S. standard."

They turned back to Shima. The chairman was still standing on the other side of the conference table. "That's a generous offer," Fritts told him, "but we have to say no."

Shima asked why.

"Your system isn't good enough to be the U.S. standard."

Shima didn't say much. To Abel "he seemed to get a washed-out kind of look on his face." The chairman turned and left the room. Abel looked over at Fritts. "Can you believe this?" he said, still whispering though they were now alone. "Do you know how much money we walked away from?"

"I know," Fritts said with a sigh. "I know."[1]

While Shima was trying to bludgeon Washington policymakers, Paik was making real progress in San Diego. Eight months of trial and error had moved him close to producing an acceptable digital high-definition picture that would fit into a 6-megahertz cable channel—or, with certain alterations, a 6-megahertz broadcast TV channel. The Holy Grail was within reach. At the same time, along the way Paik had created a digital algorithm that compressed a *conventional* TV broadcast so that four programs could be broadcast over a single cable channel. That hadn't been particularly difficult; the algorithm followed the same principles as the one he was creating for HDTV. So all his theoretical work for televisions of the future had also produced a product that could be marketed soon.

High-definition television might be years away, but cable operators were looking for inexpensive ways to increase their channel capacities

---

[1] In an interview with the Tokyo bureau chief for the *New York Times,* Shima confirmed that during one meeting with Fritts and Abel he offered to waive licensing fees if the United States adopted the Muse system. But Shima said he did not remember making the second offer—to split royalty income with the NAB. Fritts and Abel described the encounter in detail during several separate interviews. Shima said, "I don't remember at all what happened, but I don't remember saying that. I don't think I ever said that." Perhaps Abel and Fritts told the story, he added, because "they really don't have good reasons to say why they wouldn't use NHK's Muse system."

Two years after that second visit to Washington, Shima was forced to resign his position following a controversy over false testimony to parliament. Testifying about where he had been when an NHK broadcast satellite exploded just after launch in Florida, Shima said he was visiting the New Jersey offices of General Electric, the manufacturer of the satellite. But a few weeks later Japanese newspapers reported that he had actually been in Los Angeles. Some of the press reports said he had been traveling with a female companion, an allegation Shima vigorously denied. As he resigned, Shima said, "This is a major crisis for a public broadcaster, and as the responsible person I deeply apologize." In June of 1996, Shima died.

right now. Once Paik and the others turned this computer-simulated system into an actual product, cable operators would be able to buy a General Instrument digital encoder. GI's Jerrold division could make a digital decoder for the cable box on top of the customer's television set. With everything in place, a cable operator running an ordinary fifty-channel system could quadruple his capacity overnight and offer his customers two hundred cable channels without significant additional capital investment.

"This is our future," Heller predicted. Until now, the VideoCipher Division had carried out all this digital research in secret. But early in 1990, GI began showing off its new digital cable system at industry trade shows. It looked like a hit.

At the Sarnoff Shrine in Princeton, meanwhile, Jim Carnes had just announced the new partnership with Philips. Now the lab was hard at work on ACTV. Then at a cable TV convention, Sarnoff engineers saw a demonstration of the General Instrument digital cable system—several conventional television shows over a single cable channel.

Now this is pretty interesting, Carnes thought. He and others at Sarnoff had certainly known of General Instrument, but nobody in the TV business considered GI to be particularly big in advanced research. Jerrold, the division that made cable boxes, was seen as a plodding, predictable enterprise. The company's new VideoCipher Division, out there in San Diego, was best known for the huge mess they'd made out of the HBO scrambling contract. But a digital cable system—this was something else. It didn't take a rocket scientist to figure out that any algorithm capable of compressing a conventional television signal to one-quarter its normal size might also be useful for HDTV. Maybe we had better go have a talk with these people, Carnes decided.

When Carnes and Joe Donahue arrived in San Diego, in March of 1990, the leaders of the VideoCipher Division were conflicted. They still wanted a partner; GI wasn't even in the TV business. But they weren't predisposed to trust, or even to like, these men from the Sarnoff Shrine. As Larry Dunham shook Joe Donahue's hand in the soft-lit lobby of their suburban office complex, he recalled their last meeting just a few years before with a twinge of anger. When Dunham and the others had been negotiating their original contract with HBO, the cable company had insisted that a second manufacturer be found to make some of the

VideoCipher II boxes. This was a standard arrangement—undertaken so that production would not stop if one factory encountered problems. The VideoCipher Division had approached RCA, among other companies, and one afternoon Dunham had found himself in Joe Donahue's office in Indianapolis. They talked for a while, Dunham recalled, and then with a perfectly straight face Donahue made RCA's offer: "We'll pay you twenty-five cents for every box we make."

Dunham couldn't believe what he heard. "Twenty-five cents?" he said. "We've already invested $5 *million* in this project!"

Donahue had just looked at him. "When you leave here and get back on the plane," he said, with a slow, sly tone and a supercilious RCA smile, "we'll spend $5 million on advertising before you land." RCA hadn't gotten the contract. But four years later, here was Donahue again, exploring the idea of taking on the VideoCipher Division as a partner. Well, by god, Dunham decided, we're not going to roll over for them. We won't show them everything we have—for by now Woo Paik had almost finished his 6-megahertz digital HDTV system. But down in the Advanced Development lab, nobody gave the General's men even a hint of how far Paik had really come. They demonstrated just what they'd shown at the trade shows: digitized conventional television.

Carnes and Donahue weren't saying much about the little bit of digital research in their own labs, either. Before coming out here, Carnes had concluded, "If we go in there and say we know how to do this and we know how to do that, it's really going to turn them off." Besides, the Sarnoff engineers had only theories, nothing to show—though, not surprisingly, they still figured they were far ahead of everyone else. But for all Heller and Dunham knew, ACTV was all these people had.

Both sides circled warily, neither wanting to give anything away. Finally Donahue came out with the pitch: We think there will be a migration to a digital standard sometime in the coming years, and we'd like you to join our consortium. When the time comes, we'll have a shoot-out, and we'll decide which digital system to use at that time—yours or ours.

But you don't even *have* a digital system, Heller was thinking. We have one, and you want us to join your consortium so you can have full access to all our work? And then you might not even use it? Forget it.

Dunham was having a hard time figuring out what the Sarnoff men really wanted. The Sarnoff consortium was having political problems, he knew. Everybody thought it was too European. Maybe Sarnoff just wanted another American name to put on the letterhead. At the same time, he realized: Hey, maybe we're the only ones with a digital system. These guys certainly don't have anything. Maybe we're way ahead.

Dunham and Heller offered to join if Sarnoff and its partners agreed now to use their digital system.

We can't do that, Donahue said. We'll have a competition. If your system wins, then we will use it. Best man wins.

This is crazy, Heller and Dunham both thought. They don't have anything. They're out here fishing. "I don't think we're interested," they told Donahue. Thank you for coming out. Let us think things over, and maybe we'll get back to you.

Carnes and Donahue couldn't believe it. This pissant little company out here in an office park in *San Diego,* of all places, is actually sending away the Titans of RCA? "I guess they didn't see that we had anything of value for them," Carnes recalled later. "But I sort of thought, in this consortium we represent 40 percent of the consumer electronics industry in this country. I sort of thought, We're the lab that invented color television. I kind of thought, well, maybe they might want to talk to us."

Actually, Dunham had not entirely dismissed the idea of joining the Sarnoff consortium. He had begun to realize that he had a hot product but no way to market it. If we're going to make anything out of Woo's invention, he concluded, sooner or later we're going to have to team up with a manufacturer. We've got to find a way to increase our bargaining power with the consortium. Somehow we've got to make them commit to using *our* digital system.

Jim Carnes and the other men from the Sarnoff Shrine weren't the only ones who had been impressed by the VideoCipher Division's demonstration at the trade shows. Joe Flaherty of CBS had been struck by it, too, even if he did think fully digital TV would arrive about the time someone had invented an antigravity machine. A few weeks after Carnes and Donahue left San Diego, Flaherty showed up for a visit. That wasn't surprising; as the chief technology officer at CBS, he was always trying to keep up with the latest research. Because of his position, Wiley had also recruited him to be a senior volunteer adviser in the HDTV race.

A balding, roly-poly Irishman, Flaherty grew up in Kansas, the son of a radio engineer. On the wall of his spacious executive-row office on the fifty-first floor of Broadcast House in New York hung a large black-and-white photo taken in 1937. It showed Flaherty's dad standing in a field, a satisfied smile on his face as he leaned one high-top leather boot against a massive concrete anchor for a new radio tower. When TV came along in the '40s, Flaherty and his dad built one of the nation's first amateur stations and got a license to operate it. After young Joe Flaherty earned his engineering degree, in 1955, he started work for General Sarnoff at NBC. Two years later he was laid off, so he moved up the street to CBS. There he remained, climbing gradually up the ladder through the engineering department until he was named senior vice president for technology.

Flaherty had been following the development of HDTV since the mid-1970s. He took something of a paternal interest. While he was the chief of engineering for CBS, he helped push the network from film to videotape ahead of everybody else. By the late 1970s, the transition was almost complete. The only holdout was Hollywood: the studios that produced the network's prime-time westerns and adventure shows didn't want to switch from film to tape. They didn't think the resolution of conventional television was good enough, so they refused to stop using 35 millimeter film, which offered superior clarity.

Since then, Flaherty had been on a hunt for a higher-resolution television system. As a result, he had been the most important American backer of NHK's effort to have Muse accepted worldwide. That had made him quite a controversial character, but by 1990 if anyone in the broadcasting community was an established expert on high-definition television, it was Joe Flaherty. He was generally recognized as the dean of the nation's professional broadcast engineers, as well as Wiley's most important adviser.

Dunham and Heller didn't view Flaherty with the same distrust they had held for the Sarnoff characters. Flaherty was almost a legend in the business, and in person he was a comfortable, easy man (unless you crossed him; he had quite a temper). So they decided to show him everything, including their digital, high-definition television system for cable TV. The system wasn't completely ready; there had been no real pressure to finish by a certain date. But Paik could at least offer a basic demonstration.

Flaherty was a cool, careful character. But as Paik described how he had first produced a digital system for satellite subscribers and then continued squeezing the signal until it would work for cable TV, Flaherty had a hard time keeping his excitement in check. He knew full well that with only small modifications a digital signal for cable TV could also be sent over the air. Here it was, the impossible dream: digital television! And as Ed Krauss ran the computer simulation—snippets of the movie sequences and other scenic views—Flaherty could contain himself no longer. "Are you considering this for high-definition television?" he asked. In other words, had they considered entering Wiley's race?

Paik smiled to himself, but Heller was taken off guard. He stammered, "Well, uh, we've talked about that, but, uh, we haven't really made any decisions." Flaherty could see that they probably hadn't talked about Wiley's race at all. They don't have any idea what they have here, he decided. They really don't understand what they've done. "You better decide pretty fast," he told them. "The application deadline's coming up." That was news to Heller, and even to Paik.

Back at his office in New York the next day, Flaherty didn't even try to contain his enthusiasm as he told George Vradenburg about what he had seen in San Diego. Vradenburg was the general counsel for CBS and another principal on Dick Wiley's Advisory Committee. He began to see a business opportunity for the network here. If this digital system was really as good as Flaherty said (and who would know better?), maybe CBS and General Instrument could become partners.

Vradenburg called Dunham: Why don't you send Woo Paik back here to put on a demonstration for some of our people, let us see what you've got?

Dunham liked the idea. Even if they didn't join up with CBS, these discussions might increase their bargaining position with the Sarnoff consortium. A few days later, Paik assembled his tapes (including some he hadn't shown Flaherty), and flew to New York on May 15, 1990.

The demonstration was brief: A smiling, smooth-faced Japanese model wearing colorful traditional clothing glided across a set. Traffic poured over a crowded city bridge at rush hour. Wildflowers swayed in a slight breeze on a sunny afternoon. All of it lasted less than a minute. But that was more than enough; the CBS executives were rapt as the scenes played across the screen. This wasn't just beautiful high-definition television, it was *fully digital.*

"Unbelievable," Vradenburg declared. "Crystal clear. I've never seen anything like it."

"Impressive," said Flaherty, and the others echoed that.

"Is this something we should pursue?" Paik asked, uncertainty in his voice.

"Absolutely," Vradenburg said. "Certainly," Flaherty added. Then Flaherty told Paik that the application deadline for the HDTV race was June 1. Just two weeks away! Paik and his assistants looked at each other. They packed up their equipment and caught a plane back to San Diego that evening.

A few days later—Friday, May 25—Vradenburg had to be in Los Angeles for meetings, so Dunham and Heller drove up to see him. You really ought to join the HDTV race, Vradenburg told them. (By then, CBS's business leaders had vetoed the notion of a partnership with GI; it didn't seem to fit their larger business interests.)

The whole idea of entering the race struck Dunham as fanciful. Hell, he thought, we aren't even in the broadcast TV business. Still, as they drove back to San Diego that evening, Dunham and Heller began to realize that Woo Paik really had created something extraordinary. Nobody else has this—not CBS, not the Sarnoff consortium, not the Japanese. *Nobody!* If we do file for the race on June 1, no one will be able to come in after us, and this will increase our leverage with the Sarnoff group. We can negotiate with them again from a position of strength.

Monday morning, Dunham called General Instrument's top management in New York and described what he was planning. He could tell that they thought the whole idea was crazy. But he explained that the entry fee—$175,000—was a small investment and entering would increase the bargaining leverage with the Sarnoff consortium. Dunham got permission to proceed. On May 28, three days before the entry deadline, he told Paik, who was secretly delighted. With just a couple of days left, Paik scrambled to write a short paper describing his HDTV system. He had to provide enough detail so that the Advisory Committee could see that it wasn't a fantasy—without giving too much away. Paik and Heller decided to name their system DigiCipher, the logical heir to their bastard first child, the VideoCipher system for satellite subscribers. Maybe with this new digital invention they could leave all those old problems behind.

Dick Wiley hadn't known anything about General Instrument until he came to New York for a speech in mid-May and George Vradenburg took him aside to whisper something in his ear: This company in San Diego seems to have come up with a *digital* system. Wiley was skeptical, but he said, "Tell them they better hurry up and get their application in." So when Quincy Rodgers, General Instrument's Washington representative, called him on Thursday, May 31, Wiley wasn't really surprised.

"Dick," Rodgers said, "we have something we'd like to show you." He explained, though Wiley wasn't sure he believed it. Still, he told Rodgers he'd come by in the morning. Dunham and Heller flew to Washington that evening, and Rodgers drew a cashier's check for $175,000.

The next morning, a steamy hot Washington day, Wiley walked down to GI's offices with his assistant, Lex Felker, the former FCC official. GI's Washington office was part of a rent-an-executive-suite complex three blocks away. Everyone sat around a conference table in a miniature law library as Heller and Paik presented some charts, explained what they had, and handed over the check. While Felker listened, he was thinking, These are some smart guys! Wiley went over the rules and requirements of his race and asked Heller to show up for "precertification" interviews with Advisory Committee engineers in a few weeks. He didn't say much beyond that, and the meeting was brief, almost perfunctory. But when it was over, Wiley was beginning to believe that maybe, just maybe, these men had achieved what everyone said could not be done: They'd broken the sound barrier for the first time, proved Fermat's last theorem . . . As he walked back to the office, Wiley turned to Felker and asked, "Can it be? Can it really be?"

9 *Just who* are *these guys?*

Soon after their meeting with Wiley, the men from General Instrument walked the three short blocks over to FCC headquarters for an appointment they'd scheduled with Chairman Sikes late Friday morning. Wiley had called Sikes the day before to give him a "heads up," and like Wiley, the chairman had been skeptical. He'd been listening to technogossip for years and had never heard a word about GI. Still, when Heller and Dunham told him they had entered the race, paid the filing fee, and were proposing a fully digital system, Sikes abandoned any reticence. He grabbed the arms of his chair and exclaimed, "That's just what I *said* would happen!"

At the MIT Media Lab a few days later, the Grand Vizier greeted the news that his prophecy had come to pass with accustomed grace. "I used to feel like a lone voice in the woods," Negroponte sighed. "Now, suddenly, what we were talking about is becoming real."

While the digital prophet and his apostles accepted GI's announcement with self-satisfied warmth, the other contestants in Wiley's race were far from pleased. Generally, first word came on Monday, when the weekly trade journals arrived.

"The complexion of the FCC's drive to set a standard for high-definition television changed dramatically last Friday when General Instrument, New York, announced that it is proposing an all-digital HDTV broadcast system," *Broadcasting* magazine offered. "According to some observers, DigiCipher will deliver enough quality in such a spectrum-efficient manner that it will far outpace analog and hybrid analog/digital systems currently proposed for FCC adoption." But, the magazine went on, "All, including GI, caution that the system has been developed only to a computer-simulation stage. Whether it can live up to its theoretical potential is expected to be known sometime after February 1991, when GI plans to begin testing its first hardware."

Several contestants seized upon only that last phrase, choosing to believe that GI hadn't really done it. Until this moment, Zenith had been the consensus front-runner, the all but declared winner. Just two weeks before, the company had fired those pompous poseurs from AT&T, telling them in essence, "We've already won." Now, as Wayne Luplow read about GI's entry, he took it with "a fair amount of disbelief."

In New York, Keiichi Kubota's first thought was that it couldn't be true. NHK had far more to lose than anyone else; Japan was actually producing Muse equipment. *Broadcasting* magazine had called him for his reaction, and "based on our experience," he said, "it is very difficult to realize an all-digital HDTV broadcast system." That evening he called his supervisors (it was morning in Tokyo) and gave them the news. The general conclusion was incredulity. Some NHK executives told Kubota that GI "must have made a mistake," while others exclaimed, "They're lying!" Even if GI was telling the truth, Kubota thought, the digital system wasn't likely to present a picture any better than Muse's.

At MIT, Jae Lim was stunned. All these months, and his old friend Woo Paik hadn't said a word. Lim was still struggling with Bill Schreiber's half-digital system. He was short of money, and he had already told Wiley he would be late paying his filing fee. What's more, he still couldn't make the damned thing work. So Lim believed he knew how many obstacles awaited Paik as he tried to turn his computer simulation into an actual product. That could take years and cost many millions of dollars. When Ed Andrews, a reporter for the *New York Times,* called, Lim suggested that the United States might very well want to adopt a conventional, analog system for the time being (his own, perhaps?) and then move to a digital standard later, when the technology had improved. No one was arguing against the idea "that the world is going digital," Lim said. "The question is, when are we going to have it at a price acceptable to consumers? And in the meantime, should we stick to a totally inferior system?" Andrews took this to mean that Lim didn't think Paik was likely to produce a marketable product anytime soon.

Only the leaders of the Sarnoff Shrine immediately accepted the announcement for the bad news it really was. Carne's first thought: They had it all along. When we were out there in San Diego, they already had a fully digital system, and they didn't tell us. Now, Carnes knew, "we really screwed up—bad." At the same time, though, Carnes was finding it easier to make command decisions and deal with problems like this. Just as GI was making its announcement, he was promoted. Now he was the president of Sarnoff labs. Because of Glenn Reitmeier's work and continual nagging, Carnes had already been inclined to work on a digital system. So the new president called Wiley and asked, Do you think we should drop ACTV and try to come up with a digital entry? Wiley didn't hesitate even a moment. Given his view of ACTV, his answer probably would have been the same even if GI had never shown up. "Yes I do," he said.

Carnes had a clear-eyed view, but his consortium partners and others in the broadcasting industry still couldn't seem to let go of ACTV. To the broadcasters, Advanced Compatible Television was *cheap.* And to the American leaders of Thomson and Philips, the allure of double sales—an ACTV system in every home now, a true high-definition system a few years later—remained overpowering. So these former RCA executives clung to the hope that they might outmuscle everyone and get their way once again.

In the Glenview Bunker, Wayne Luplow had pooh-poohed General Instrument's announcement at first. But Jerry Pearlman, Zenith's president, saw the problem clearly. It was the same one that had dogged his company from the earliest days: holding on to comfortable, profitable technology even as the rest of the world moved on. Pearlman had been raised in that tradition; it was a part of him now.

"It's not important to be first," he liked to say. "As I have said many times, we weren't first in radio, but we got the biggest market share. We weren't first in black-and-white, but we got the biggest share. And we weren't first in color, but we got the biggest share."

Nonetheless, this time was different. Zenith's hybrid system, the presumed winner just two weeks ago, looked stillborn today. "In this business now," Pearlman had to concede, "no one holds a lead in technology for very long." So he accepted the hard fact that Zenith had to create a fully digital system of its own. But he still faced the quandary he'd confronted a year before: Zenith just didn't have the staff. Debt payments were piling up; a serious liquidity crisis loomed. There was *no way* he could go out and hire two dozen new engineers. So what choice did he have? Hard as it might be, Pearlman had to call AT&T once again.

At Bell Labs, Eric Petajan smiled to himself, thinking: Zenith has come crawling back. This time, though, he wanted to make one thing clear: If we go back into business with Zenith, we're going to be *full partners.* No longer will AT&T serve as a mere subcontractor to those penny-pinchers in Glenview.

As it turned out, Petajan and the others needn't have worried. When Zenith explained how the labor was to be divided, Petajan quickly realized that Zenith was asking Bell Labs to build the *whole thing!* Under the Zenith proposal, AT&T was to build the coder and the decoder—virtually the entire system—while Zenith would be responsible for the transmission system, the device that sent the TV signal over the air. As a very rough comparison, AT&T was building a computer, while Zenith was building a modem, the small box that allows a computer to communicate over a telephone line. Of course, a digital modem for broadcasting a high-definition television signal was a far larger and more complex device. Still, AT&T was going to be responsible for the vast majority of the work. That was fine with Petajan. As he saw it, the new division of responsibility and the huge job ahead "brought a kind of formalism to the project that took the emotion out of dealing with Zenith." And with Bell Labs in charge, the ill-fated motion estimator was going to be one of the first devices drawn into the schematic diagram.

At first Petajan and his little team were jubilant. Soon, however, a sobering thought came over them all: We're taking on one hell of a big project here, and time is short. According to the Advisory Committee's schedule, we have to have a system ready for testing in just over a year, even though right now we have *nothing*—no staff, not even a drawing. In fact, among the thousands and thousands of engineers working for AT&T around the world, no one had any specific experience in broadcast television engineering. Nobody. Petajan figured he'd have to find people working in related fields—picturephones or computer programming—who could apply their knowledge to his project. He visited AT&T facilities across the nation and gathered the engineering staffs in conference rooms or dining halls to present the challenge: "We're entering the race to create the next generation of television for America," he would say. We are going to build a digital television, the best one in the world. We're not in competition with just the other companies in this country; we're up against the Japanese. This is going to involve really long days, crazy hours, impossible deadlines, but this is a very high-visibility project. It'll be really good for the company—and for the country—if we win.

Petajan knew "a self-selecting process" would take over here. Only a certain sort of engineer—hungry, unsettled, ambitious—would volunteer. That was just the sort of person he wanted. In some locations he would stand on a chair while he made the pitch. Looking over the audience, he would see the fire in some eyes. These were the ones who came forward. By early October he had recruited ten people, and they had set out a rough system diagram—ten or so discrete parts that needed to be designed and built. They divided up the work; each engineer took charge of one part. And they approached the project with an overriding goal: Maybe GI had already created a digital system, but "ours is going to be novel," as Petajan put it. "We're going to include features nobody else has."

But as one engineer after another looked at the part he was supposed to design, each of them came to a disquieting realization: Every one of these parts is really two pieces—maybe even five pieces, or ten. This is going to be a far bigger project than we realized. We're going to need more people—many more. And look at the calendar . . . four months have already passed.

Jerry Heller and Larry Dunham had guessed that GI might face incredulity from some of its new competitors, but they never expected the withering scorn they got from the broadcast engineers who served as the judges in Dick Wiley's race. They might have: these were the very men who had been saying for years that digital, high-definition television defied the laws of physics. "A person tends to be skeptical; digital's impossible. That has been the dictum," said Charlie Rhodes, a former engineer for Philips who was now the chief scientist at the Advanced Television Test Center taking shape in Alexandria. Many of the engineers had staked their reputations on that maxim. Then along came these unknown characters from San Diego. Nobody in the East Coast television-engineering community had ever even *heard* of them before.

And the prevailing question came to be: Just who *are* these guys? They aren't even *in* the TV business!

In July, six weeks after they'd entered the race, Heller and Paik had to present themselves to these very Advisory Committee engineers for precertification, an aggressive grilling intended to determine whether the proposal had any real chance of working. The other contestants had already been through this two years before, during Hell Week.

"We had a lot of flat-earth people out there back then," Wiley said. "We had to weed some of them out." One of the earlier contestants had offered a system purporting to use a form of digital compression that, according to one of the interviewing engineers, seemed "to defy the laws of physics." Eventually that company dropped out. So as GI arrived for its own interrogation, several of the engineers were thinking they'd already heard "a lot of fast and loose stuff," as Peter Fannon, the test center's director, put it.

The meeting was in a wood-paneled conference room at the National Association of Broadcasters, which should have been a hint of what lay ahead. The inquisitors were arranged around a horseshoe-shaped table, and as Heller and Paik walked in, their jurors stared at them with blank expressions, thinking, in Fannon's words: "Who do you think you are, anyway? Do you mean to say you really think you can do this?" After brief introductions, Paik took the stand, a chair placed right in the center of the horseshoe. He was surrounded. The questions came fast, and though the tone was civil, it was not difficult to detect the snarl just beneath. An engineer on one side of the table posed a question, and as Paik turned to answer in his heavily accented English someone on the opposite side of the table growled, "Huh? What was that you said?" Paik spun around and repeated his answer, only to get the same dismissive stares from the engineers on that side. So it went for half an hour.

The jurors worked through many of the engineering hurdles, though it was clear that they were not persuaded. Then they seized on one issue: signal fade. A conventional television signal has a range of, say, thirty-five miles. Actually, however, some people farther out can still receive a picture and sound, though the signal will grow progressively fuzzier with every additional mile from the transmitter. In the business, this is called "graceful degradation." But a digital signal is different. There is no graceful degradation. Homes up to about forty-five miles away will receive a perfect signal, with no ghosts or snow or impairments of any kind. Finally, at some fixed point, the signal simply stops. One home will get a studio-quality picture, while the house next door is unable to pick up anything at all. That, the Advisory Committee engineers concluded, simply would not do. You mean to tell us there's no graceful degradation, one of them asked, his words heavy with scorn. The signal just falls off a cliff?

That's true, Paik said, but you have to understand that the digital signal does not fall off until after it has already reached *farther* than a conventional signal is able to go in any usable form. The engineers weren't buying it. That won't work, they grumbled as they shook their heads. Nobody will *ever* accept this "cliff effect." Paik didn't want to be combative, but he was thinking, This is really stupid! Maybe what I should do is introduce some random noise into the signal at thirty-five miles out so these people will see "graceful degradation."

Heller was next, and the questioning directed at him seemed even more aggressive. The veneer of civility was gone. But he never abandoned his cool, detached manner, even when the jurors found a new point of attack: channel acquisition. They'd read GI's entry papers and deduced that with the proposed digital system it might take a little longer to change channels. On a conventional TV, channel acquisition is nearly instantaneous. Almost as soon as a viewer lifts his finger from the change-channel button on the remote control, the new channel is on the screen and the picture is locked in. But GI's digital TV took a little longer. Are you telling me the acquisition time is going to be one-half to three-quarters of a second? one of the engineers asked, head tilted, eyes wide.

"Yes," Heller said.

"Well, that'll never fly. It'll never fly."

By then Heller had heard enough. He was angry, but he kept his voice even as he told the group, "Look, you have to look at digital as a whole. It has advantages and disadvantages." Maybe channel acquisition is slower at this stage of development, he went on. But with digital TV you're going to lose all the picture impairments—ghosts, noise, and snow—that bedevil conventional, analog broadcasts. You'll have all the flexibility that comes with digital signals, make televisions interoperable with computers and the other tools of this new, digital age. To himself Heller added, You'll also draw this backward, calcified industry into the modern era.

The engineers were having none of it. They looked at Heller with expressions of wanton disbelief. In the end, though, as much as these jurors might have liked to send Heller and Paik packing, they didn't really know enough about GI's proposed system to say for certain that it could not work. They had little choice but to precertify GI, although this didn't mean they couldn't send Heller and Paik back to California when they finally brought their equipment in for testing—or perhaps even before.

General Instrument had entered when the race was already in high gear. Dick Wiley's labyrinthine network of committees, subcommittees, working parties, subgroups, and every other manner of bureaucratic substructure was holding meetings almost every day. They were examining and debating all the questions that would have to be answered before a new television system could be introduced to America—along with others that, in all likelihood, no one would ever ask:

How will the testing be conducted at the Advanced Television Test Center? What will be used as test material? Will field tests be needed to complement the laboratory analyses? Who's going to evaluate all the data? Once a winner is selected, how much will these new sets cost? When should manufacturers be brought in to look at the prototypes? What will TV stations have to spend to install new transmitters? How many stations will have to build new TV towers? How will cable companies be affected? How easily will the new American system coexist with high-definition systems under development in Europe and Japan? What about the VCR industry? The list went on and on.

The contestants attended most of these meetings. They had no choice, unless they wanted to find that a decision had been made putting them at a disadvantage. This stultifying series of meetings had helped drive Bill Schreiber out of the race; he hadn't had the time or patience. Now GI had to start sending someone, even though some of the subcommittees and working parties were run by the very same engineers who'd been at the precertification meeting. Heller was an executive vice president; he couldn't spend all his time in Washington. Paik was hard at work in the lab again. A working prototype of his HDTV system didn't have to be ready for testing until next year, so as soon as he returned to San Diego Paik put the project aside and resumed work on the digital transmission system for conventional cable television, the one they had been showing at trade shows. This was a product GI could market right now, and Paik wanted to finish it, then turn back to HDTV.

With Paik and Heller otherwise employed, the task of attending Dick Wiley's subcommittee meetings fell to Quincy Rodgers, General Instrument's man in Washington. Rodgers was a competent, affable lawyer-lobbyist in his fifties, with a fair knowledge of the engineering issues that were important to the cable television world he had been representing all these years. He'd been with GI for decades but had come to know Dunham, Heller, and Paik only after General Instrument bought the VideoCipher Division. The HBO piracy problem had required a fair bit of explanation in Washington. Much of that had fallen to Rodgers. In those days, he noted, "almost all of our equipment had quality problems," so the VideoCipher wars were the normal grist for his mill. "When things are going well, it's bad for me," he liked to say. "But when things are going badly, it's good for me." The problems gave him work, the chance to show his value to the company.

But Wiley's Advisory Committee and all these meetings were an entirely different matter. They were boring, impenetrable, and everyone wanted to ask him dozens of questions about GI's digital system and the company's larger plans. Now that a fully digital system was in the race, many of the previous assumptions would have to be rethought. Someone asked, for instance, how much a digital TV would cost. Rodgers had no answer: "I'll get back to you," he said. He'd send questions out to San Diego by fax. The problem was that nobody in San Diego had the answers either. Rodgers and a part-time assistant, one GI officer remarked, were "like the cavalry at the outpost fighting the Indians—calling for reinforcements that never came."

Soon Heller and Dunham realized they'd have to hire someone to be the company's point man for HDTV, and their search settled on Bob Rast, an engineer with deep roots in the broadcast and cable TV world. He grew up in the Washington D.C. area, studied engineering at the universities of Maryland and Pennsylvania, and spent his formative years working for RCA, where he became friends with Jim Carnes. As a young engineer at the Sarnoff Shrine, he had made a name for himself by taking control of a floundering engineering project and, as the undeclared team leader, pushing it to completion. He'd been awarded ten patents and won one of the company's prestigious David Sarnoff Achievement Awards on his way to a promotion into management.

Rast left RCA in 1981, several years before the company's fall from grace. After that, he headed research and development for a cable company, and in 1990 he was a cable television consultant based in Denver, working on a project to assess the potential impact of high-definition television on the cable industry. He already knew Jerry Pearlman, Joe Donahue, Peter Bingham, and the other principals in the race. In fact, at RCA he had worked for several of them.

Rast had met some of the General Instrument officers at the trade shows where they were showing off their digital system for cable, and he wanted to get a closer look at GI's work. Before he could get around to it, GI had entered the HDTV race, and a couple of weeks later Heller called Rast to ask if he was interested in a job, saying that Jim Carnes had recommended him. Rast was excited by the offer. He'd never had particular respect for GI, but this new division in San Diego had actually invented digital television.

Rast was tall and trim with sandy hair and a modest mustache. He had a sharp wit and an easy laugh; he was fun to be around. Beneath the amiable exterior, Rast was a warrior, a guerrilla willing to do whatever it took to win. The HDTV race was well suited to a man of his temperament. Soon he would become its most visible and active player.

About the time Dunham offered Rast a job, control of General Instrument came into play on Wall Street. Forstmann Little & Company, a leveraged-buyout group, was offering to purchase GI for $1.6 billion, and GI's management was interested. The sale would make them rich. All hiring froze while this deal was being negotiated. Finally, at the end of August, the buyout agreement was settled. Rast was hired, and he moved his family to San Diego.

At first, he didn't feel particularly secure among all these high-priced engineers. "Jerry's from MIT, Woo's MIT, everybody's MIT," he recalled with a laugh. "These guys came out of this highly secret government kind of work. They think they can do anything—*anything!* And here I am, the freak—University of Maryland." But "I grew up in RCA," he added. There, "I kept meeting all these people in the elevators . . . the father of this, the father of that. I've already been intimidated by the masters." Besides, he quickly learned that a race of this sort was equal parts engineering and politics, and as far as the politics was concerned these MIT engineers "don't know what the hell they're doing!"

As Rast saw it, the VideoCipher Division sat out there in a San

Diego office park, completely isolated from the rest of the industry and from the people who were in charge of the race. What was worse, the HDTV project had serious engineering problems. Paik's high-definition compression system looked sound, in theory. But Woo hadn't even begun to work on a transmission system, and nobody at GI had any background in that. Adding to Rast's worries, "they don't seem to understand anything about manufacturing. They have big problems with quality control. And their HDTV system, it's all theory. Here I am telling everybody how great this HDTV system is, and it's just vaporware. They have nothing to show." Worst of all: "Woo isn't even working on it anymore!"

Maybe Paik wasn't working on it now, but by the fall of 1990 word of his invention crept from the trade press to the mainstream papers, then in December to the front page of the *New York Times:*

In a remarkable technological turnabout, American companies have moved from laggards to leaders in the worldwide competition to develop the next generation of television.

The advances have occurred despite both the late entry into the race by American companies and the Bush Administration's refusal to support substantial financing for research. Instead, the Government relied on a high-stakes competition among private companies that was orchestrated by the Federal Communications Commission.

"The conventional wisdom is that we are a poor third in HDTV," said Alfred C. Sikes, chairman of the Federal Communications Commission. "It's my view that U.S. companies are now on the leading edge."

Not all the companies were willing to admit it, though. Petajan was still recruiting engineers for the AT&T-Zenith digital project, and his partners in the Glenview Bunker weren't stepping out on a limb for him. Asked in the fall whether Zenith would offer a fully digital system, company spokesman John Taylor said, "It's possible, but there are serious technical challenges that need to be addressed."

At the Sarnoff Shrine, the engineers were more optimistic. Glenn Reitmeier had been promoting a digital approach for more than a year now, and in mid-November the Sarnoff consortium announced that it would push ACTV to the background and aggressively try to develop a fully digital approach. After all, if those people out in *San Diego* could do it. . . .

"The whole world is going digital," admitted Joe Donahue of Thomson. He even dropped the mantra he had been confidently espousing these last few years: ACTV now, digital later—"a convenient two-step evolution for broadcasters."

Keiichi Kubota and his supervisors at NHK were still having a hard time believing they might have been beaten. Kubota saw one of GI's computer simulations in the fall, and it looked so good that he was certain "it can't be right. They made a mistake with the software." Later he called Wiley, who urged him to move NHK toward a new, digital system. Kubota told him: We have too much invested in Muse, too many installed sets. It's not right for us now. Wiley was suspicious. "I thought there was something else going on there," he said. "Something else deep down they weren't telling me." Still, he could see that the decision had been made, so he told Kubota, "OK, it will be a good test: analog versus digital."

At MIT, Jae Lim finally flopped over into the digital camp—at least in theory. "It can be done," he said in November. "We have been making rapid progress in the past two to three months. Lab studies show, in fairly impressive detail, that the technology is here." Maybe, but Lim still didn't have the money. The FCC had announced that laboratory tests of all the proposed HDTV systems were to begin in April 1991, and by December 31, 1990, the contestants had to advance part of the testing fee—about $200,000. Lim didn't have it. He'd tried everything now. So finally he concluded: I need a partner. He didn't take long to settle on Woo Paik, his old college friend.

General Instrument was now under new management, with a brand-new chief executive officer, Donald Rumsfeld. He had held several senior positions in the Nixon and Ford administrations, including secretary of defense. For the last few years, Rumsfeld had been chairman of a large pharmaceutical company in Chicago, so he moved General Instrument's headquarters there from New York. But when he took over GI, in the autumn of 1990, Rumsfeld found a company in shambles. GI owned a diverse range of businesses that didn't seem to have much to do with each other. Why, he wondered, did a telecommunications business also own a company that sold and serviced racetrack betting systems? At the same time, GI's core business, the cable TV division, was "declining rapidly, heading south," as Rumsfeld put it. As for the VideoCipher Division, Rumsfeld realized that its piracy problem was critical. "If you're in video security and you can't protect your signal, well. . . ." He frowned.

One area, at least, did excite him: digital television. Rumsfeld was an experienced, no-nonsense manager, a difficult man to impress. Still, he immediately saw the potential of Woo Paik's invention. "Digital compression has unambiguously moved into the position of being the next generation of technology," he said, adding with a satisfied laugh, "You don't see a generational change in technology very often. And we are the pioneers. It's difficult even to contemplate what being a pioneer in digital compression for HDTV offers us."

In San Diego, Paik turned back to his HDTV project in January 1991, but Heller and Rast were worried now that he was in over his head. How on earth was he going to build a transmission system? "I started asking questions, and people didn't have answers," Rast said. "We're faking it on transmission. We're making the pictures, but transmission's weak, and I'm asking, 'Where are the skills going to come from?' " Heller agreed. "It's too much of a risk to have Woo do both compression and transmission," he said, though Heller also knew that "Woo didn't share this view."

In fact, Paik was offended. Even though "I couldn't say strongly I could do it, because I didn't have a finished product" or a clear path toward getting one, Paik also didn't think it was "a good idea to create compression and transmission separately. It's like separating the heart and the body and hoping they work when you put them back together." Besides, Paik had faced uncertain challenges many times before. Who'd have guessed that he could create digital television to begin with?

Regardless, Heller and Rast decided they needed to bring in a transmission expert. Larry Dunham agreed—especially because he still believed that GI would be better off with a partner in this venture, and by now it was clear that the Sarnoff consortium had set off for its own fate. So Heller and Rast set out to find a partner, and they settled on Motorola. In early September 1991 they met with a large group of Motorola executives at their offices in Chicago. The scheduling was tight, and Heller and Rast arrived late. When they entered the conference room, the Motorola executives were already testy.

Rast took the lead. He gave a presentation with an overhead projector describing their digital TV project. When he finished he saw that the crowd was still unfriendly. What about transmission? one of the Motorola men asked.

That's exactly why we're here, Rast told him. We need help with that.

The Motorola men just kept peppering him with hostile questions —finding vulnerabilities, poking holes. If you can measure up to our standards, they seemed to be saying, then we will help you. If we could do that, Rast was thinking, then we wouldn't have come here in the first place. When they left, Rast could easily see that "we made no headway. They weren't impressed with us at all."

Back in San Diego, Heller fell back on the companies he already knew, the San Diego–based aerospace contractors he had worked with for decades. He picked a satellite communications firm named ComStream. The company had an engineer with expertise in the transmission field, so Heller and Dunham began partnership negotiations. Paik didn't like the idea one bit. This HDTV system was his child. He didn't want to share it with anybody. We don't need them, he kept telling Heller. We're ahead of them. But Paik could see that Heller thought "this guy was far better than any of us. Jerry would not stop, he would not stop." Finally, Heller arranged a meeting between Paik and the ComStream engineer, but Paik was something less than warm and welcoming, and the meeting did not go well. When Heller heard about that later, he thought Paik had deliberately turned the engineer off. Paik had a different interpretation. The ComStream engineer, he said, "discovered how well we were doing and became concerned that there was really not much for him to do."

In the end, the deal fell through. By now it was late autumn of 1991, and the leaders of the VideoCipher Division believed they were in trouble. Heller was upset. "Jerry assures me that, in the end, Woo will not be able to deliver, that the job is too much for him," Rast wrote in his diary. Dunham, he observed, "seems shocked" by the state of affairs. Dunham wanted to meet with Rast to discuss it, but before a meeting could be arranged Dunham abruptly resigned, a victim of the piracy wars and personality conflicts with Rumsfeld, the company's new president. The situation was spiraling out of control.

It was just about this time that Paik got a call from Cambridge. Jae

Lim wanted to know if GI would take him on as a partner. When Paik told the others about Lim's call, one by one each of them smiled.

Nobody at GI held even the faintest hope that Jae Lim could help them build a transmission system. Nobody really believed he was actually capable of building a complete digital television system, either. But that wasn't the point, for Lim had something even more valuable—the last test slot.

Through the fall, Wiley had been negotiating the order of testing with the contestants. Nobody wanted to go first, and the arguments had gone on for weeks. Finally Wiley had simply assigned the order based on how advanced each contestant's work seemed to be. Sarnoff's ACTV system was up first. The men of the Sarnoff Shrine had lost all interest in Advanced Compatible Television, but they wanted to test it anyway. Broadcasters had put a lot of money into ACTV, and some of them still held on to the fantasy that it might be selected. Sarnoff was scheduled to bring it in for testing in April 1991.

NHK was up next, two or three months later; Narrow Muse was nearly finished, too. After that came GI. Paik had begun work on his digital system before anyone else, so it seemed only fair that his digital system should be the first tested.

Next came Zenith-AT&T. After that, the Sarnoff consortium's digital entry. Then finally, more than a year after the start of testing, Jae Lim was to come forward with his system. Lim didn't have a digital system or any real capability to build one, and everyone knew it. But in the fall of 1990, neither Zenith-AT&T nor the Sarnoff consortium had much more than theories, either. Lim had as much right to say he could create a system as anyone else. GI saw the potential in this immediately. Woo's is the first digital system up, they said to themselves. If Woo can't create a transmission system, or if he encounters some other problem, then we'll have a second chance with a joint GI-MIT system more than a year later. This is insurance; it might save us.

Rast was dispatched to negotiate a deal with Lim, and they met at a rotating bar atop a Cambridge hotel. This was not Rast's sort of place, and he was rolling his eyes even before he sat down. They ordered their meals, and Lim started to make his case: If you like Woo, he said in his earnest, preachy manner, bring me in as a partner and you'll have seven more just like him, my graduate students. They have the same training Woo has. You'll also be bringing in the reputation of MIT, and MIT's technology will be of great value to you. And don't forget, he finally said, "I have the last test slot."

"Ten, twenty times," Rast said later, "he tells me this, going 'round and 'round in the rotating restaurant."

Finally Rast asked him, So tell me what you want? Lim was not shy about answering: If GI and MIT were to build a second system together, and if that system won, then Lim and MIT would get 50 percent of the profits. Rast just looked back at him, trying not to let his jaw drop. Fifty percent? he was thinking. This guy's got almost no technology to offer. We're going to be the ones who build this system, whatever it is, because he can't do it. We'll end up paying for all of it. And he wants 50 percent? No way!

"Jae," he said, "we are so far apart I am not even going to tell you how far apart we are, because it will offend you. And I don't want to offend you." Lim pressed, so Rast spread his arms wide: "You're here," he said. And then he held his fingers an inch or two apart: "And we're here."

But you'll get my graduate students, the MIT reputation, Lim told him again. And *I have the last test slot.* Finally, though, Lim could see he was making no headway, and he managed to coax GI's offer out of Rast. It discouraged him. He was not about to accept this offer, at least not yet. He didn't say so, but he had other possible partners. So when the dinner ended, nothing had been concluded.

Lim liked the idea of working with GI. "They have lots of MIT graduates," he said, "and that feels comfortable." But the HDTV race was a high-profile enterprise now, and the *Boston Globe* had published a story noting that he was looking for a partner. He'd taken calls from several interested parties, and like Rast and Heller he was negotiating with Motorola. That company seemed serious, and Lim was enthusiastic about working with them. "They have lots of MIT people, too," he said. "They're a strong company." Lim was pinning all his hopes on a deal with them. With Motorola as a partner—with all that money and expertise—maybe he really could build the winning system, beat Woo Paik and everybody else. Maybe he could *save America* after all.

Back in San Diego, nobody was terribly worried about the disappointing talks between Rast and Lim. If Lim was going to throw up absurd demands, well, the VideoCipher Division could do without him. Then on December 10, Motorola gave Lim its decision: Sorry, but we're not going to make a deal with you. Building a digital system is going to be a huge task. We can't free up an engineering staff for this. Motorola probably had a fairly clear idea of just how big an undertaking this was. Lim didn't know it, but Rast had showed them with that slide show in the conference room a couple of months earlier.

Lim was deeply disappointed, and he was also quite worried. The deadline for paying the $200,000 testing fee was December 31, just three weeks away. He had to find a partner. If not, on January 1 he would lose his test slot, and along with it any real value as a potential partner. He would be out of the race. His shot at saving America would vanish. General Instrument, he thought. Somehow I've just got to get their attention.

In late December, Lim phoned San Diego again, and word of that call rocketed through the offices like the crack of artillery fire. "He's negotiating with Scientific Atlanta. He's trying to make a deal with *Scientific Atlanta!*"

Scientific Atlanta was the second largest maker of cable TV boxes, and General Instrument's most ferocious competitor—"Darth Vader," as Rast put it. Just mentioning the company's name was like waving a red cape in front of a bull. Lim let the VideoCipher Division know that he was in "very serious discussions" with the Atlanta firm. He was vague about who had called whom. "I can't remember whether Scientific Atlanta approached me or vice versa," he said in an airy tone. To GI it didn't really matter. Quickly the word came down from the top: Get him out here. Make a deal. Do it now.

So Lim arrived in San Diego on Christmas Eve and spent Christmas Day at the Marriott Courtyard Hotel, just down the street from GI's offices. In a rapid series of discussions over the next couple of days, both sides modified their positions: Lim scaled back some of his demands and GI offered him a little more money while also promising to pay his test fees and even to provide occasional cash grants to his program at MIT. Finally they reached a deal in principle, and by the first of January GI had paid Lim's test fee; his test slot was now secure. A few days later, Heller flew to Cambridge to work out the last details. He'd been away a long time now. Heller had spent so many years in California that he'd forgotten what winters in Cambridge could be like, so when he climbed out of the cab in front of Jae Lim's office he stepped onto a sheet of ice, broke through it, and soaked his shoe and sock up to the ankle. Heller grunted with disgust.

That night over dinner at a local seafood restaurant, they hammered out the full arrangements. GI would build two complete systems for testing, Woo Paik's and Jae Lim's. One would be called the General Instrument system, the other the MIT-GI system. Lim and Paik would decide the technical parameters of their joint system, and on paper, Lim would have the final say. In truth, however, Paik would wield the most influence. It was Paik's lab and GI's money, after all. Lim did get an important concession, though: since his background was in audio engineering, Lim would be in charge of the audio system.

Lim balked again when Heller suggested that Lim would receive a higher payment if his system won. If you have a financial incentive to favor your system, Lim argued, then you'll put more energy into that one. Heller agreed; Lim's share of the profits would be the same no matter which system won. They seemed to be finished, and Heller relaxed. But Lim had one more condition. He was in this race to *save America,* after all. He was worried that GI might be the subject of another takeover one day.

"GI is strong," he explained later, "but maybe not strong enough. They're only a one-billion-dollar company. But Sony, Matsushita, are much larger and stronger." Even without a takeover, it was more than possible that GI might decide to merge with one or another of the other contestants sometime later, just as Heller was doing now. If that happened, god forbid, Jae Lim's system might fall into the hands of *a Japanese company*! There was no way he would ever permit that. So Lim issued his final demand: GI could not make any partnership agreement with a foreign company without getting his permission first.

Heller wasn't particularly concerned about that. OK, he said. What's the harm?

But of all the agreements he and Lim made that evening, there was none his company would grow to regret more.

At Zenith, Luplow was incensed by GI's pact with Jae Lim. From the Bunker in Glenview, he saw right through it.

"This is crazy," he said. "All they've done is buy themselves a retest." That seemed even more obvious after the shoot-out to determine what kind of system Paik and Lim would build. Early in 1991, Lim flew out to San Diego, and down in Paik's lab they ran computer simulations to compare Lim's proposed compression algorithm with Paik's. Paik still hadn't built anything; he'd spent all of 1990 working on his digital system for conventional cable TV. He turned back to HDTV only in January 1991, with GI's date at the Test Center less than a year away.

For this little in-house competition, Paik was both a contestant and a judge. Not surprisingly, he chose his own system. (Later, even Lim admitted that Paik's was better.) Lim proposed a few technical add-ons. But the truth was that the original GI system and this new MIT-GI system were going to be quite similar. So Luplow called foul. Why should GI get two chances to test essentially the same system? He asked Wiley to intercede.

By now Wiley was accustomed to these flare-ups; his children always seemed to be squabbling. Normally he looked for some sort of accommodation to smooth things over. This time, he simply asked Heller and Rast to give him something to show Luplow that the two systems were different in some way. Heller and Rast thought about it, and finally they came up with what seemed the perfect solution: proscan. It could solve a lot of problems.

Fifty years before, when General Sarnoff's engineers and others on the National Television Standards Committee designed the country's television standard, the technology of the time did not allow them to fit even the prescribed 525-line TV picture into a 6-megahertz channel, that figurative 6-inch pipe. So they devised a rudimentary compression technology, called interlacing, that took advantage of broadcast television's extraordinarily fast transmission speed.

TVs receive sixty picture frames every second, and at that speed the picture changes faster than the eye can perceive. In fact, even if the rate were cut in half, to thirty frames a second, the changes would still come too fast for anyone to notice. So the founding engineers decided to send half of each frame every thirtieth of a second followed immediately by the other half. This cut the size of the signal precisely in half so it could fit into the channel. With that, "interlacing" was born. Now every broadcast TV signal worldwide is interlaced. Before leaving the TV tower, television signals are sliced into dozens of horizontal strips. First, the even-numbered strips are transmitted, then the odd-numbered strips. In the home, these alternating strips of picture are recombined, or "interlaced," inside the TV set to make a complete frame again.

For more than thirty years interlacing worked well enough and hardly anybody complained, even though the process did harm the picture in small ways. The most significant problem was picture shimmer. Sewing together all of these alternating strips thirty times a second causes the picture to sparkle, shimmer, or "boil" ever so slightly. Few viewers notice, however; TV pictures have always shimmered, so most people never knew they weren't supposed to. The other major problem is that the alternating strips don't always match up perfectly. As a result, horizontal lines on the screen—the painted lines on a basketball court, the stripes on a football referee's jersey—sometimes appear to be jagged. The problem is particularly pronounced with text. Small letters are hard to read; they look as if they have been chopped up. But hardly anyone complained about any of these "artifacts," as the flaws are called. TV had always been that way.

That started to change in the early 1980s, as millions of Americans began buying personal computers. Many of the computers came with interlaced monitors, which were cheap. The interlace shimmer—not particularly bothersome on television—was quite irritating on computer screens, which are typically viewed from only about eighteen inches away. The jagged horizontal-line problem made tiny letters and numbers messy and difficult to read. So the computer industry adopted "progressive scan" monitors for all but the least expensive computers. Salesmen began promoting "Non Interlaced" monitors as an essential part of any top-of-the-line computer system.

With progressive scanning—"proscan"—an entire frame is presented at one time, "painted" onto the screen from top to bottom. Proscan removes picture shimmer and leaves the text easily readable. But progressive scanning also eliminates the compression that interlacing provided. The signal is back to being twice as wide, since all of it is being sent at once. Computer manufacturers didn't care about that; they weren't broadcasting the signal anywhere. Broadcasters, on the other hand, swore by interlacing. All their production procedures depended on it, all their equipment was made for it, and hardly any of their viewers were asking for change. For years, the interlace-progressive question had remained an obscure technical issue. But once this arcane debate spilled over into the HDTV race, it grew almost overnight into an explosive interindustry argument that in time threatened to consume the contest.

Almost as soon as General Instrument announced that it had invented a digital high-definition television system, the computer manufacturers began to see that the new TVs could serve as computers, too. No one knew precisely what that meant, but computer executives began to imagine huge new markets if, one day, TVs and computers could be combined. So they started demanding that the new televisions use progressive scanning. That pitted them against the broadcasters, who wanted nothing to do with proscan. In 1991 the arguments were beginning to turn ugly.

Zenith and AT&T announced that their new system would use computer-style progressive scanning. That meant they could not compress the signal as much, and therefore could not provide a picture as sharp as the others. But they argued that eliminating the interlace shimmer and blur more than compensated.

GI's system was interlaced, and Sarnoff planned to produce an interlaced system, too. So Rast and Heller got an idea: Why don't we make the MIT-GI machine a progressive scan model? That will differentiate it from our original entry; we'll be able to grab this second test slot and hedge our bets. We don't know how the interlace-proscan debate is going to turn out. We'll be covered no matter what happens.

At MIT, Nicholas Negroponte and others had already begun arguing loudly that HDTV ought to use progressive scanning. Jae Lim heartily agreed, so GI offered the idea to Wiley, who figured progressive scanning might be enough to demonstrate that the MIT-GI entry was different from the GI entry. It seemed to quiet Luplow.

One squabble was resolved. But this little argument over GI's second entry was a picnic compared to the lying, cheating, and puerile quarreling that lay ahead.

3

Tests

10 Golden Eyes and the Hoover test

The Advanced Television Test Center was born of cynicism, and it endured a difficult childhood. In a small and unassuming postmodern office building in the Washington suburb of Alexandria, some of the broadcast industry's best engineers suffered warring board members, broadcast industry obduracy, technical disputes, and the last-minute change that forced them to prepare for evaluating digital systems no one had ever seen before, much less tested. But at last, after several delays, in July of 1991 the test center was ready. Testing was about to begin, and the people who ran the place were taking their work very seriously indeed. They wanted no one to argue that the testing was biased or unfair. They were determined not to give losers evidence for any lawsuit challenging the selection process. "We wanted to be bullet-proof from a legal point of view," said Mark Richer, a PBS executive who watched over the testing on behalf of Wiley's Advisory Committee. One of his duties was to ensure that the test regimen was precise and unvarying, riddled with safeguards. They were determined to make no mistakes. The test center was making history.

But when the first contestant finally arrived to be judged and graded, the moment hardly felt historic, for the tractor-trailer truck that pulled up to the test-bay loading dock held the only working version of Sarnoff's Advanced Compatible Television. Jim Carnes knew full well that ACTV had not even the tiniest chance of winning. By now all of the contestants save the Japanese had announced their intention to build fully digital systems, though none of them, aside from General Instrument, was fully confident that they could do it. In fact, even as the Sarnoff truck pulled into the test center driveway, most of the Shrine's engineers were back at Princeton working feverishly on their new digital entry.

But Carnes also knew that broadcasters had pinned so much hope (and money) on ACTV that he couldn't just discard it. The less informed among them still dreamed that it could win. So Carnes had asked Wiley: Can we test ACTV and simply put the test results aside? Wiley agreed. Sarnoff, like all the contestants, was paying the bulk of the test costs, and the test center might appreciate a dry run of sorts. Peter Fannon, the test center director, went along with this idea, even though he'd been dismissive of ACTV all along.

A week ahead of ACTV's move-in date, Fannon showed the Sarnoff engineers around and explained how the eight weeks of testing would go. The test center was laid out on two levels, all of it attractively furnished with bright, modern furniture and exuding an airy sense of confidence and importance. Downstairs were the contestant-equipment bays; upstairs were a reception area, the viewing room, and offices. As the Sarnoff group was led into Charlie Rhodes's office, the chief scientist showed them the three old radios, from the '20s and '30s, that sat on a credenza behind his desk. All were in working order. The first, a large black Bakelite number, bore a nameplate that said "General Instrument, New York." The second was a wooden model with a handle on the top—a "portable" that weighed forty pounds; the manufacturer was RCA laboratories. And the final one, a brown box, "was Zenith's first model," Mr. Rhodes told the visitors. "Gives you a sense of perspective, huh?"

One floor below, Fannon took the Sarnoff team past the equipment bays. The test center had two of them so that one contestant could unload and hook up while another was finishing his tests. Each bay had wide, roll-up doors at the back leading to a loading dock. Once the Sarnoff technicians had unloaded the ACTV equipment, they were to hook it to trunk cables that hung from racks in the ceiling. Contestants, Fannon told them, had several days to connect their equipment and make sure it was working, but once that was done they would not be allowed back into the equipment room. Nobody could make any adjustments or changes to the equipment while the tests were running. This was a rigid rule. Nor would the Sarnoff engineers be allowed to wander through other parts of the lab. They would be restricted to one mean little room with a desk, a phone, a few chairs, and a single TV monitor that showed what the testers were seeing upstairs. The Sarnoff team also noted that every door in the test center had a heavy keypad lock. Test tapes were held in a fireproof, climate-controlled vault.

On the other side of a thick wall was the test bed—racks and racks of test equipment. Fannon led the Sarnoff people past all of it: On the left were five equipment racks with blank fronts. Exactly 3.03 miles of coiled coaxial cable TV cable was inside. Part of the test signal was sent down this long line, delaying it slightly so the engineers could simulate "multipath distortions"—ghosts that develop when part of a TV signal bounces off a building, say, and then arrives at the receiving antenna a second later than the rest of the signal. Another, kilometer-long run of cable rolled up on a spool sat off to one side, for cable TV compatibility tests. To the right, a computer ran a sophisticated program, written by the FCC, that provided interference similar to the problems a new HDTV channel might encounter if it were broadcast next to conventional channels—the HDTV signal on channel 2, say, and a conventional station on channel 4.

"We're going to run your system in every single television-station assignment it might be given in North America to see how it interacts with other channels," Fannon explained.

Four more equipment racks held devices intended to simulate typical sources of interference affecting over-the-air television. Sarnoff's system was going to encounter simulated interference from airplanes flying overhead, mobile radios driving past, lightning, electric starters. Finally, near the bottom of one rack, was a single switch.

What's that? one of the Sarnoff engineers wanted to know. "The Hoover test," Fannon replied, deadpan. "It's a vacuum cleaner motor."

Upstairs, a panel of expert viewers would assemble in the viewing room. These people were called "Golden Eyes," because they were experienced TV engineers trained to spot problems with the picture that normal viewers might not notice. They would sit in plush swivel chairs so they could swing around to see all the television sets arrayed along two walls. In the front, dominating the room, sat a 65-inch rear-projection HDTV screen made by Hitachi, though someone had placed black tape over the nameplate. Along the left wall were twenty-four conventional TV sets, all different makes, models, and sizes. "We have sets from every manufacturer who sells in the U.S.," Fannon said.

Wiley's Advisory Committee had hired a New York production company to tape a series of high-definition segments that would demonstrate a system's abilities in a variety of circumstances—fast-motion sequences, high-contrast scenes, and others. As each of these was played, the Golden Eyes would look for flaws and note them on a chart. After that, they'd serve as judges for the interference tests. They would watch the conventional monitors while the ACTV system was placed in dozens of simulated station assignments. In each simulated slot, technicians down in the lab would turn up the signal power bit by bit, extending the signal's coverage area, until the Golden Eyes began to see interference popping up on the adjacent, conventional channel that was being simulated on all of the regular television sets. Each Golden Eye would hold a small buzzer and stab the button at the first sign of a problem.

When it was all over, after about eight weeks of testing, the Golden Eyes would turn in their results. The test center would grade the results, write a report, and give it to Wiley.

The first problem was the Golden Eyes. Some of them were color-blind. As soon as they arrived, the test center had given them visual acuity tests, and Fannon could hardly believe it when some failed. One man hadn't even known that he had no color perception, and finding replacements proved no easy feat. Few people had the training to do this—and the freedom to spend days in a room watching TV.

Then, when the testing began, the test center engineers started noticing certain problems with their equipment. One piece of software actually tended to *reduce* the resolution of the test signal. The test center's management came to be grateful that they were testing only ACTV. As Fannon wrote in a memo to Wiley, ACTV "cannot come close to the resolution necessary to see the flaw." Still, the problem had to be fixed before the real HDTV systems came in.

But the big problem lay with ACTV itself. For all the hoopla behind its creation, for all the glare and fumes that had poured forth from the old RCA publicity machine, ACTV just didn't work very well. The results of one test sequence looked so bad that the Sarnoff engineers asked if they could go back into the equipment room and install a filter to erase a nagging flaw that kept appearing on the screens. Absolutely not, Fannon told them. That would represent a change in the system, and changing the system during the test was prohibited. At another point in the tests, ACTV projected such objectionable interference onto adjacent TV channels that Carnes asked the test center's leaders if the interference tests could be rerun. While they considered that, Sarnoff ran some tests of its own back at the Shrine and apparently got the same miserable results. Sarnoff dropped the request.

As the weeks passed, the results grew ever more dismal. ACTV "just didn't work well," Fannon explained later. "It didn't look good at all." But the curious thing to him was that the Sarnoff people didn't really seem to care. In fact, halfway through the test sequence most of the engineers returned to Princeton and went to work on their new, digital system, leaving behind a junior engineer, who sat down in the contestant room working on designs for Sarnoff's new system, holding the papers on his lap.

When the ACTV tests were over, on September 15, 1991, Advanced Compatible Television was sent off to its grave. Sarnoff packed up the ACTV equipment, drove back to Princeton, and stuck the boxes in a storeroom at the Shrine. The test center stuffed the scoring sheets in a box. To this day, no one has ever looked at them.

Even as ACTV was crashing through its last tests, a platoon of Japanese engineers was moving equipment into bay No. 2, a few feet away on the other side of a cinder-block wall. Keiichi Kubota and his men had several days to set up, but as the deadline neared they fell into a panic. The prototype wasn't working right; it seemed to be plagued with bugs from top to bottom.

For more than three years, NHK had been scrambling to create Narrow Muse for the American market, and they had barely finished in time. Finally, in the late summer of 1991, they had packed the fragile prototype equipment onto an airplane. Now at last here they were at the test center, and when they looked inside their equipment racks they were dismayed to find that some of the circuit boards had been damaged in shipping. Dozens of delicate chips had been knocked partway out of their sockets. They'd never be able to hunt all of them down in time. Kubota knew full well that "we can't ask for a delay." Not now. Not with Sikes and Wiley on the warpath. What on earth was he going to do?

Over at the FCC, Al Sikes had been growing impatient. More than five years had passed since that morning back at the Commerce Department when he was stunned to learn that the United States was backing

"Japanese hegemony over advanced television." Now here it was mid-1991, and the nation still seemed years away from deploying an HDTV system of its own. Every time Sikes looked up, it seemed, Wiley was telling him about another delay. Months earlier, he had set a deadline—1993—for completing the race, choosing a winner, setting an official new broadcast standard and loosing manufacturers to begin producing high-definition sets. But the way things were dragging along, Sikes was worried that they'd miss the deadline by months, maybe even years.

I want to see tangible progress, he warned during one of Wiley's Advisory Committee meetings in April 1991. I want to see this race move forward so we can be sure we'll meet our deadline. Wiley was a loyal soldier, and he turned up the heat on his troops. No more delays, he ordered. Let's get this moving. Move those systems into the test center and get them out. No excuses.

The engineers at Zenith, Bell Labs, Philips, and the Sarnoff Shrine heard the call. Soon they doubled their shifts. Every one of them was behind schedule—way behind. None was certain, or even confident, they could build these new systems at all, much less get them to the test center by their deadline. As Carlo Basile, a senior Philips engineer, said later with just a hint of hyperbole, "I was entering the most miserable period in my life."

The engineers who ran the test center were no happier. Wiley was pushing Peter Fannon, and Fannon was pushing all of them. Workdays stretched to ten hours, then twelve, then fourteen. They couldn't even keep track of the overtime any longer, and toward the end of the ACTV tests they threatened mutiny. Finally, the lab closed for a week in the middle of the ACTV tests. *Another* delay. And when NHK arrived two weeks later, the building fairly throbbed from the pressure. Down in equipment bay No. 2, Kubota felt it pressing against him, even as his engineers were saying that they couldn't get the system up and running in time. It wasn't just a question of pushing chips back into sockets. Each board would have to be tested and retested. There were days of work left here. They wouldn't be ready.

Is this what all our work has come to? Kubota wondered. Twenty-five years of research, and here we are, the only analog system left in the race, while all the digital systems remain unproved, unfinished, and perhaps even unworkable. Now NHK is going to be disqualified? Before Kubota could allow that to happen, at least he had to ask. He was "very nervous about this." But he swallowed hard and climbed the steps to have a talk with Fannon. We need more time, he said. Our system was damaged in shipping. We need an extra day. Just one day. Fannon had to consult with Wiley, whose view was that within reason everyone should get a fair chance to compete. In the end, it was agreed: Take the extra day. Get your system running again. As Kubota told his men, he thought: Now I am in their debt.

The NHK engineers worked without a break, and soon they had identified all the out-of-place chips. With that extra day, they got the system ready. Deeply grateful to have been given the additional time, Kubota told Fannon, We're ready to start. So Kubota and his assistants shut themselves into the contestant room, and as the Japanese watched, the high-definition images taped by the New York production company played over the monitor.

A hand reached into the frame and turned down a kerosene hurricane lamp. A worker wearing a gray jumpsuit crossed a waiting room and affixed a sign to an office door. A young man rose from a sofa and walked over to a bookshelf. None of the scenes held any intrinsic interest—they were boring, in fact—but each was designed to test various aspects of a system's performance, and right away Kubota could see that his company had dropped the ball. "It just didn't look very good," he said. Kubota had left the Narrow Muse project the year before, when he was assigned to the United States. At that time, a great deal of work remained to be done. A year later, there in the mean little contestant room, Kubota saw the truth: The prototype hadn't been improved at all since he left Tokyo—since about the time GI announced it had a digital entry.

"Once General Instrument announced, our people didn't really think we could win," he realized. "They lost interest." Kubota looked at the floor, "very, very disappointed."

Fannon, for his part, was shocked. Early in the first day of the NHK testing, he walked from his office to the viewing room and stood behind the Golden Eyes as they looked at the initial test images on the big, 65-inch high-definition screen. The pictures looked terrible! Narrow Muse wasn't nearly as good as NHK's original system. In the hallway, he turned to one of his assistants. "What's wrong with this?" he asked. "It looks like shit!"

Every Friday, the test center engineers met with the contestants to tell them how the week of testing had gone. From the beginning, Kubota had known that Narrow Muse was performing poorly. Then during week two, video noise began popping up along the left edge of the screen. It looked as though someone were scratching at the picture. At week's end he learned that his troubles were even deeper.

Sitting around a large table in the conference room, the center's engineers explained that Narrow Muse had demonstrated serious adjacent-channel interference. When Narrow Muse was played on channel 9, say, it threw heavy interference onto channel 7. During the testing, the Golden Eyes had been stabbing the buzzer almost as soon as channel 7 had been turned on. There were also other problems, which the engineers described as "phase noise modulation, channel-change and ghost-canceling delays, and picture artifacts." Despite the soft tones and neutral language, the bottom line was clear: Kubota's system was flunking.

Kubota knew he could fix some of these problems if only he were allowed back into the equipment bay to make a couple of adjustments, change some wires around. His engineers wanted desperately to do it, but Kubota told them: It's forbidden. That is the rule. Besides, Fannon had already granted Kubota one big favor—that extra day during the set-up period. He didn't dare ask for another favor now. This was the end. Kubota felt as though he couldn't breathe.

Twenty-five years of research; hundreds of millions of dollars spent. We beat the world. Just a couple of years ago, nobody else even *had* a high-definition television system. Now here we are in Washington, and we are failing the tests. Has all the work, this long, exciting odyssey, been for nothing? As bad as Kubota felt, he could not bring himself to ask Fannon to let them go back and fix the system. With a certain understatement, he later explained, "I was very, very sad."

11 Building it is your problem

Wayne Luplow's men had finished developing the transmission system for the new Zenith-AT&T machine right on time, and long before this point they were supposed to have begun integrating it with the complex device the Bell Labs boys were creating. Now their date at the Advanced Television Test Center was just a couple of months away. But every time one of the Zenith engineers called Murray Hill to find out when AT&T would be ready, Arun Netravali, Eric Petajan, or one of the others would say, "Two more weeks, just two more weeks." Three months had passed since the first time they'd heard that. They had come to call it "the sliding schedule." Now it was clear to everyone in the Glenview Bunker that AT&T was in trouble.

They're in way over their heads, the Zenith engineers were saying among themselves. *We're* the ones who know about television. They should have listened to us.

But it was too late. Once again, the two camps were barely on speaking terms. Zenith could only sit back and wait—angry, resentful, and scared. The company had fallen into even deeper financial trouble. The 1990 annual report had shown a loss of $52.3 million, and late in that year the company had reported to the Securities and Exchange Commission that it was facing a serious liquidity crisis. Bankruptcy seemed possible. Nonetheless, the 1990 annual report had also offered a promise: HDTV would "allow us to rise above the tight margins in the color television business and should provide excellent opportunities to restore profitability." A few months after the report came out, Woo Paik's invention had knocked Zenith's entry out of the race. Zenith had lost $51 million the next year, and it was clear that layoffs loomed ahead in 1992. HDTV still might save them, but their fate was in the hands of those arrogant dilettantes at AT&T. The men in the Bunker could see that those people had no idea what they were doing, but there wasn't a damned thing anyone at Zenith could do.

By the beginning of 1991, Petajan had recruited almost twenty AT&T engineers for his project. Most of them knew next to nothing about television. "All of us could at least spell 'video,' " as one of them, Jim Lauranchuk, put it. Nonetheless, in the early days the work was exhilarating. They had set out a proposed system design brimming with fancy bells and whistles. Sure, all of the contestants were building digital HDTVs, but Bell Labs' machine was going to be the smartest, *the best.* Kim Matthews, a quiet, bearded young man who'd joined the project early to help design the compression algorithm, quickly discovered that he and the others were also working on "the most visible project at AT&T." Senior executives always seemed to be stopping by. Reporters trooped through every few weeks. The HDTV team even got its picture in the company's annual report.

By late winter 1991, however, the glory had faded as they fell further and further behind schedule. One by one, the bells and whistles were dropped from the system design, and Petajan's team began to see that they would be lucky just to produce a decent TV picture within the allotted time. The system was due at the test center in September. They were supposed to marry their work with Zenith's transmission box a couple of months ahead of that. It seemed hopeless.

In February, Petajan changed the work schedule: Everyone was to start coming in at nine o'clock in the morning and stay at it until ten or eleven at night, and they were going to start working on Saturdays, too. They ordered pizza every night, until several of them complained that they were gaining weight. Then they tried sandwiches. Finally, Bell Labs brought in a caterer. After dinner each night, as they turned back to their circuit boards, every one of them was thinking "this schedule's going to slip," as engineer John Mailhot put it. "It's just got to slip."

Occasionally one of Luplow's men would fly to New Jersey to have a look at the work in progress and check on some portion of it that bore on Zenith's part of the design. Usually the Zenith engineers hated those trips. When Dennis Mutzabaugh walked into the lab, he felt as if the Bell Labs boys who looked up at him were thinking, You guys don't know anything. Just hook up your equipment, keep your mouth shut, and get out of the way. What he saw was a dozen steel equipment racks laid out in a rough semicircle and thick tangles of cable snaking between them. A two-ton air conditioner roared in the background, trying to keep this welter of electronics cool. One or two engineers sat on stools in front of each rack; beside each of them tools and test equipment had been laid out on carts so the engineers could easily roll the gear around the machine. Listening to the chatter, Mutzabaugh began to see the problem: "Their approach seems to be: 'We want to prove our philosophy,' rather than build a workable product. They're research people, mostly interested in proving a point."

Longtime Zenith engineer Ray Hauge said, "They weren't thinking about the whole system, how they'd connect with the next board. They just wanted to prove that *their* board was going to do what it was supposed to do." AT&T's engineers heartily disagreed with that notion. At the same time, though, the Zenith engineers said they kept spotting fundamental errors. Bell Labs was trying to run a TV signal down a hundred and fifty feet of cable, one of them noted, even though television images degrade with every passing foot in a wire. "They just didn't know," Mutzabaugh said. But he couldn't bring himself to tell them. The partners were estranged, hostile once again. Mutzabaugh was convinced that "if our marriage breaks up, AT&T will survive, but I don't know about Zenith. If our transmission system works, and we fall down because of their compression system, it'll really gall me."

Zenith engineer Tim Laud had attended AT&T's kickoff meeting almost a year before, and he'd listened as the Bell Labs boys talked about their plans. "No white wires!" they had said then. In a prototype design, a white wire is used to correct a mistake. "We don't have time to build the system and then correct it," Laud heard them say at the start. "We have to build it right the first time." A year later Laud went back to Murray Hill. Looking over the welter of cables strung between the equipment racks, he counted at least two hundred white wires.

Then Laud started pulling at some of the cables; he was trying to determine where exactly Zenith would plug its transmission system into AT&T's compression coder. The Bell engineers pointed to some wires and plugs, and Laud puzzled over them for an hour or more. Nothing looked right. At first he couldn't figure it out. Then finally it hit him: "They built the interface connector upside down!" Back in Glenview,

Laud reported this, and his supervisors knew what they had to do. At this point, they said, it's hopeless to ask AT&T to fix it. We'll just put ours in upside down, too.

Arun Netravali, the senior project manager at Bell Labs, had his own complaints. He talked with Luplow or somebody else at Zenith almost every day, and each time it seemed they had a new demand. For example, "They told us we need to have a more ambitious motion estimation system. They kept saying you have to have it for sports." That was galling enough, considering how dismissive Zenith had been of motion estimation during the first round. But if he questioned requests like this one, he said, the Zenith people would tell him, "Look, we know television. You don't. And this is what we have to do." As Zenith's leaders piled on demands, Netravali said he also got another clear message: The Zenith people were telling him, "Building it is your problem." So as the weeks and months passed, the tone of the phone calls grew ever more rancorous and resentful. With all this interference, the Bell Labs boys were thinking, it's no wonder we're falling behind.

As the summer of 1991 turned to fall, Jerry Pearlman, Zenith's president, grew ever more worried with every passing day. Various delays at the test center had pushed Zenith-AT&T's start date back from September 1991 to a point closer to New Year's. But even this new, borrowed time was passing with no appreciable progress at Murray Hill. As he drove to work at the Bunker each morning, he passed roadside Christmas tree lots. Now Pearlman was calling Bell Labs nearly every day. Their time was almost up. "We were supposed to have a month to tune the system," he said. "Then it was a couple of weeks. Then days." The way this was going, it could turn out to be "hours—then nothing."

In San Diego all the while, Woo Paik had been fighting his colleagues' doubts about his ability to build a transmission system. He settled the debate by doing it. In January 1991 he finished his work on the digital converter for conventional cable TV, turned back to his HDTV project, and worked at it steadily for nine or ten months. By the time GI's date at the test center came up, Paik had finished the prototype, complete with a perfectly valid transmission system, and he had built a second, backup system, too.

The world's first fully digital, high-definition television system was ready for testing.

On November 12, Peter Fannon, the test center director, wrote Bob Rast, saying, "This will confirm our understanding that you plan to move the GI DigiCipher system into the Test Center on Tuesday, November 26. We will have a video and photo crew to record your move-in and setup, and you are welcome to ask the crew to take such pictures as you may like for your records and use. Should you wish to have a 'photo opportunity' in your ATV equipment room with representatives of the press once you are set up, this can be scheduled."

On November 26, the GI truck pulled up to the test center loading dock in Alexandria and unloaded just two racks of equipment. GI technicians carried them into the equipment bay and quickly hooked them to the overhead cables. The truck driver then went on to GI's offices in downtown Washington, and the backup system was set up in a small room there. Rast was so confident that Paik's digital prototype would work—and work well—that he planned to run a continual demonstration in the Washington office, even as the other prototype was being tested. NHK had come to the test center with half-a-dozen racks of equipment, nearly two dozen engineers, and an intense, diligent manner that never let up. GI showed up with two racks, three engineers, and a Southern California approach: Hey, no problem, this will be fine.

Nobody at the test center had really known what to expect when they hooked the world's first fully digital, high-definition television to their test equipment. Is this going to be like testing a jet engine in a lab set up for propeller craft? Are they going to blow us away? What they got was one of the cleanest television signals they had ever seen. Hardly anyone ever noticed all the flaws in conventional TV, but when suddenly they were removed and the resolution was doubled at the same time, the high-definition images were mesmerizing.

"Hot damn!" Fannon declared as he stood in the viewing room watching as the first test images played across the big screen. "Now we're getting somewhere. *This* is HDTV!" Over the following days, GI's system did demonstrate problems. Overall, though, the Golden Eyes and other test center personnel were stunned to see how easily GI's digital signal flowed through their equipment and handled the various obstacles thrown its way.

The most worrisome early difficulty came when the digital signal was broadcast alongside a traditional analog TV signal. Down in the test lab, the engineers reported that their equipment was showing that the GI system was throwing significant interference into the conventional

TV channel. Up in the viewing room, however, the Golden Eyes couldn't see a thing. The regular TV picture looked just like it always had. Finally the engineers tried a different approach. They turned the HDTV signal on and then off, to see if the Golden Eyes noted any difference in the conventional signal. They did. The interference had been hard to see because it looked so much like the normal "boiling" endemic to all interlaced TV broadcasts. Even digital's *flaws* were better than those on conventional TV. While GI's system did cause this interference on adjacent TV channels, by and large it performed remarkably well. In fact, it even seemed to have the ability to self-correct!

Toward the end of the test period, a troubling "green shift" appeared in the test signal: the system's color accuracy slipped. By this time, GI generally kept only one or two junior engineers at the lab, and during the scheduled meeting at week's end, test center engineers showed them the problem. Just like Keiichi Kubota a few weeks earlier, the GI engineers knew they could fix it if they could get back into the equipment room and readjust the dynamic-range controls. At a similar point a few weeks earlier, Kubota had agonized, but GI's engineers felt no such torment. When the meeting ended, one of them walked down to the equipment room, pulled a circuit board out of the rack, found the faulty setting, and adjusted it. As he finished, he noted what he'd done in the daily log and then went back to his hotel. When the tests got under way the next day, the test center engineers were amazed. The green shift was gone! Wow, several of them thought, this is one fine machine.

Charlie Rhodes, the test center's chief scientist, wasn't so sanguine. Something suspicious is going on, he realized. Colors don't just change like that on their own. As GI's tests drew toward their close, Rhodes decided he had better take a closer look.

Monday, November 25, 1991 was an official holiday in Japan. By government decree it was "Hi-Vision Day," the grand debut for regular daily broadcasts of "Hi-Vision," NHK's HDTV system, known as Muse in the United States. The date, 11/25, was not chosen by accident; 1,125 was the number of lines on the HDTV screen. In Washington that very week, General Instrument engineers were hooking the world's first all-digital high-definition television system into the Advanced Television

Test Center's equipment room. But in Tokyo, the first HDTV system, the grandfather to DigiCipher, started near full-time broadcasts. Beginning that morning, NHK transmitted Muse programming by satellite nationwide for eight hours every day. The nation celebrated with speeches, demonstrations, and proud pronouncements. "This is a curtain-raiser for the Hi-Vision era," boasted Akio Tanii, president of the Matsushita Electric Industrial Company. Across the country, Japanese crowded in front of huge high-definition monitors set up in public buildings and department stores.

Still, the moment was bittersweet. As the *New York Times* Tokyo bureau chief observed in his account of the day: "Although they toasted the occasion, the nation's broadcasters and electronics executives spent the day wondering whether, after two decades of work and billions of dollars in investments, they were backing an aging television technology that might burn out before its first commercial break." Worse still, even the Japanese who were impressed with the pictures they saw on department store monitors were unlikely to walk over to a salesman and buy one. The sets were selling for $30,000 each, on average, and that was only the first expense. No one was likely to buy an HDTV without a high-definition VCR to go with it, and NEC had introduced a fancy new one just in time for Hi-Vision Day. Its price: $115,000.

A few years before, the Japanese government had predicted that almost a hundred and fifty thousand Japanese homes would have HDTVs by now. The sad truth was that manufacturers had sold barely two thousand sets. The manufacturers seemed unwilling to lower prices to encourage greater sales, preferring instead to take the huge profits from the few sets they were selling. "The price is coming down a little more slowly than we hoped," Haruo Kurakata, a spokesman for NHK, observed with considerable understatement.

As if the prices weren't problem enough, there wasn't much high-definition programming available for the new service. Nobody else in the world was producing any. NHK had to fill all this airtime with its own material. The network set up separate crews with their own high-definition equipment, making the broadcast of this programming an extraordinarily expensive enterprise. On Hi-Vision Day, the Muse broadcast opened with a startlingly clear and realistic tape of a rocket blasting off. But the programming seemed to fizzle after that. None of the nation's favorite shows were yet available in high-definition, so toward day's end, NHK broadcast the finals of an important sumo tournament. Most Japanese may not have seen any larger meaning, but it happened that a wrestler named Konishiki won. He was American.

General Instrument left the test center positively triumphant. They had passed almost every test. Woo Paik's digital system had shown some interference problems, but that was fixable. The real news, as GI saw it, was that they had the only working digital, high-definition television in the world. Zenith and AT&T couldn't claim that, even though they were carrying their gear into equipment bay No. 1 as GI packed to move out of bay No. 2. AT&T's "sliding schedule" had slid right up to moving day. The agreed timetable giving the partners three months to join the system's halves had indeed slipped until finally they had "weeks, days, hours . . . nothing," as Pearlman had warned. Zenith and AT&T began final assembly—hooking Zenith's part up to AT&T's, sorting out cables and connections, debugging the inevitable unexpected problems—right there in the test center equipment bay. Bob Rast watched with amusement, looking for advantage. These people are crippled, he thought. They might not make it. And of course, the fewer competitors the better. This is a race, after all. Maybe if we watch closely, maybe we can move in for the kill.

But in the meantime, Rast began telling others at GI, it's time to show off what we have. We ought to put on a flashy demonstration.

For the last few weeks, GI had been running a demonstration using Paik's backup prototype over at the Washington office. But the rooms were cramped. This hadn't been very effective, so Rast and the others decided to put on a truly dazzling demonstration and invite the other contestants. They've got nothing to show, Rast said with a grin, "so, gee, they'll end up having to say how great *our* system is."

For several days, GI's leaders talked about where to stage this demonstration and get the highest impact. Maybe the Air & Space Museum? The White House? They recalled the Muse demonstration five years earlier, in January 1987. None of them had seen it, but by now everyone at GI knew the history. NHK's demonstration in the FCC meeting room and at the Capitol had been the first act in this long drama. Without it, they probably wouldn't be here now.

We've got to do it in the Capitol, they concluded. It's fitting, it's right. The Japanese had demonstrated their system in a Senate office building just across the street from the Capitol building itself. Rast and Quincy Rodgers figured they would go one better. GI won permission to run a demonstration in an attic gallery right next to the Senate chamber itself.

"We're going to shake up the world," Rast said with a cunning smile.

Back at the test center, Charlie Rhodes was investigating the GI mystery. How exactly had that green shift just gone away overnight? "That doesn't happen all by itself," he said. Rhodes pored over the test data until finally he discovered what had happened. Actually, it was right there in GI's own activity logs: that young engineer had gone back into the equipment room and adjusted the machine. No wonder the system had done so well the next day. GI had cheated! And when Rhodes told Fannon, "the roof fell in," as Fannon put it. "We have a rule," he angrily declared when he got Rast on the phone. "It was clearly explained to you." The system has to be tested exactly as delivered, and you *changed the system*!

Rast said he hadn't known anything about the tinkering. Let me check and I'll get back to you, he said, then set out to find out what happened. Paik told him he knew nothing about it. He and Paik asked the young engineer who had been left behind to manage the testing.

Yeah, I readjusted the dynamic range, he said. Rhodes said we could.

Rast called Fannon back and repeated that, pointing out that the engineer had noted what he had done in the daily activity log. On the other end of the conference call, Fannon and Rhodes exploded: We gave you permission? That's nonsense!

Rast pulled the receiver away from his ear. "It sounds like I raped someone's mother!" he said. But he stuck to his version: Rhodes told us we could do it.

The argument went on and on. The dispute was never resolved, and finally Fannon simply decided to write that into the test report. Rast closed the issue by repeating his contention that if they had cheated they wouldn't have written this into the logs. "We've had a big misunderstanding."

Explaining it later, Rast liked to say, "There was a massive miscommunication. I think Charlie Rhodes has a memory problem." The altercation was duly noted in the test report, but nothing ever came of it. At the test center, though, General Instrument would never be viewed the same way again—though at that particular moment Rast didn't really care. The demonstration at the Capitol was less than a week away.

A couple of days before the big show, Quincy Rodgers helped Paik get the equipment into the Senate gallery and then up a narrow, twisting staircase to a loft just above. The Senate chamber was on the other side of the wall. The ceiling here was high, good for air circulation. But Paik was still concerned; the equipment was sensitive to heat. Can't we stick an air conditioner out there? he asked, pointing to a small window that looked toward the Supreme Court.

"No, Woo," Quincy said with a bemused smile. We can't stick an air conditioner out the side of the United States Capitol.

Two days later, Monday evening, March 23, 1992, the guests began arriving. The invitation list was not as long as NHK's had been five years before. The room was small, and so the audience was a select group of about fifty people. In the front row sat Al Sikes and the other FCC commissioners. Behind them were Wiley and other leaders of his Advisory Committee, a scattering of senators and congressmen, NAB officials, and assorted others. GI had little trouble getting all of them to come. This was a demonstration of an American success story, after all, and by 8:00 P.M. the guests had settled into their seats.

Quincy Rodgers stood in the back. What a great turnout, he thought. At the same time, though, like every GI employee there he kept thinking, "God, I hope the damned thing works." Donald Rumsfeld, GI's president, had been ruminating for days about what to do if the system didn't work. What would they say; how would they explain? "I mean, really, we have no backup" equipment readily available, he told the others at GI. "It might fail. Really, this is just an R&D project...."

The lights dimmed, and everyone looked at the 65-inch, wide-screen Hitachi high-definition projection television at the front of the chamber. GI's demonstration tape was to be broadcast here from WETA, a PBS station in suburban Virginia about seven miles away. The screen clicked on just as the regular PBS programming was interrupted with an announcement: "This station will be going off the air for about twenty minutes so we may conduct a historic test, the first over-the-air broadcast of digital, high-definition television." With that the screen went blank.

Rodgers sucked in his breath and waited. Suddenly the screen clicked on, bright as day. Static and noise. That was it. Then just as quickly the screen fell dark again.

"Oh god," Rodgers muttered to himself, staring at the blank screen. Nothing happened. Minutes seemed to pass. Rodgers started to sweat. The air felt deadly still. The silence seemed to roar. Then from upstairs in the loft Rodgers heard a voice babbling away in Korean. It was Woo Paik, and his tone was high-pitched, panicky. A moment later Paik came racing down the stairs, talking rapidly into a cellular phone. All eyes followed as he adjusted something behind the monitor, then bounded back up the steps, still jabbering in Korean into the phone.

Moments later the screen came back on, though the picture was not high-definition. The demonstration tape was playing now. A square-shaped, conventional TV signal filled only the center of the wide screen. This TV within a TV showed stars in the sky as a woman's voice intoned, "For most of the time, the sky was silent, full of silence. Then in November 1920 that silence was shattered forever [by the first radio broadcasts] and the world would never be the same." Through the following eight minutes, she gave a brief history of broadcasting, then introduced Harry Blackstone Jr., a silk-tongued magician. He threw his cape over an old TV set in the foreground, then pulled it off to reveal a new, wide-screen TV.

"Now it's time for you to listen to and look at the world of tomorrow, today," Blackstone said as he glanced down at this new-looking TV. An American flag popped onto the screen as Blackstone announced, "This is HDTV!" Slowly, as a drumroll played, the camera closed in on the wide-screen TV. Then with a flash the big Hitachi monitor filled out with a full, high-definition image of the same American flag, transformed suddenly from a fuzzy, conventional-TV view to a clear, perfect, high-definition image. The effect startled everyone in the room. Several people gasped.

"God, would you look at that!" exclaimed one of the FCC commissioners in the front row, his words caught on tape GI was recording. The demonstration continued with clips from *Top Gun;* F-18 fighters roared off of carrier decks and the engine noise roared through stereo speakers, just as it had when NHK showed similar footage from the same movie five years before. Then came a rodeo, aerial views of the Grand Canyon, flowers, tropical fish. All of it flowed by so velvety smooth and clear that even the cynical government officials in the room—knowing full well that "this was a political event," as FCC Commissioner Ervin Duggan said later—couldn't help but smile as they watched. And when the demonstration ended fifteen minutes later, the applause was long and loud, obviously heartfelt. Most in the audience rose to their feet. Here was an unambiguous American achievement. Everyone could be proud. The Yellow Peril had been vanquished. At last, *America* was in the lead. GI was flying now.

The press coverage was generous, not to say fawning. Most of the people who had attended the demonstration gushed. Sikes said he was "thrilled."

"What has been accomplished in just a few years is amazing," bubbled Margita White, president of the Association for Maximum Service Television. Wiley called GI's system "a giant step," and from then on his privately held, disparaging view of the contestants in his race began to change. "HDTV has this mosaic quality, like a fabric that flows," he started saying. "I love it."

Not surprisingly, some of GI's competitors were not so generous. This first demonstration doesn't mean much, Joe Donahue of Thomson said dismissively. "NHK demonstrated its *analog* Muse system first. Does that mean they're going to win now?"

Wayne Luplow of Zenith was scornful. "The transmission was only seven miles," he said. "If that's as far as they can go, they'll leave out all the people in the suburbs." But the truth was, General Instrument's demonstration stood as the smallest of Zenith's problems just now.

12 It broke my heart

As bemused Advanced Television Test Center employees looked on, Zenith and AT&T scrambled in equipment bay No. 1 for two long, difficult weeks, trying to meld the two halves of their system into a coherent whole. This was supposed to be a leisurely time for hooking their machine to the test center trunk lines and running some preliminary signals back and forth, just to see if everything worked. But as Fannon, Rhodes, and the others could clearly see, Zenith and AT&T were still *building* their machine, working around the clock to get it finished in time for the first day of testing. Two days after the tests were supposed to have begun, Fannon warned, "If you're not ready by noon, I'm going to have to call Wiley and declare a non-start." Well, they weren't ready by noon. But without saying anything, Fannon gave them the rest of the day, and the partners finally finished late that night.

Testing started three days late, and even then nobody from either company really knew if the system would work. At the last, they'd shoved parts into sockets, hooked cables to boards, not entirely certain the connections were right. When the first day of testing dawned, all of them crossed their fingers. Zenith and AT&T each kept three-man teams at the test center, including Rich Citta of Zenith along with Eric Petajan and Arun Netravali from Bell Labs. Fannon advised them that they would have to find a way to make up the lost three days. But then worse news came almost immediately.

Petajan, Citta, Netravali, and a couple of others were locked into the tiny contestant room, watching the monitor as test center technicians ran the test material through the system in one long burst—several dozen short sequences. This was just to see if everything was working properly; nobody was taking measurements yet.

"Oh, shit!" Netravali exclaimed, as a bold white vertical line appeared on the screen. "What the hell is that?" There was *no way* the system could win with a flaw like that. So they called Fannon immediately. The system is broken, they said. They needed time to track down the problem. Fannon had been here before; all the systems tested so far had faced problems, though none quite so visible. He was firm: under the rules, testing was to continue anyway. It did.

That weekend Citta and the others on his team found the culprit, a circuit board in Zenith's part of the system. On Sunday at 3:00 P.M. Chicago time, Ray Hauge, one of the engineers who had stayed behind, got a call at home. We've got a bad chip on one of our boards, he was told. We've got to replace it. Get into the office—*now*. Hauge lived in the country, an hour out of town, but he hurried in and immediately set to studying the problem on a "shadow system" Zenith had left behind at the Glenview Bunker. He found it fast enough. An integrated circuit was programmed incorrectly.

Hauge wasn't surprised. As he saw it, they'd worked so fast that it was "like trying to tie your shoelaces while you're running in a race." The solution here was simple enough: "Pop out the old chip and put in a new one with the correct program," he said. That's exactly what Hauge did on the shadow system, and the problem seemed to go away, as best as he could tell. Zenith had a shadow system for only its part of the machine, so some of this was guesswork. Still, late that evening he sent a replacement chip by air express to Citta at the test center.

Monday morning, Citta's team plugged in the new chip while the Bell Labs boys looked on, privately gloating to themselves. Their biggest problem so far hadn't come from AT&T's sliding schedule or the hurried final assembly in the test center equipment bay. No, after all those screaming phone calls, all that seething resentment beamed at them from the Bunker in Glenview, "in the end it was *Zenith* that made the mistake," Petajan observed with obvious satisfaction. "None of our stuff failed to work properly."

During the NHK test six months earlier, Kubota hadn't dared to fix some of his system's problems because he'd known that would have been a violation of the rules. A few weeks later, GI had shown no such reticence. Now Zenith, too, had made an illegal repair, but Citta didn't try to hide it. He told Fannon exactly what had been done. With no hesitation Fannon told him: You've got to take that new chip out and put the old one back in. You can't make changes to the system while testing is under way. Citta and Luplow couldn't believe it. We made a *mistake,* they argued. We put the wrong *chip* in. It's just a *mistake,* not a design failure. If we'd put in the *correct* chip, as planned, this never would have happened. This isn't a design error. It's an *implementation error.*

Right there Zenith introduced what came to be the most abused phrase of the HDTV race. It was almost if an aircraft manufacturer had explained, "Our prototype aircraft didn't crash on landing, killing everyone on board, because we designed it incorrectly. No, somebody just forgot to put tires on the wheels. It was an oversight, an *implementation error.*" Fannon listened to this argument and concluded that he couldn't decide it on his own. He wrote a turgid memo to Wiley: "The proponent reported that the source of the 'cyclical artifact' that occurs under certain conditions has been tracked to a particular integrated circuit (IC). The proponent said that the IC was not programmed correctly and, therefore, the advanced television system has not been operating 'correctly' (i.e., as designed and certified). . . . Addressing this matter may be more complex than similar situations in the past."

A few days later, Wayne Luplow faxed a letter of his own to Mark Richer, the PBS executive who oversaw testing for Wiley's Advisory Committee. During testing, he wrote, "an unexpected vertical bar of interference was noted." This was "caused by the insertion of incorrectly programmed ICs which we can trace only to 'human error.' " We want permission to change chips, he said, to correct this *implementation error.* Richer, too, bucked it up to the Advisory Committee chairman, and Wiley quickly decided what ought to be done. The Advisory Committee, in his view, had not been established to disqualify contestants. Wherever he could, within the rules, he wanted to cut some slack, keep everybody in the race, because Wiley believed that was the only way to produce the best possible HDTV standard for the country. So on April 8, a month after the Zenith tests began, Wiley sent a letter by fax to every member of his Advisory Committee. Based on the recommendations of his engineers, he wrote, "I have concluded that a retake of some of the tests that have been run on the Zenith/AT&T system is in order." He explained the situation, and added: If you disagree, let me know by tomorrow. Then Wiley called the other contestants.

He explained the situation to Keiichi Kubota, who didn't like it one bit. If NHK had been allowed to fix its problem, Narrow Muse would have performed much better. Why should Zenith be allowed? But NHK didn't want to cause trouble. We cannot support this, Kubota said, but we will not oppose it either.

Then, on April 10, Jim Carnes picked up the phone at the Shrine and listened as Wiley explained the Zenith problem. We've got to do a retest, Wiley said, and Carnes couldn't help but smile to himself. He had been trying to manage the panic spreading like a virus through his own lab. Their test date was coming up fast, and they still couldn't get their prototype to offer even a hint of life. We'll never make it, his engineers had been telling him. Now Wiley was asking if the Sarnoff consortium would agree to let Zenith stay at the test center two extra weeks. Normally in situations like this, Carnes would have conferred with his partners before answering. This time, though, he didn't hesitate even a beat. "Sure, Dick," he said with a Cheshire grin. "Give them all the time they need. That's the best thing for the country." Then he called his engineers and told them the good news: We've got two more weeks. *Two more weeks!*

Finally Wiley called GI, and Bob Rast was not nearly so agreeable. This just doesn't smell right, he was thinking. Zenith is "pulling a fast one." Rast told Wiley, "I think they may have a design flaw in their system, and now they are trying to fix it by saying it's an *implementation error.*" Wiley took the complaint under advisement. But the more Rast thought about this, the angrier he got. "They're misrepresenting what happened, I just know it," he said. "This is bullshit!" He took it up with Rumsfeld, GI's president.

This is outrageous, Rumsfeld told Rast. "They can't do this. You lay down rules. We played by the rules. These guys at Zenith didn't. Now they're moving the finish line in the middle of the race." So Rumsfeld told Rast: Let's not let them get away with this.

Rast pressed. Finally Wiley made a Solomonic decision: In the interest of keeping on schedule, let's continue testing with the old, flawed chip. Then at the end, we'll rerun the affected tests with the new chip. Meantime, we'll continue discussing this and then hold a meeting to decide which of these test tapes will be judged.

Mark Richer, the chairman of the subcommittee in charge of testing, was a smart, savvy, and sardonic young man, and he too was skeptical of Zenith's explanation, though as an Advisory Committee leader he generally kept his views to himself. The decision over which tape would be considered fell to his committee, so he scheduled the meeting for late April. Rast had been through his own inquisition over the problem during GI's tests a few weeks before, and he prepared a list of questions for this one. Now he was ready to throw accusations at someone else. "Were the errors design-related or implementation-related? Please provide a rationale for the answer," he planned to ask. "What is the date of the computer file that was used to program the improper parts? What is the date of the computer file that was used to program the proposed replacement parts?" If this were truly an "implementation error," then Zenith's people would have to demonstrate that they had written the program for the integrated circuit *before* the system was sent to the test center, and then simply failed to install it.

Richer convened his subcommittee on Wednesday, April 29. Representatives of all the contestants were there, but Rast took the floor. He acted as prosecutor. By now he knew precisely what to ask: What are the date codes of the replacement ICs? The Zenith people mumbled their responses, but in the end it was clear: "The date occurred *after* you discovered the problem in the test center," Rast declared, and no one disputed this. So in fact there had been no "implementation error." Zenith had made a mistake in the system's design and had discovered it only during testing. Then they had rewritten the computer code to correct it. Richer concluded that Zenith had cheated, and he was angry. He was convinced that the new test results, recorded with the replacement chips, should be thrown out. "Zenith has now made clear that the program for the new chips was in fact written after they discovered this problem during testing," he wrote in a fax to Wiley at 6:40 the next morning. "General Instrument pointed out that the Advisory Committee leadership were not aware of this when the issue was discussed before. It is unclear to me that the system *ever* worked properly."

Wiley read the fax. But by now he was weary of the whole thing. He'd seen all the memos. He'd taken all the screaming calls. To him this was a technical dispute among engineers, largely unfathomable to the rest of the world. We've already discussed it, he concluded. It's probably not the best thing for the country to test an obviously flawed system when we have the correction in hand. Let's just let it go. Accept the tapes made with the new chips.

Richer couldn't believe it, and neither could Rast. He was livid. Zenith "pulled a fast one, and they got away with it!" Rast declared. "The process doesn't have integrity anymore."

"Zenith might win through cheating," Richer realized. "This isn't right." A few months later, however, Richer came to believe that simple justice may have prevailed because, "in the end, even with the new chip in place, Zenith didn't do significantly better" than they had with the first chip.

The Sarnoff consortium's digital HDTV program began as a warm and earnest exercise in international cooperation. But as the construction project labored along, it fell into a desperate, fevered struggle that seemed doomed to fail. General Sarnoff's heirs careened toward disqualification.

No one saw that at the start, when Sarnoff and Philips divided up the work, then set out to build their new machine. The engineers at Philips Laboratories, forty miles north in Briarcliff Manor, New York, would build the audio system and part of both the encoder and decoder. At the Shrine, Sarnoff engineers worked on the transmission system and the other parts of the coders. Like Netravali and Petajan had done at Bell Labs a few miles away, Jim Carnes and Glenn Reitmeier rounded up as many engineers as they could find. "This is your chance to make history" and carry on the General's proud tradition, Carnes told the recruits. Many of them rolled their eyes as Carnes spoke; he'd said exactly the same thing about ACTV just a couple of years before. Still, Carnes got many of the Shrine's best people, and he moved them out of the main laboratory complex to an outbuilding in the back called the Field Lab. It was just beyond the garages that had once housed the General's limousines. The building hadn't been used for at least ten years, and when engineer Bruce Anderson unlocked the doors and walked into the main room, he looked around at a musty, cluttered place with several large, archaic pieces of equipment pushed up against the walls. He wiped the dust off of one of them, and his eyes widened: These were VideoDisc presses. These old monsters had made the movie disks—the failed product of the '70s that helped bring RCA's demise. "These things are just like records," the old RCA marketeers had boasted with puff-chested bluster a decade before. "And people buy *lots* of records." But of course they'd been wrong.

Blanks were still loaded in the presses, the controls were still set, and boxes of blank disks still sat alongside—all frozen in time, as if the pressmen had simply left for lunch in the middle of a production run and never come back.

Now some of their descendants batted the old disk blanks back and forth across the floor as if they were hockey pucks. Then they shoved the old machines aside to make room for the new one they were about to start building. Some wondered whether this new product would be any more successful. One engineer invented a slogan that played off a recent movie title and also stood as a modern-day statement of the old RCA hubris. "Field Lab of Dreams," a sign on the front door read. "If you build it, they will come." In the early days of 1991, dream they did. Just like the Bell Labs boys just up the road, the General's men set out to build the best HDTV system there ever was. They had a year; the system wasn't due at the test center until early 1992, and they began with unfettered enthusiasm. Then, like their competitors to the north, Sarnoff began falling behind. By summer they were working late every night, eating dinner in the company cafeteria and assuring themselves, just as the Bell Labs boys had, that the schedule's going to slip. It's just got to slip.

At Philips Labs, the work was going no faster. Carlo Basile was promoted to Research Department head at about the time the project started. He was a fiery young man who grew up in Queens and had moved to wealthy, quiet Briarcliff Manor only recently and with clear reluctance. The HDTV project didn't improve his mood. When he took his first look at the work he was supposed to direct, Basile was "astounded, and depressed." That night he unloaded to his wife. "I don't think we can do this," he complained, throwing up his hands. "As an engineer," Basile liked to say, "if I have a solution, then it's just a question of putting together a plan and implementing it. But here I don't have a solution." Back at the labs, the others saw his agitation and tried to reassure him. Look, they said, this is a calculated risk. And the calculation is that the timetable will be pushed back. There will be delays. The test center won't be ready for us in February.

Philips, too, was working under the abiding faith that the schedule would slip, it's just got to slip.

By late fall 1991, both labs were pushing their men as hard as they could. The engineers starting working in two shifts, twenty-four hours a day, seven days a week. Long before, they'd stopped dreaming of building the best and fanciest system in the world. As Bruce Anderson now put it, the job was "Just build something that works." Some nights, Sarnoff engineers finished their shifts, brushed their teeth, and rolled onto cots in a tiny back room next to the furnace, then rolled out again a few hours later. Sarnoff cafeteria workers carried food to them across the grounds so they could eat while they worked. Wives and children felt abandoned. When one team supervisor finally found a moment to go home, his young daughter looked up at him imploringly and asked, "Now that you and Mommy are separated, are you going to get a divorce?"

Generally, Carnes knew, this sort of fast-track production of a prototype "isn't really an option for a research lab." His men weren't used to it. But he continued telling them, over and over, "You're making history here." More important, he would say, "if we come in just thirty days late, we might as well not come in at all."

Early in 1992, Carlo Basile saw that his men couldn't do much more on their own; it was time to move their part of the system down to Princeton and try to make it work with the other half. A week later the complete system, fourteen vertical equipment racks, was laid out in a rough square in the center of the old VideoDisc pressroom. Philips and Sarnoff engineers hooked cables back and forth, leaving a spaghettilike snarl in the center of the square. But when they turned it on, nothing came out. Not even a cough. They were now into the extra two weeks they had gained because of the Zenith retests—"going so fast that we had to design boards without even knowing what the next and last boards on the line would be doing," complained Liston Abbott of Sarnoff. Another Sarnoff engineer, Rocco Brescia, turned to the guy next to him one day as both of them were finishing work on their circuit boards and said, "By the way, your board's gotta have the frame storer that will do the deinterlacing." The other engineer looked back at him with alarm. "I thought *your* board was going to do it," he said.

Though Rast lost the battle with Zenith over the so-called implementation error, he quickly launched another one over Sarnoff. Rast had started his career at the Sarnoff labs, and he still had a few friends at the Shrine. Other people in his network had their own connections there. Some of these people told him the truth: They're in serious trouble. "There's no positive information coming out from there," Rast realized. "They're in real trouble, so let's not do anything that inadvertently *helps* them."

That turned Rast's attention back to Zenith. When the regular test period ended, Zenith would get two extra weeks to retest certain sequences with the new chips in place. Two extra weeks—a positive *gift* for Sarnoff. From what Rast was hearing about the work at the Shrine, he could see that "these guys are dead. They're *dead!* And now because of these retests, the coroner's giving them a stay."

Rast called the people he knew at Zenith, and he also spoke to Robert Graves, an AT&T engineer-lobbyist. He tried to put the bad feelings from the recent confrontation behind them, arguing that the greater good required them to cooperate now. Do you really need those retests? he asked. Just how serious is this problem? What are we really talking about here? Is it really that big a deal? Because you've really got to get out of the test center. Get out of the test center and get Sarnoff in there as soon as possible. They're wounded. Come on, now. We can kill them!

Graves, not surprisingly, had no sympathy for this argument, nor did the people at Zenith. Both companies were far more interested in getting the best possible test score. As for the Sarnoff group's problems, "if two weeks makes the difference between a marvelous system and going down the tubes, then forget it," Graves said. "That's not good for the nation." The retests proceeded.

The ducks were coming back to the General's ponds, the tulips outside the Shrine's front door had bloomed and fallen away. The date at the test center was just a few weeks away. But when Carnes had a look at the work in his lab, his only thought was: "This is scary. In three months the line on the graph hasn't moved at all."

The machine just sat there, dead, and neither he nor anybody else at the Shrine was sure what to do. The goal of their earlier projects, including ACTV, had been simply to take the existing technology and make it better. Carnes had approached those jobs with a straightforward philosophy: You start with a picture at a certain level, and you see problems. You go back at it, debug the problems, work steadily, and as you do the picture gets better and better. For particularly vexing problems along the way, you seek out some of the older veterans—the fathers of television, at work elsewhere in the building. They come over and have a look, chin in hand, and suggest other approaches. Finally, "When the deadline comes, whatever you've got, that's the picture you deliver." End of project.

As Carnes was learning to his dismay, digital TV didn't work that way. With digital, he said, "you have the same number of problems.

But we have nothing, no picture at all, nothing to look at and figure out how to improve. And there are no old guys sitting around anymore telling us what's wrong." The time was slipping quickly past, and with just days to go "we can't even get a stable checkerboard"—a standard test pattern—"through the system." We're not going to make it, Carnes decided. We've got to go see Wiley.

On May 19, three weeks after Wiley had closed out the debate over Zenith's "implementation error" and just two weeks from the day the Sarnoff system was due at the test center, Carnes, Joe Donahue, Peter Bingham, and Robert Hynes of NBC met at Wiley's law firm and assembled around the heavy walnut table in the conference room just across the hall from Wiley's corner office. Photos of turn-of-the-century Washington hung from all four walls, and Wiley's staff had set out pastries and coffee for the guests.

We need an extension, Donahue announced.

How long? Wiley asked.

Four weeks.

Wiley sat up straight. He couldn't believe what he'd heard. Sure, he'd known they were going to ask for more time. But a *month*? That's impossible.

Four weeks? Wiley repeated back to them. Four weeks? You know I can't do that. But then he offered a compromise. As usual, he'd thought this through the day before, though he had expected a shorter delay. Even with this absurd request on the table, he still made the offer he'd formulated in advance. Look, he said, I can get the test center to work overtime, squeeze the testing into a shorter period if you show up late. But you've got to get there sooner than four weeks, because any testing we don't have time to do in the allotted period won't get done at all. And that will count against you.

The Sarnoff consortium had no choice. They agreed. The next day, May 20, they sent Wiley a letter signed by all four men confirming the agreement: "It is clear today that we will not be able to meet the June 3 'start of test' date. While all of the major subsystems are working well and promise to deliver winning performance, we require additional time to complete the system integration phase of our work." That was a rosy interpretation. But this was Sarnoff, after all. A press release was in order.

"Our system is producing excellent HDTV pictures and sound during in-house laboratory tests," the release quoted Carnes as saying. In the Field Lab, some of the engineers found that more than a little amusing. "If we're getting excellent pictures, I haven't seen them," said Joel Zdepski. "Maybe a little two-second flash of something. That's all."

As a courtesy, the consortium faxed a draft of the press release to Wiley. He insisted on editing it, adding a couple of phrases to ensure that no one would believe he was cutting these people any slack. They accepted his changes, sent out the release, and then turned all their energies back to the Field Lab.

Friday, June 12, 1992, was "a good day to work here," Rocco Brescia observed. That morning, more than a week after they were due at the test center, the engineers decided they had done as much as they could. Their machine had at last produced a reasonable high-definition picture, at least for brief periods. Several bugs still had to be resolved, and they'd had no time to make spare copies of several circuit boards. Nonetheless, the moment had come to ship it out.

That morning the team gathered in the Field Lab for the last time. The system was powered down now; the cooling blowers were off, and the room was eerily quiet. Several on the team felt oddly sad. They lined up in front of their machine for a photo, and somebody stuck a pair of rabbit ears on top of the closest racks. Afterward, they stood back to watch, hearts filled with pride and nostalgia as the packers crated their baby and loaded the boxes onto the truck. It filled up a twenty-foot tractor trailer. After the truck left, they held a barbecue, then went home for their first real sleep in almost a year. A handful of engineers accompanied the truck to Washington, and after the equipment was unpacked, debugging continued right there in the test bay, just as it had with the Zenith-AT&T teams a few weeks earlier.

Peter Fannon and the other test center officials were not surprised by that. But they were stunned at the size of this system. *Fourteen racks!* General Instrument had come in with just two, and their system had worked pretty well. No wonder these people had so much trouble getting it built on time. And when the engineers turned this monster on, Fannon wrote to Wiley, the power transformer in the system's equipment room began "running hot, and the air conditioning is also overloaded by the heat output of the system."

A week later, the engineers were still scrambling to get the machine working—to no avail. Several more days passed. Finally, at 2:00 A.M.

on June 22, they found the last bug. They were ready. Testing began the next morning, two weeks late. Under their agreement with Wiley, the most important tests would be performed first. Testing would continue on Saturdays, and the contestant would pay for the overtime. Some of the less critical tests would be dropped when the time ran out.

Like all the other contestants, the Sarnoff and Philips engineers found that they'd made an error. A filter intended for the transmitter had mistakenly been inserted into the receiver, and that was screwing up the picture. The Sarnoff group was able to demonstrate that they did in fact have an implementation error, and a change of filters was allowed. After that, testing proceeded apace. Then, as agreed weeks before, the test center gave the consortium's leaders a list of the tests that probably would not be performed because the system had been delivered late. Now, however, the partners didn't like that plan anymore.

Mustering all the RCA hubris still coursing through their veins, they decided that the test center should just go ahead and perform *all* the tests, regardless of the earlier agreement with Wiley. An NBC executive began calling members of the National Association of Broadcasters' board of directors with a request: Could the NAB board please pass a resolution calling on the Advisory Committee to give us all of our tests? Carnes and another NBC executive both called Wiley, then wrote letters urging him to change the rules. The consortium believes "that complete testing is the way to ensure that the best ATV system is selected for the United States," Carnes wrote. The full consortium sent another letter, offering curious logic and a veiled threat. Full testing should be allowed "to preserve fairness to all proponents" and "to insulate the selection process from potential legal challenge," it said.

Wiley found all of this outrageous. Preserve fairness? You got there late. Everybody else got there on time. Is that fair? What happened to our agreement? He told NBC he resented this "end run" to the NAB, and he denied the request outright. The consortium kept beating on him, though, and in the end Wiley relented: All right, all right. You can have three extra days. Just three days for a few more tests. Case closed.

Jae Lim wasn't having much trouble assembling his MIT-GI machine. The bulk of it was already built. Woo Paik had constructed it in 1991, and this year GI was demonstrating it across the country. There were differences, though. Along with a few technical innovations of Lim's design, the MIT-GI machine was to have progressive scanning. The pictures would be transmitted as a unit instead of in the alternating strips of conventional, interlaced television; this was the distinction that had allowed GI and MIT to present their second entry as a new and different system. It also proved to be Lim's Achilles' heel, and it led to one, last childish squabble assuring that the testing period ended on a petty, puerile note.

Lim, the final contestant, was due at the test center in August of 1992, and construction was proceeding on schedule. But he couldn't test his machine because he didn't have any progressive-scan tapes. A progressive-scan television can display interlaced images; they are converted to progressive format inside the set. Still, to show the system's full capabilities, Lim needed some tape footage—*material,* it was called—filmed with a progressive-scan camera. The problem was that there were only two such cameras in the United States. Zenith and AT&T owned both of them. That was not surprising, since their system also used progressive scanning. So Lim asked Zenith if he could have a copy of some of the company's progressive-scan tapes or borrow a camera to tape some of his own.

Sure, Zenith responded. We'll rent you the camera for two weeks, at a cost of $250,000. Or we'll make a copy of our progressive material. That will cost you $750,000. "We consider these payments quite nominal," Wayne Luplow wrote. But when his letter arrived, Lim, Rast, and Wiley erupted: This is extortion! Lim and Rast declared, and Wiley believed Zenith was acting "in bad faith." Writing back to Luplow, he said, "I must say that I find your proposals both disappointing and unrealistic. The charges you suggest can hardly be called 'quite nominal.' "

This was May 27, and the test date loomed just eleven weeks away. Irritated because his contestants were acting like children again, Wiley put out the word: I want this resolved *soon.* Meanwhile, Lim started to panic: How can I have any chance of winning if I can't even test my system? How can I possibly *save America?*

On June 10, Robert Graves of AT&T started trying to negotiate a compromise. He told Paul Misener, a young lawyer not long out of Princeton who served as Wiley's Advisory Committee assistant, that Zenith had not consulted with AT&T before making the $750,000 offer, even though Luplow's letter had said that the offer came from "the top management at Zenith and AT&T." Graves said his company was upset about that, and personally he was "dismayed and disappointed." He promised to knock the price down "below six figures." At the same time, though, he also made it clear that he wasn't happy about it. He told Quincy Rodgers, GI's Washington representative, "It seems increasingly bizarre to me that we are being asked to provide equipment and materials to our competitor at this late date." We're interested in cooperating, Graves added, but it's "crystal clear that GI and MIT should not expect us to make use of equipment or materials available without compensation."

As soon as Graves entered the debate, Rast suggested a new price to Luplow: $10,000. Five days later, Luplow wrote back, "We must say that we do not consider your offer of only $10,000 for the use of the camera plus additional equipment to be a constructive step toward resolving the situation." And he countered with an offer that kept Graves's promise to knock the price down below six figures, though just barely. "We are prepared to offer use of the camera for a one-week period for the sum of $95,000," Luplow wrote. He suggested two possible dates. This just made Rast and Lim angrier, and by now all parties were seething. On June 18, Rast wrote back to Luplow, "The gap between us on this appears unbridgeable." The price is too high, he said, and one of the proposed dates "occurs after the system has been shipped to the Test Center." Lim, meanwhile, was beginning to despair. Time is running out, he complained to Wiley. It looks like I'm not going to get the material in time. This is hurting my chances.

Finally at the end of June, Peter Fannon offered to loan Lim two fifteen-minute progressive-scan tape sequences that the test center held in its own archives. The cost of packing and shipping and protecting this valuable commodity was $5,000.

Great, Lim said. But I think the Advisory Committee should pay the $5,000, not me.

Wiley cracked. "This is absolutely out!" he thundered. "This irritates me beyond belief." The proposal fizzled, leaving Lim, the supposed victim, in Wiley's doghouse, too.

On August 20, 1992, Lim and a small group of GI engineers moved the MIT-GI system into the test center for the final round of testing. They unloaded two racks of equipment. The logos of both General Instrument and the Massachusetts Institute of Technology had been stenciled in white on the smoked-glass doors of each rack. Hookup proceeded smoothly, and testing began on August 28, right on time. Lim had rented a small condominium nearby, and he walked over to the test center every day, often with a couple of his graduate students. The GI engineers generally left him there on his own, even though they, not he, had actually built this machine.

Right away, Lim saw that his system was performing poorly. "This isn't working," he told Fannon, pacing back and forth, his voice soaked with despair. "And I don't know why." In certain of the test sequences, the screen was bursting with video "noise"—sparkles all over the screen. Lim blamed this on his failure to get progressive-scan test material: If only I'd been able to properly test the system first, he lamented. But he also faulted GI's engineers, saying they had made an error while writing one of the system's computer programs. "The problem may have gone unnoticed due to the lack of material" for his machine, he reasoned in a letter to Mark Richer, chairman of the subcommittee that oversaw testing. I'm a victim, Lim concluded. As he complained to the test center's management, he alternately lathered blame on Zenith and on GI. Testing proceeded, and the problems only grew worse. In one tape sequence—the one showing a young man rising from a sofa and walking over to a bookshelf—part of the picture seemed to be boiling. This was no small problem; people would never put up with that on their TV sets.

Lim's system was failing as he watched. "It broke my heart," he said. And as best he could tell, the difficulties were related to the progressive-scan sequences. Zenith and AT&T, the other progressive-scan system, had encountered the same problems—though Lim didn't know it then, because the test results were being kept secret. During their tests weeks earlier, in the scene in which a hand turns down a hurricane lamp, the dark area around the lamp had erupted with fireworks-like sparkles.

Moving into the test center a couple of weeks before, Lim had been utterly convinced that he had the very best high-definition television system in the world. He could win this race. He could *save America*! Now his entry was stumbling, falling, and Jae Lim feared that he might actually be eliminated from the race. As he watched, the flawed images coursed through his system into the test equipment, then back out onto the monitor in the little contestant's room downstairs—ugly, mangled pictures that seemed to reach out to him from the screen. Lim winced with each new failure. Sometimes he wanted to turn away. But there Lim sat alone—humiliated, heartbroken, as his dream collapsed.

Finally the testing was over, and Jae Lim was hardly the only contestant who found himself slinking out of the Advanced Television Test Center. All the systems had problems—Zenith's and MIT's more than the others. Nonetheless, the testing had made a larger, far more important point. Maybe none of the entries was quite ready. Not one of them worked perfectly. But all those months of testing had found no fatal, irremediable flaw. Digital, high-definition television wasn't just possible. It was here. Even with its cynical origins, Dick Wiley's race seemed on the verge of producing a wonder for the world.

In truth, however, this race was far from over. The remaining technical obstacles would prove to be the smallest of the players' concerns, for none of them had yet even heard from their most formidable foes.

13 *We'll build the sets*

By the autumn of 1992, high-definition television seemed unstoppable. Manufacturers were getting ready, and the latest projections from one of Wiley's subcommittees showed that the first large, rear-projection HDTVs would probably sell for about $3,000. Prices would fall rapidly after about 1 percent of the population, the nation's technophiles, bought those first sets. After that, sales would climb an ever rising curve for the next ten or twelve years. Or so everyone hoped and believed. Network engineering executives, meanwhile, were issuing confident promises to broadcast high-definition programming during the very first season.

"At the earliest possible moment to do it in a way that reaches even a few TVs, we will do it," promised Michael Sherlock, the NBC vice president.

"Fox will get in right away," averred George Vradenburg, who had left CBS to become a vice president at the new Fox network. "We'll have a full season of programming right away."

Joe Flaherty, the CBS vice president who served on Wiley's Advisory Committee, reminded everybody that switching to high-definition programming would not be a decade-long process, like switching from black-and-white to color television. There was one simple reason: the networks already had vast libraries of high-definition programming in hand. All the movies and most of the prime-time programs produced in Hollywood were recorded on 35 millimeter film, and that was already a high-definition format. Film offered almost four times the resolution of conventional 525-line television—even though nobody could tell by watching this material on the present generation of TV sets. "More than half of our nighttime material is already on film," Sherlock said, echoing Flaherty's refrain.

Even the broadcasters' most hated competitors were recognizing the inevitability and joining the parade. Motorola, the power behind the Land Mobile lobby, announced that it was willing to form an alliance with General Instrument to provide integrated circuits and related equipment for GI's high-definition television products. As you can see, Motorola said in a letter to Wiley, "our interest in advanced television extends beyond Land Mobile."

George Bush climbed aboard the bandwagon, too. Two years before, he had fired HDTV-advocate Craig Fields, setting off a squall in Congress and on newspaper editorial pages nationwide. Now, just weeks before the 1992 elections, with Bill Clinton gaining ground, the president offered a new view: he praised Wiley's race, saying it was a great example of "industry-driven initiative."

Most important, the FCC, at Chairman Sikes's behest, issued an extraordinary, far-reaching rule that made adoption of HDTV mandatory for broadcasters and homeowners alike. Under this new order, once the Advisory Committee and the FCC chose a winner in 1993, every TV station in the land would have two years to apply for the temporary use of a second channel. The FCC would then loan each station a vacant UHF channel free of charge, with the obligation that it be used for the transition to HDTV. Within three years of that, each station would have to build a digital, high-definition transmitter and soon after begin transmitting all their programming on the second channel simultaneously, in high definition. Finally, fifteen years later—in the year 2008—the FCC expected everyone in the nation to have bought a new, digital HDTV. That year, on a date certain, all conventional TV channels would go off the air, and the stations would have to give those channels back. Some twenty years after the debate began, Land Mobile would have a shot at the unused channels once again because, like it or not, the conversion to high-definition television would be complete.

In a roadside business park a short drive down Route 1 from the Sarnoff Shrine, Gregory DePriest presided over a large suite of offices, most of them empty just now. From the highway, the building appeared unremarkable. But these rooms were a staging area for a powerful foreign army preparing for battle in the high-definition television wars.

Six years before, DePriest helped get the HDTV race started. During that seminal meeting at NAB headquarters in 1986, he sat across the table from John Abel and watched bemused as Abel was struck by that sudden thought: HDTV . . . maybe that's it! DePriest was the one who had suggested calling NHK. He had changed jobs since then, and at the end of 1992 he was the American head of Toshiba's high-definition television research lab, a suite of empty rooms. Here, the Japanese TV manufacturer lay in wait for HDTV.

Up and down Route 1 outside Princeton, and in California, other Japanese manufacturers had opened similar facilities—"listening posts," they were called. Their purpose was to follow the HDTV race hour by hour, even minute by minute, taking account of every twist and turn so the Japanese could spring to market with high-definition televisions at the first possible moment no matter who won the race. "We'll have six research-and-development engineers on staff here," DePriest explained. Now, his engineers were studying the different systems entered as contestants in Wiley's race, testing computer simulations of each, based on published descriptions.

Down the hall from DePriest's office, Sheau-Bo Ng, a Toshiba engineer, worked at a computer console in one of the few occupied rooms. A jumble of televisions, computers, and assorted electronic equipment lay scattered around the floor behind him, some of it only half uncrated. He was studying an HDTV monitor as it displayed an endless-loop videotape of water running over some rocks. The tape had been made in the rock garden of a Buddhist temple back in Tokyo. This was a digital TV torture test. The rushing water filled almost the entire screen, and since it flowed quickly, every part of the screen showed entirely new information in every frame. This was the hardest kind of material to compress, because the compression coder being simulated in Sheau-Bo Ng's computer couldn't pick parts of the picture to repeat from one frame to the next. As a result, the system could not keep up with the unstoppable rush of data. The computer was overwhelmed, and so the picture looked indistinct, muddy. A small leafy branch off to the left of the picture was so ill-defined that one leaf blended into another. The rocks sticking out of the water looked like mud, not stone. All and all, this system, a simulation of General Instrument's, was putting on a sorry performance.

The Toshiba engineer also had computer models of the other compression algorithms, and over the last few days he had put Sarnoff's, Zenith's, and MIT's systems to the same test. All of them had failed. He was trying to find out why so he could learn what the systems were able to do and what they weren't. That way Toshiba would be ready to start manufacturing TV sets just as soon as the winner was chosen.

Toshiba was loosely allied with General Instrument, having agreed to make a digital VCR to go with GI's system. But DePriest, a friendly, open North Carolinian, made no pretense of partiality. "We don't really care who wins," he said. "Give us a standard, and we'll build the sets."

When a winner was chosen, DePriest's office suite and the others like it along Route 1 would fill up quickly. "Our in-house estimate," DePriest said, "is that it'll take a year" to start manufacturing HDTVs, once the FCC published the new standard. Already engineers and marketeers in Tokyo were getting orders, preparing to move. Soon a winner would be chosen. NHK might lose the HDTV race, but Japanese television manufacturers were poised to be the real winners of the contest.

Broadcasters had been watching high-definition television rush toward them with a concern that bordered on loathing. For months, John Abel had been unhappy, but Al Sikes's new rule sent him over the top. Under the FCC's plan, broadcasters would have no choice. Abel and others in his profession tallied the potential hazards and expenses, all the while continuing a fervent search for some means of escape. If Sikes's plan went forward, they would have to buy high-definition equipment and broadcast high-definition programming by a date certain just a few years away. All this equipment was certain to be expensive. Some broadcasters' groups had ginned up new estimates ballooning the HDTV conversion costs to well over $30 million per station. Even conservative projections showed that the new equipment would cost $1 million, $2 million per station at least. That was more than the net value of some small TV stations.

In 1987, of course, HDTV had helped pull the National Association of Broadcasters out of a terrible jam, and it had given Abel an important lobbying victory that had boosted his career. He still bragged about it occasionally. Now, however, Abel was squealing like a stuck pig. The FCC rule wasn't fair. "It's not reasonable; it's discriminatory," he complained early in 1993. "What's our incentive to do this? There is none. We can't charge higher ad rates. With HDTV it's the same ad market. So what's the incentive? The FCC has a gun to our heads!" Network executives agreed. Sure, their vice presidents for engineering, like Joe Flaherty at CBS, may have been confidently predicting that the networks would broadcast high-definition programming at the first possible moment, but in many cases these people were on their own, with little support from more senior executives, the ones responsible for spending the money.

At CBS, in fact, it appeared that Flaherty was being used as a front of sorts. As soon as the rest of the network hierarchy realized that HDTV was real, and coming at them fast, Edward Grebow, a senior vice president, called for a meeting to figure out what CBS should do. Among the questions to be discussed, he said in an internal memo distributed among CBS top management, was "Is there any viable way to delay this implementation date?" And "If HDTV is inevitable, can we get anything in return from the FCC or Congress?" While these and other questions were being debated, "we plan to continue to limit HDTV activities to those projects that Joe Flaherty can undertake on his own, with the exception of a study on the costs to implement HDTV at our owned stations and the network."

Flaherty and other network engineering executives continued to work with Wiley, but by late 1992 their bosses had dropped all reticence. Their unvarnished views poured forth at the annual "HDTV Update" conference sponsored by the Association for Maximum Service Television. Over the years, the HDTV Update had come to be an important place to gauge the evolving political thinking. This one, in October 1992 at the Westin Hotel in Washington, was called "Countdown to Consensus." In fact, it was a battle royal.

"The rush to develop HDTV may have pushed aside practical, real-world considerations," warned Daniel Burke, the president of Capital Cities/ABC, speaking to about two hundred and fifty people packed into an amphitheater-style conference room. Hotel staff scurried to bring in folding chairs for the late arrivals, who stood in the back after seating ran out. As Burke spoke, murmurs of approval spread through the crowd. The FCC should heed "the law of unintended consequences," he warned. The millions of dollars that TV stations would have to spend on the conversion to HDTV could drive some out of business, meaning that "a significant portion of the national audience could disappear from the networks' coverage capacity, perhaps forever."

Five years earlier, in 1987, one rallying cry had sustained the broadcasters through their battle against Land Mobile. If America's broadcasters were "precluded from offering HDTV as a free, over-the-air service to the nation," NAB President Eddie Fritts had said back then, it would mean *the death of local broadcasting as we know it!* Now Burke was turning this mantra on its head. If we are forced to broadcast HDTV, he asked, "could this mean the end of the universal, free, over-the-air delivery system as we know it?"

Al Sikes sat in the audience listening to the broadcasters' caterwauls and complaints, and in the back of his mind he heard other American industry leaders of the last twenty years: computer manufacturers in the early 1980s, complaining that it was beneath their dignity to sell personal computers because their expertise was in mainframes; automobile manufacturers declaring that they didn't want to build small cars because they'd *always* built big ones, and that's what they did best. When his turn at the microphone came, Sikes told the broadcasters, "If General Motors hadn't upgraded its plant, it would be dead today." And so it could go with you. "Every industry that has failed to upgrade its plant is dead or dying. Now is not the time to get weak-kneed."

But Sikes wasn't changing many minds. "We're staring down the barrel of a gun," one station owner growled later that day. An ABC vice president was even more dismissive. HDTV, he said, is liable to become "the eight-track of the '90s."

Early in 1993, the Paul Kagan Company, an education and consulting firm, held a seminar in New York for broadcasters from local TV stations across the country. Paul Kagan had been holding these seminars at the Park Lane Hotel every year since 1976, and never before had the room been filled with such angry-looking men. But then this was the first time Kagan had organized a discussion of digital, high-definition television. It was aptly entitled "Digital TV—New Business or New Problems?" Here the broadcasters' ire blossomed in full flower.

Before now, most of the people who'd come to this conference had barely followed the national debate; for them, high-definition television had remained the subject of near mystical futurespeak. Now, however, HDTV was bearing down on them like a freight train under full steam, whistle screaming, wheels screeching. Like their counterparts at the networks, these people were not at all pleased, as Philip Lombardo, the president of a company that owned several medium-size TV stations, made clear in the seminar's opening session. "I don't think there's any demand for HDTV," he snarled, and all morning long his expression, even at rest, was a wide-eyed hostile challenge. He and about seventy-five other skeptical television executives sat at small, round tables filling the conference room. On the dais were several leaders of the broadcasting industry, here to tell them why they should be *enthusiastic* about digital HDTV. After all, when television service goes digital, TVs will become computers. They'll come alive!

The moderator called on James McKinney, the former FCC official. And from the moment he opened his mouth McKinney's enthusiasm was infectious. When television transmits its signal digitally, as computers and compact discs do, broadcasters will be able to subdivide the TV signal, he said. Most of it they'll use to transmit high-definition programming—wide-screen TV with stunning resolution. But every one of you in this room, McKinney promised, will be able to use the rest of the signal, about 10 percent of the space, to transmit new services for sale to the public. Then he offered examples.

"You can send one newspaper page in 1.25 seconds," he said. It could be downloaded onto a computer-style hard disk inside the TV and then viewed anytime. "Or you can send a hundred-page newspaper—say, the *New York Times* Sunday edition—in two minutes, or the entire Manhattan telephone directory in fifteen minutes. You could send the TV guide, and it could specifically be customized" to instruct receiving sets to store certain shows for viewing later. "You could run stock quotes customized to the customer's own accounts. You could provide interactive quiz shows," so that viewers could participate from home. "You could provide user-controlled alternative camera shots for sporting events. One customer wants to watch the play from the end-zone camera, another wants to watch from the fifty-yard-line camera. You could do that.

"And how about a twenty-four-hour dedicated news and sports-data channel, where scores could be constantly updated? You could download software. You could download video games to the home. You could provide interactive educational programs. Any number of multilingual services, any number you want. Or become a music store: download music to a write-able compact disc in the home."

With digital television, McKinney said, all that and much more is possible. And to demonstrate, he turned the microphone over to another panelist, who showed a video shopping system: A female model appeared on the TV screen wearing a business suit, and with his remote control the panelist spun her around, and then pressed a button to zoom in on her lapels. He pressed another button, and she took off the jacket, so that customers could look at the blouse. (Around the room, several people snickered about the potential this might offer for another sort of service.) Finally, he said, you could enter a credit card number from the remote control and order the suit, to be delivered by Federal Express.

"As amazing as it may sound," McKinney said, taking the floor again, his voice fairly floating with wonder, "the television manufacturers tell me they can build the sets that will do all those things at *no additional cost!*"

McKinney was right, because the "digital armies" that would soon be dispatched to everyone's home care not what mission they are given. They can bring TV pictures or anything else. Even after providing a full and perfect high-definition TV picture, complete with six-channel surround sound, sizable numbers of these "troops" will be left for additional missions. As McKinney noted, only about 90 percent of the digital bitstream, at most, is required to provide a high-definition picture. The other 10 percent is free to transmit whatever else the broadcasters want to send—actually a conservative estimate, since most television programming is not full motion. Sometimes, as Woo Paik found years before, fewer than 10 percent of the troops in the digital army are carrying TV pictures. Several times each hour, for example, most TV stations broadcast a station break—a still picture of the station's call letters and logo. During that five seconds, digital compression can reduce the transmitted data to a bare trickle so that almost all the bitstream is free for other uses. In that time a TV station could transmit enough data to fill the entire Manhattan Yellow Pages.

Digital, high-definition television had the potential to give broadcasters tremendous new marketing powers, if only they had the imagination to see how they might use it. Just then, however, they didn't. At this point, only the engineers, with their talk of megabit rates and their data-packetization schemes, understood the full potential. But all that jargon meant nothing to the broadcasters. These people were marketeers; to most of them the technical talk might just as well have been couched in some remote Bantu dialect. Since digital HDTV was still just an advanced research project, not a real product quite yet, nobody in marketing had grabbed hold of it and begun to figure out what might actually sell.

In fact, in 1993, digital television was at about the same place on the marketing curve as personal computers had been in about 1980. Back then everyone knew that PCs would be a great technological advance:

someday almost everyone would have one. They would be *indispensable,* industry analysts said. At the same time, though, most potential buyers couldn't really figure out what they were actually supposed to do with PCs. There was talk of balancing the checkbook on the computer, of writing letters with a word-processing program instead of a typewriter. And, of course, computer games. But these ideas held scant appeal for most people. Nobody had thought of the "killer applications" that would make *everybody* want a personal computer. That didn't come until 1983, when Lotus 1-2-3, the wildly successful spreadsheet program, came on the market.

Now McKinney and others were proffering ideas to broadcasters, telling them how they might make money with digital television. But nobody had yet thought up the "killer applications" for digital television, and most of the broadcasters at Paul Kagan's seminar weren't confident that anyone ever would. So when the time came for questions from the audience, Philip Lombardo was not mollified. His hand shot up fast. "My problem," he said, the edge to his voice still knife-sharp, "is that nobody's saying how we're going to make money at this. If we end up at the end of the hunt here by paying $10 million for HDTV and we haven't figured that out," stations in smaller markets will be driven out of business. "And therefore I keep pressing: How are we going to make money with this?"

Testy now, McKinney told him, "Well, I've just given you fifteen or twenty different possible ways to do it."

"But I don't want to be in the data business!" Lombardo exploded. "I'm in the television business. Are you? You're telling me I gotta get out of my business and go into the business *you* think I should go into." If you go through with this, Lombardo went on, "You're going to drive local TV stations all across the country out of business, and the American people are going to lose their local television service!"

That sent McKinney over the top. Six years before, he had been witness to HDTV's ignoble birth. He'd heard the broadcasters' rallying cry back then. Now he shot back, slamming his fist on the table. "That argument is *exactly* the argument the broadcasters made in 1987 that got this whole process started!"

The HDTV contestants watched all this with a curious ambivalence. Sure, it was exciting to see Al Sikes push the process forward, and none of them was surprised at the broadcasters' furious response. Fifty years before, hadn't many radio broadcasters reacted the same way to the advent of television? Besides, the contestants had other concerns. Quite frankly, all of them were wounded. Every one of them had encountered problems at the test center, and as the test results were released one by one, each of them began to see just how severe their failings really were. Even some of the contestants who had done well enough were tripped up during the next stage of testing. Wiley's committee had also contracted with a laboratory in Ottawa to run "subjective tests," using tapes recorded at the test center. Panels of volunteers watched the tapes, not knowing which contestant had produced them, and then rated the quality of the picture.

In the end, Zenith-AT&T, MIT, and the Japanese all were told they had picture-quality problems. General Instrument's largest failure was still interference with other TV channels. The Sarnoff group had interference problems, too, and their system struggled to handle fast-motion sequences. In fact, nobody had a clean report, and most of them rationalized their failings with a common refrain: It was just an *implementation error,* not a mistake in design. Zenith's fatuous explanation for its flawed chips had spread like a virus among the others.

"We have discovered a significant implementation error in the video compression part [of our system]," Jae Lim reported in a letter to Wiley. "The error has now been corrected." General Instrument, explaining its results, generally boasted that Woo Paik's first system had put in a fine performance. As for the interference problems, the company said, "that was an implementation error, and we are correcting it." Wiley didn't take well to this explanation. "It's an attempted artifice," he said. Soon after that, Zenith stopped using the phrase, and when the test center published Zenith-AT&T's disappointing results, Zenith's team said simply that they had been too rushed. "It's goddamned difficult to schedule an invention," Jerry Pearlman said. "All of us materially underestimated the complexity of going from a simulation to hardware."

Whatever the problem, all of them set to correcting their mistakes. And in time each genuinely believed that their systems were performing far better than before. Zenith-AT&T and MIT, the two contestants offering progressive scan systems, noted that the progressive-scan camera that had been used to shoot some of the test footage had added some interference of its own. And for that reason they argued that their tests hadn't been fair. But then another idea began to take hold. It struck Jerry

Pearlman first. One summer morning he called Robert Graves of AT&T and told him: Robert, I think we need a retest.

Graves chuckled. "Where would you like to have it?" he asked. "On the moon?" But Pearlman was serious. Our system is performing much better now, he said. We were too rushed before. Wouldn't it be the best thing for the country to test these systems again, now that we know they are performing as they should? Pearlman wasn't proposing to run the entire test sequence again, just a few days of evaluation to demonstrate their improvements. Eventually the idea began to grow on Graves. Here was a *real* lobbying challenge. Besides, he and Pearlman knew full well that Zenith and AT&T had little chance in this race if their entry was judged on the test results just released. "If we are not retested," Pearlman said, "I think it means we will lose."

Wiley's Advisory Committee was due to declare a winner in just a few months. Even now his engineers were poring over the test data. Wiley had scheduled a meeting with them for February 1993—the Special Panel, he called it—to draw conclusions from the data and make the first cut. Given their system's performance, Pearlman and Graves knew, the Special Panel would knock Zenith and AT&T out of contention. So Graves began formulating the best case possible. Finally, he called Wiley. Dick, he said, I think it would be the best thing for the country to order "supplemental testing" to reexamine those parts of the system that weren't ready for the first round. "We're preparing a new television standard for the next forty or fifty years," he said. "Shouldn't we be patient for just a few more months, so we can be sure we get it right?" Besides, Graves went on, "the testing wasn't fair. We thought we had to get to the starting gate on time to be in the race. The next people"—the Sarnoff group—"got there weeks late, and they were still allowed in. If we had been given two extra weeks, we wouldn't have had all of our problems."

Pearlman called, too, making the same case, and Wiley didn't reject the idea out of hand, though he knew Sikes would never go for it. Graves was talking about another week or two per system. But considering how sluggishly this process had worked so far, Wiley guessed it would really take a year. There'd undoubtedly be months of wrangling about who would be first, what would be retested, and how. One or another of the contestants would be late, and then they'd argue about misprogrammed chips or some other nonsense. And then the test center would have to ship more tapes up to Ottawa for subjective testing. All in all, "supplemental testing" would delay things into 1994 for sure. And Sikes had been firm: He wanted to issue the standard in 1993. For now, Wiley asked the contestants to send him documents describing any improvements they had made, and he promised at least to consider the retesting request. After all, it was beginning to look like George Bush might lose the election. If he did, Sikes would leave the FCC. Who knew how the next chairman might feel about Sikes's deadline?

While they waited for Wiley to make up his mind, Zenith-AT&T and the Sarnoff group rushed to put on their own public demonstrations, just as GI had after leaving the test center. The demos seemed to have two goals: Zenith and AT&T, as well as Sarnoff, wanted to outdo the GI extravaganza while also demonstrating that they had fixed the problems uncovered during testing.

Zenith invited everyone who mattered—essentially the same group of guests who'd been at GI's demonstration at the Capitol—to the Glenview Bunker in late May. That proved to be a tactical error. Most of the guests, including nearly all the important ones, lived in Washington. While a congressman or an FCC commissioner could easily find an hour or two for a trip across town, many of them couldn't get away for an overnight trip to Chicago. Dick Wiley was among those who could not attend. Still, the guests who did show up were promised the first *long-distance* HDTV transmission. Wayne Luplow had openly disparaged GI's demonstration—the transmission was "only seven miles," he'd said. "If that's as far as they can go, they'll leave out all the people in the suburbs." Well, Zenith planned to broadcast their signal all the way from Milwaukee to Glenview, a full seventy-five miles. WMVT, a public television station in Milwaukee that broadcast on UHF channel 36, loaned Zenith its facilities for a late-night test, and near midnight the guests assembled in front of the big-screen projection TV in Zenith's demo room.

After Zenith's leaders introduced the show, the lights dimmed and all eyes turned to the big screen. Seventy-five miles away in Milwaukee at that very moment, Zenith's technicians were trying to control their panic. They'd been sending test signals for hours, and everything had been working fine. But now, at the critical moment, the signal was fading down to zero. Frenzied, they scrambled to figure out what was wrong. They tweaked knobs, shot fevered glances at their scopes and monitors.

Finally a desperate technician, Steve Heinz, reached over and gave one equipment box a sharp slap. Suddenly the picture roared back. The signal surged over the airwaves and into the demonstration room. The technicians smiled at each other as relief washed over them like a cool breeze.

Like GI's before, Zenith's show was a grand success. The government officials and writers from the trade press all left mightily impressed. Zenith mailed press releases and commemorative paperweights to all those who had been unable to attend.

A few months later, the Sarnoff group held its own demonstration, but they learned from Zenith's error. They staged it at a Washington hotel and promoted the event with all the swagger and strut the old RCA publicity machine could muster. The consortium also added a new wrinkle. They loaded a van with an HDTV receiver and then drove east toward the Chesapeake Bay to see how far they could go before the signal faded out—seventy miles, it turned out. And at that distance, the consortium boosters boasted, a conventional TV brought along in the van for comparison was picking up *no signal at all.*

That claim cried out from the press releases that followed: At seventy miles, our HDTV picture was perfect, while "the conventional NTSC signal could not be picked up, even though its transmission power and tower height were much greater than those of the HDTV transmission." Powerful stuff. Still, this third demonstration left some people yawning. By now they'd seen it all before. And besides, Ed Williams, a test center engineer, was suspicious of the consortium's last claim.

There was *no* conventional TV signal at seventy miles? he thought. That can't be right. There has to be *some* sort of signal, even if it's weak. Otherwise, he said, "the laws of physics don't apply." Finally Williams discovered that the consortium had been using one of those TV sets that turn the picture off automatically if the reception falls below a certain level. The set showed a blank blue screen instead. "So the picture was there," Williams said with a wry smile. "They just didn't show it."

General Instrument didn't watch all this quietly. While Zenith-AT&T and the Sarnoff group showed that they had working systems, GI demonstrated the first cable broadcast of HDTV and then the first satellite transmission. These shows at conventions and trade shows got some attention—though GI, too, found that the power of demonstrations was diminishing. Nonetheless, GI's leaders still believed they held the lead. The others were just trying to catch up. Their smug self-satisfaction didn't last long, however; they were about to encounter a dangerous new enemy.

The last most people in the HDTV race had heard from Nicholas Negroponte had been a self-satisfied pronouncement: "I used to feel like a lone voice in the woods," he had said when General Instrument announced its fully digital system in the summer of 1990. "Suddenly, what we were talking about is becoming real."

Now digital TV was a fact, not a futurist's prophecy, and in Negroponte's world the present was always lacking. It had to be. So by January 1992 he was singing another song. "There's no evidence that anybody out there wants high-definition," he said. "I'm sorry the industry spent so much money on it." Instead, he opined, the public would want sophisticated computers that also happened to show TV pictures. "They'll have a 'tell me more' button," he said. Viewers would be able to push it and get additional information about whatever they saw on the screen.

This was not a particularly far-fetched concept. In fact, given the flexibility of digital TV, Negroponte's vision would be easily realized. And so by the beginning of 1993 the Grand Vizier's position had evolved toward a target no one could possibly hit. "My problem is the reckless nationalism that surrounds the television business," he said one January day, perched on a sofa in his Media Lab aerie. His staff waited on him as he spoke, bringing him a box lunch, handing him messages and memos through the afternoon. "The government should abandon this race, this 'bake-off,' " he said, his voice laced with scorn. "The government should take the lead in proposing a *world* system, with the U.S., the Europeans, and the Japanese. Thank God all these systems are flawed. We have a chance here for *world harmony.* I can't tell you how passionate I am about this!"

World harmony. The United States, Europe, and Japan all abandoning their research, undercutting their own industries, and sitting down to agree on one high-definition television system that could be adopted around the globe. All of them would forget their nationalist aspirations, stop worrying about their own political and economic needs for the sake of the greater good. America would abandon its lead, forsake the corporate profits it might bring. Now *there* was a goal.

"Sure, it's an uphill battle," Negroponte admitted with a shrug. And with that on the table, was there any real chance that the players could *ever* catch up with his prophecies? Now he was safe. The Media Lab was insulated from entanglement with the real world. Down in the bowels of his empire, Negroponte had already purged all evidence that the lab had ever been a player in the HDTV race. William Schreiber was long gone. He was a senior professor, based in his old department now, and spent most of his time posting Schreibergrams to addresses around the globe. And Jae Lim, Schreiber's heir, had never been a part of the Media Lab. He lived in a private world, wrapped in his own certain sense of rectitude. All that was left as far as HDTV was concerned were the lab's sociologists and political scientists who had been engaged in related work. At the beginning, the leader of this group was Russ Neuman, a political scientist who had seemed intent on puncturing the HDTV balloon. While Washington lurched through its drunken binge in 1989, Neuman had directed the shopping mall audience-reaction survey, the one suggesting that the public had little real interest in HDTV.

Over the following years, Neuman and several other people working with him in the Media Lab had also come to be active political advisers of sorts for some of the government and congressional leaders who managed the HDTV race. They carried the MIT badge, a powerful validator for many politicians in Washington, but these people from Cambridge were rabble-rousers first and foremost. Then, with the introduction of digital television, they began arguing that the nation's new digital, high-definition television standard should be built with "open architecture." To them this meant that the new TVs should be designed so that other manufacturers—computer companies and anyone else who might come along—could adapt them to their needs. It started as a classic '60s-liberal campus quest—"free access by the masses to this new technology," as Suzanne Neil, one member of Neuman's team, put it. "That grew, by association with others, into concern for open architecture in the set—interoperability."

A laudable goal, certainly, but all of this lobbying and troublemaking made Negroponte uncomfortable. "The problem with the Media Lab is that it has to be much more focused on pure, classical research," Neil observed. This mandate discouraged examinations of "real-world questions," she said—including their "open architecture" campaign. As a result, Negroponte "put up with us, but we were not good for him." Neil and the others realized they couldn't remain supplicants in the

Grand Vizier's court forever. They needed a new sponsor. They found one in the most unlikely of men—Michael Liebhold, a functionary from Apple Computer, who was the company's corporate liaison with the Media Lab. Apple was one of Negroponte's corporate sponsors, so Liebhold attended occasional meetings at the lab.

Negroponte didn't think much of Liebhold at first. The professor went by to see him once while visiting Apple's West Coast offices and seemed to view him as a character not significant enough to merit vizierial attention. "He had a little cubicle half the size of this one," he said, gesturing with distaste toward the walls of a small office where he sat for a discussion. At the time of Negroponte's visit, Liebhold was researching Apple's concept of "paperback movies"—films on compact discs that could be played in a computer. "He had all these discs around him," the Grand Vizier recalled. "He didn't seem to have much support at Apple. He was weak."

Liebhold was not an engineer. "My academic background," he explained, "is in film, television, and English literature. I'm self-taught." But he was a bulldog—a lightning rod for fights. He loved to attack easy assumptions, even if that laid him open for scorn. He seemed to thrive on being contrary. On the East Coast, it helped that he looked like a Northern Californian, and he made no attempt to change his appearance. He favored jeans with a sport coat of uncertain origin and an open shirt collar. He wore his thinning hair at a modest length, but over his neck it fell into a slight, fanlike tail. And he was a sardonic fellow, whose manner could turn on a dime to a tableau of scornful disbelief. "You mean to tell me that . . . ," he would say, turning his head in profile and squinting theatrically as he repeated the offending assertion.

When Neuman's soldiers spotted Liebhold at Media Lab meetings, they quickly decided he was their man. "On our own as cranky academics we weren't getting attention," said one of them, Lee McKnight. "We needed people in the computer industry. Initially Mike was not aware. He was doing paperback movies, that kind of thing. But he came to our meetings, and he was drawn into our guerrilla activities."

"We were lobbying Washington," Neil explained. "But it was like trying to battle mercury. Our power was extremely limited. And we were looking at who would make sense to help us." As far as Neil could tell, Liebhold had no particular background or knowledge in their field. "But he was a fast study," she said with a grin.

So they began coaching him, encouraging him, and soon even

Negroponte was impressed. "Out of the blue, he came to one of our meetings, and he was wonderful," the professor recalled. "He said all the right things." As the Media Lab guerrillas coaxed and advised, Liebhold began arguing that the computer industry needed to be brought into the Advisory Committee process, since digital HDTVs actually were computers. "I view advanced television as a terribly important mechanism for harmonizing image communications," Liebhold started saying. "Many of the stakeholders are bound to be dramatically impacted by the design of the HDTV standard, but they aren't being included in the process—namely, people from education and from the scientific, technical, and computing industries." With pronouncements like this, Liebhold was becoming the voice of the computer industry, now that it was awakened to the importance of *digital* HDTV.

From these seeds of discontent and common interest a new player was born: MIT's Program on Digital, Open High-Resolution Systems. Through Liebhold, Russ Neuman's soldiers had managed to convince Apple to pitch in some money so they could continue their underground war after their residence at the Media Lab had grown to be untenable. Apple agreed. So did another computer manufacturer, Digital Equipment Corporation, along with a couple of other clients. Without Neuman (who moved on to nearby Tufts University), Neil, McKnight, and a third colleague, Richard Solomon, moved into their new offices, a few blocks away from the Media Lab, and began a more focused campaign on behalf of the computer industry. From here Liebhold and the "cranky academics" opened a guerrilla war against Wiley and his Advisory Committee. With Liebhold as their front man, they began lobbying Congress to let representatives of the computer industry onto the Advisory Committee.

Wiley was not immediately receptive. "I have to remind my friends in the computer industry that this is a *broadcast* standard," he liked to say. But Liebhold pressed his views on Representative Ed Markey, the chairman of the telecommunications subcommittee, and then on Al Sikes over at the FCC. All this began to drive Wiley absolutely crazy. Here was a man attacking the very integrity of his Advisory Committee, a certain red flag for the chairman. Not only that, Liebhold was sniping at him *from the outside,* making end runs around him.

"Mike Liebhold, I'll be honest, he gets to me," Wiley said.

But Al Sikes liked him. "Liebhold, I remember, was testifying at one of our FCC meetings," he said. "And it was the most riveting introduction. He spoke to us up on the bar and said the computer industry is being precluded from any active participation in the high-definition television, Advisory-Committee process." Looking down at Liebhold from the dais, Sikes told him, "You should not be precluded." A few days later, Sikes set up a meeting in his office with Liebhold and some of Wiley's lieutenants. Wiley sent Jim McKinney, who now directed a broadcast industry technical group. Liebhold offered his complaints in his usual outraged, sneering manner. "He was opinionated, aggressive," Sikes said. "But I didn't mind that."

McKinney did mind, and he went on the attack. "How dare you bother the chairman with this!" he shot back at Liebhold. Like Wiley, he found Liebhold offensive. But Sikes accused McKinney of "verbally abusing" Liebhold. "Jim, I invited him here," the chairman said. "You are being rude."

Liebhold had caught Sikes's ear. A short time later, Sikes ordered Wiley to give the computer industry a greater role. So Wiley appointed Robert Sanderson, the manager of Eastman Kodak's Image Telecommunications Center, to one of his subgroups charged with seeing that high-definition televisions would coexist with cable TV systems and satellites. Sanderson was to ensure compatibility with computers, too. Wiley also ordered a new area of evaluation for the competing systems: the ability to be interoperable with computers. This had to be a paper exercise since the new condition was added after testing was almost complete. Still, it tripped up General Instrument.

When Woo Paik began designing GI's system he didn't spend even five minutes thinking about computer interoperability. Creating the world's first digital TV was challenge enough. But the other contestants had started work on their digital systems later, and each of them had searched for new features to make their entries better—or at least different. The Sarnoff group had made the greatest move toward computer interoperability. They'd added *packetization,* an important element of interoperability that made the computer manufacturers smile and assured that Sarnoff passed the new test. But now all of a sudden the golden boys from San Diego had a problem. For the first time, their system was flunking.

"I'd like to see how GI is going to do packetization," Jerry Pearlman of Zenith said with a wicked grin early in 1993. And in San Diego, Paik complained that Bob Rast hadn't kept him adequately informed of the changing rules. If he'd known beforehand, maybe he could have made some changes before the computer interoperability review. Maybe he could have added packetization.

Jim Carnes described packetization best. He found himself explaining it over and over as he gave presentations promoting the features of the Sarnoff group's system. In his analogy, the high-definition television signal wasn't like an army on the march. It was a train steaming from the station to the viewer's home—lots of trains, really. Each train represented a portion of the signal and included a locomotive pulling, say, thirty-two cars.

Well, with Woo Paik's system every train looked the same. When it arrived at the TV set, twenty-five of the cars, more or less, carried TV picture information. Three or four cars carried the audio signal. And the remaining two or three carried additional data and administrative information.

The Sarnoff system was different. One boxcar on each train carried a label, and that label told the TV set what all the cars behind it were carrying. That way, one train could contain TV picture and audio information, just as with Woo Paik's system. But the next train could also carry anything else, because digital signals are infinitely flexible.

One train could carry stock-market quotations, or computer spreadsheet data, or the morning's newspaper, or a TV traffic report customized to show the viewer's route to work—even a beeper service or a cellular phone line. These individual trainloads of data were called "packets," hence the term "packetization." With packetization, the TV signal could be used to carry anything at all. As long as the digital televisions in people's homes were designed to read these boxcar labels, they could come equipped with a wide range of options that would enable owners to print out the newspaper, store sections of the Yellow Pages on a hard disk, call up video E-mail, or view the special traffic report. Some TVs might have a floppy-disk drive, so that data could easily be moved from the TV to a computer and then back again.

With packetization, digital high-definition television moved a long way toward interoperability with computers. Still, even if *all* the contestant systems had offered packetization, Mike Liebhold would not have been satisfied. Far from it. He wanted to turn the fruit cart upside down. And while Wiley found Liebhold irritating now, soon he would grow to loathe the man as he had loathed no one else in his long career.

14 *We're not playing anymore*

"Don't tell anybody," John Taylor, Zenith's spokesman, whispered, "but we're getting ready for a demonstration in Washington." At the same moment in San Diego, General Instrument was packing up Woo Paik's creation, preparing to send it to Washington once again—though Bob Rast also asked, "Please keep this to yourself." And the Sarnoff consortium was thinking about putting on another show of its own, though none of them wanted competitors to know. All the contestants had demonstrated their systems just a few months before. But now, in January 1993, they were jockeying, lobbying, agitating once again—getting ready for Dick Wiley's "Special Panel" meeting, the predecision gathering of engineers.

The full Advisory Committee was scheduled to choose the winner in the spring, but Wiley wanted to hold this predecision session in February. He realized that if all that dense, complex data were turned over to the full Advisory Committee, "we'd have people like [CBS chairman] Larry Tisch sitting there" listening to the mind-numbing detail from the test results. He couldn't expect senior network executives to endure several days of that. "That's when I got the idea for the Special Panel, to sift through all the test reports."

The Special Panel members were supposed to study the test data and make the first cut at a selection. Nobody knew exactly what that meant, though the best guess was that they would eliminate one or two contestants. Needless to say, that made everyone nervous, and the gossipy guesses about who was in and who was out dominated every discussion. NHK and MIT seemed almost certain to be eliminated, by most assessments, and a strong question mark hung over Zenith-AT&T, too.

Wiley hadn't ruled out a second round of "supplemental testing,"

Jerry Pearlman's idea. But he did make it clear that there would be no new tests until the Special Panel had done its work. At Al Sikes's urging, he wanted to hurry the process along, even though with Bill Clinton's election the chairman was a lame duck. So in a letter to all the contestants dated December 4, 1992, Wiley wrote that "as chairman of the Advisory Committee, I have decided that we should maintain our February meeting schedule and that it will not be feasible to conduct any further system testing prior to these meetings." He added, "this judgment does not foreclose the possibility that the Advisory Committee might decide to conduct additional tests, focusing on system improvements, *subsequent* to the February meetings." He also warned that "one or more of the systems" might have been eliminated from the competition by then. Right away, all of the contestants, especially the weakest of them, began worrying that they might be out of the race before they could be tested again. Hence the new round of public showings. Wiley had asked each contestant to provide documentation of any improvements made since testing. Zenith wanted to make it plain that the improvements were real—and they wanted to do that in Washington this time, so that everyone of importance could attend. GI, stung by the computer-compatibility review, needed to show that Woo Paik had added packetization, and Jae Lim wanted to demonstrate that his system's flaws were mere "implementation errors."

The Grand Hotel, on M Street near Georgetown, was only a few years old; it had last been the home to a major news event in 1991, when the Palestinian delegates to the first Middle East peace talks had used it as their headquarters. Now, in February 1993, just a week before the Special Panel was to convene, Woo Paik installed his system in a luxury suite on the fourth floor.

A white piano stood to one side of the living room, and the sound system's subwoofer (the speaker providing the deep bass) sat on the floor under it, housed in its own white cabinet. Five more speakers, in custom-built white columns with crown and shoe molding to match the suite's woodwork, were positioned around the room. And three rows of chairs, about twenty seats in all, were set up in front of the large-screen, rear-projection monitor at the center of the room.

GI had mailed invitations to hundreds of people around town—

government officials, industry executives, members of the press, among others—and Quincy Rodgers's office took appointments for two dozen half-hour demonstrations spread over three days. This time, they hadn't invited just the key players in the competition. With Donald Rumsfeld's help, they'd sent invitations to a wide range of the city's big names—Robert Strauss, David Brinkley, Vice President Gore—people who had no real role in the process but might help GI by speaking favorably of what they had seen. "We are demonstrating the DigiCipher interlace system, incorporating improvements approved by the FCC's Advisory Committee since we so successfully passed the Advanced Television Test Center testing procedure a year ago," the invitations boasted. "These include Dolby AC-3 sound, packetization, and various transmission improvements. And our progressive system will demonstrate significantly improved picture quality."

These were the themes Bob Rast hit hard during the show. A few weeks before, he'd hired a local film crew to tape portions of Bill Clinton's inauguration with high-definition cameras, and as that tape played Rast told each new audience, "This is the first swearing-in recorded in digital high-definition." But from a political point of view, the show's most important element came later, when the tape offered a few minutes of simulated computer images. As Rast talked about the wonders of GI's new packetization scheme, the demonstration tape showed a computer on the TV screen running a Windows-like program. The underlying message was clear: Move aside, Mike Liebhold. We're fully computer-compatible now.

The show ended with a short, almost perfunctory, demonstration of Jae Lim's system, using a progressive-scan tape from the recent World Cup soccer tournament. Rast's tone was desultory, and in fact this footage didn't look particularly good. The previous tape had been sharper, clearer, more vibrant. Lim's images were flat and boring in comparison.

Bob Graves, the AT&T executive, sat in the audience for one of these shows and groaned as he watched the World Cup tape. The Zenith-AT&T system used progressive scanning, too, and both these "proscan" systems were fighting for survival. In his view, the proscan material was "abysmal." Lim stood in the suite's anteroom during the demonstrations and greeted visitors as they arrived and left. Graves walked right up to him and said, "You just did the worst possible thing for the proscan proponents. That material was terrible."

"I know," Lim said, and he hung his head. "But I couldn't be at the meeting where the material was selected, so I can't complain."

All in all, Lim thought GI was doing him no favors. In fact, he half suspected that they were trying to abandon him. He'd heard what Rumsfeld said. GI's president had flown in a few days earlier to preview the demonstration. He hadn't liked it. "Get rid of the MIT part," he'd told Rast. All that talk about progressive scanning—"If you're a guy like Bob Strauss coming in here, that's not going to mean anything to you." Rast managed to talk Rumsfeld out of dropping "the MIT part." Still, Lim could see that GI was emphasizing Paik's system over his. Already, he'd been upset by his test results. Lim was so certain he'd been wronged that he raised $100,000 to pay for a second round of viewing tests at the subjective-test center in Ottawa; they were run after he had fixed his "implementation errors." Rast had tried to talk him out of this, arguing that people wouldn't take the second set of results seriously. Still, Lim stood in the hotel suite anteroom now, handing out the "unofficial subjective-test results" to Graves and everyone else who passed by. Not surprisingly, they showed that Lim's system excelled by every measure. As Lim considered his situation, he confided to one visitor, "When I worked out that contract"—with Jerry Heller, two years earlier—"I should have let it say that GI would get *more* money if I win."

A few blocks up M Street, at AT&T's Washington office, Zenith and AT&T were putting on their own demonstration at the same time. Zenith's and GI's invitations had arrived in mailboxes within twenty-four hours of each other. Compared to GI's show, Zenith's was dramatic. Zenith had been in broadcast-standard battles before; over the years, the company had staged dozens of demonstration of new TVs and other innovations. So Wayne Luplow and his men knew what bells to ring and which whistles to blow.

Unlike GI, Zenith was staging a real broadcast, with the cooperation of a local UHF station. As the audience watched an excerpt from an Australian movie showing a yellow biplane soaring over verdant hills, Zenith engineers at the TV station turned the transmitter's power down, down, down. In the demo room, another engineer called out the declining numbers in a tone of ever greater amazement—as if to say, How on earth could we possibly be receiving a perfect TV picture at power levels *this* low? Finally the level was down to a mere trickle—one-sixth of the power used for a conventional TV broadcast—and the picture began to break up, until it simply froze. Then, before the audience could catch its breath, Luplow came forward and stood under the spotlight, holding something behind his back. "I want to show you the special HDTV antenna we are using for this," he said with a wicked grin, and from behind his back he pulled a simple UHF "bow tie" antenna, the kind that people had been clipping to their rabbit ears for decades. That sent a ripple of warm laughter through the crowd.

During one of the Zenith-AT&T demonstrations, Quincy Rodgers, GI's lobbyist, sat in the front row next to an FCC staffer and carefully pointed out to her where the system might falter. "Look at the faces in the crowd," he whispered, as they watched a tape of a Chicago Bulls basketball game. "That's where digital systems fall down."

But this one didn't. Zenith's demonstration simply outclassed General Instrument's. "GI just doesn't get it," said Greg DePriest, the Toshiba executive, after he had seen both. And DePriest's company was supposed to be *allied* with GI. Still, no one could say whether the demonstration would be enough to save Zenith-AT&T.

At first, the Sarnoff group worried that if the others were putting on demos, they should stage one, too. In the end, though, they decided not to do it. The consortium had demonstrated their system in Washington just a few months earlier. And by the beginning of February, as the Special Panel meeting loomed, members of the Sarnoff group were saying they didn't need to do anything more. They had no improvements to show because they hadn't needed to make any.

"We didn't suggest one single system improvement," Glenn Reitmeier asserted (though Wiley's December 4 letter had noted that "all five proponents have submitted improvements to their systems"). "We're claiming no flaws in the system." As for the corrections other contestants had offered, "they're just catching up with us, so if it ends up being a close-call decision, I think it's a no-brainer." Sarnoff's system should win.

Why was that? Well, Reitmeier and most of the others in his group were the General's heirs. Could anyone possibly be better qualified? "I encourage people to look at the track record of the different companies," he suggested. Then came a new claim: Didn't General Sarnoff push America to accept color television by manufacturing the sets when no one else would? So, shouldn't the reverse be true today? "If another company is chosen," Reitmeier said, "then no one is going to make the TVs."

"Without us," Joe Donahue added with a thin smile, "I don't think HDTV can make it in America." Sarnoff's position was clear: We have no improvements. We don't want "supplemental testing." Just pick us to win. Otherwise this whole effort will fail.

Behind all the RCA bluster, however, lay a larger unease. The truth was that in 1993 they were the only ones who saw themselves as heirs of RCA's greatness. Everyone else disparaged them as "the Euros," and the consortium's members knew that could count against them in the end. The Advisory Committee had been established in part because Washington feared Japanese dominance of HDTV. Was it realistic to think that the United States government would hand the HDTV crown to Europe instead?

"The consortium, it's basically Thomson and Philips," Jerry Heller said, sitting beside his desk in San Diego, fingertips pressed gently together. "Sarnoff is just a hired hand, and NBC is invisible. And the important thing is, do we want to let this technology get out of the U.S.? The last thing we want is for some foreign company to be telling us to change our television standard, or to move manufacturing out of the U.S." Zenith officials offered similar warnings. "I'm really afraid of what it will mean if the Europeans win," Jerry Pearlman would say. Jae Lim, of course, wholeheartedly agreed.

The General's men could hardly contain their anger over these remarks. At the Shrine, Jim Carnes's face darkened. "The concept that we are Europeans is a canard!" he snorted. "We're *more* American than the others." Reitmeier, sitting at the General's Queen Anne dining table for lunch, gripped the arms of his antique chair and pushed himself halfway to his feet as he exclaimed, "That *really* gets me, when they call us 'Euros.'"

But the Advanced Television Research Consortium, as the group liked to call itself, was controlled by Europeans. It now included Compression Labs, a small California firm brought in late in the process that specialized in digital compression. NBC was just an adviser to the group, and in theory so was Compression Labs, though the company was seldom if ever in evidence at consortium meetings. This led others to believe, fairly or not, that Compression Labs had been added just to get another American firm on the letterhead. The Sarnoff Research Center had started the work back in the mid-1980s and remained the founding member of the consortium. Now, however, the General's men were indeed "hired hands," as Heller had put it in his own unflattering way. The Shrine's engineers were contract workers hired by Thomson. The real power, and all the money behind the entry, came from Philips, the Dutch firm, and Thomson, an arm of the French government—a fact all the other contestants knew only too well.

One time, at a conference in February, Glenn Reitmeier was standing in a hallway, telling a small group about Sarnoff's work when a representative from one of the other companies, standing behind him, started making frog noises: "*Ribbit, ribbit.*" Reitmeier closed his eyes, grimaced, and clenched his fists. In a more contemplative moment, Carnes acknowledged that "things would have been very different for us if Thomson and Philips had been American companies. It's like"—and then he mugged a bit, stooping over as he pretended to be carrying two heavy suitcases—his two foreign employers.

Thomson and Philips were key participants in Europe's own high-definition television research program. They and several other companies were designing an analog HDTV system known as HD-MAC, which the rest of the world had long ago concluded was a dud. The governments of the European Community had launched this research project back in 1986, when they were trying to stop the Japanese from imposing their HDTV standard on the world. In the years since, the EC had continued to subsidize this work, giving the companies pursuing the research more than $1 billion. With all that free money and the implicit promise that Europe would adopt HD-MAC once the work was finished, Thomson, Philips, and the other continental corporations were loath to abandon the project—even long after it became clear that HD-MAC was obsolete.

In the United States, some of the contestants loved to whisper that the Sarnoff consortium's entry was the bastard child of HD-MAC, though there were few real technical similarities. Meanwhile, the consortium's leaders pointed out that nobody in the race could boast absolute patriotic purity.

"There are *no* American companies in this business," Joe Donahue liked to say. That wasn't exactly true; still, Zenith's assembly plants were in Mexico. "And who's going to own Zenith when they finally go bankrupt?" Donahue would ask. "I'll bet it will be a Japanese company. And GI—they make everything in the Far East."

("We make our decoders in Puerto Rico," Heller countered, sheepishly.)

Thomson and Philips, meanwhile, made their televisions in the American plants they'd bought from RCA and General Electric. While the "Euro" allegations flew fast and thick as the date of the Special Panel meeting approached, the issue was not as clear as any of the competitors liked to suggest.

Here at last, in February 1993—more than six years after the NHK demonstration at the Capitol that spawned the race—Dick Wiley's Advisory Committee was set to make some decisions. The Special Panel gathered at a Sheraton hotel in Tyson's Corner, a suburban shopping strip outside Washington. The hotel was pleasant, though certainly not luxurious, but nobody seemed to notice the accommodations. The stakes were too high. The final week of lobbying had been intense—demonstrations, interviews, press releases, private pleadings. Now, drinking coffee at the Sheraton at 9:00 A.M. on Monday, the contestants felt like candidates on election morning. They had done all they could do. Now they awaited the returns. The meeting room was an amphitheater seating fifty or sixty people in semicircular rows of desks and chairs, each desk with its own microphone. The rows narrowed as they descended to a stage at the base. There, behind a table, Robert Hopkins, the Special Panel chairman, would preside.

Hopkins was executive director of the Advanced Television Systems Committee, a broadcast industry group whose principal mission was to publish official documents setting out new technical standards for the world of television. An engineer, he'd worked for RCA for many years, including a long period at Sarnoff. On the Special Panel meeting's first morning, he stood just inside one entrance to the amphitheater reminiscing with Jim Carnes about the old days at the Shrine. Today the two of them were about to take part in the first cut at approving a new digital television standard for America. As both of them considered that, they recollected the days not so long ago when they walked the halls at the Shrine with slide rules hanging from holsters on their belts. Their conversation now was easy and comfortable. Still, the tension in the air was palpable—for Carnes, of course, but also for Hopkins. He saw this meeting as an important responsibility, and Wiley reinforced that in his brief remarks when the four-day event opened at 9:30.

"Like most of you, I used to have a day job before HDTV came along," he joked. "This has been a long process, and all of you have worked hard. But in my view this is truly a historic meeting, reminiscent of the [National Television Standards Committee] meeting fifty years ago."

Hopkins agreed. He was an unsentimental, businesslike man—some might say pedantic at times, though always pleasant and helpful. At the same time, however, he brooked no nonsense, and he held clear notions about how his Special Panel meeting was to be conducted. He wasn't going to let the contestants push him around. No sirree! Years before, Wiley's subcommittees and working groups had begun work with no clear rules of order. The contestants had attended most of the meetings, and it had never been decided exactly what role they should play. Soon they began taking over many of the sessions, dominating the discussions with ardent advocacy of their own interests. Hopkins wanted to end that right here and now. A week before the Special Panel convened, he vowed, forming his hand into a tight fist, "I *absolutely will not allow* a proponent to take over the meeting. I will not allow it! The chairman has a button to cut off all the other microphones in the room, and I will use it!"

As the meeting opened, the contestants were unaware of Hopkins's oath and probably wouldn't have cared much if they had known. All thoughts were mired in the politics of the moment: who was up, who was down. The only point on which everyone seemed to agree was that NHK, the sole analog system left in the race, was likely to be eliminated.

On the first morning, Keiichi Kubota of NHK sat among the other contestants at the center of the amphitheater, near the top. When the time came, late in the morning, for each contestant to give an opening statement, Kubota was up first and gave his best shot. "We believe being analog is an advantage," he said in a calm, even tone. "It's a proven technology. It can be sent using the same transmitters and the same antennas as [conventional television]. We spent an enormous amount of time developing the Muse algorithm, and we have never experienced a failure."

All that was well and good, but Kubota also knew that if Narrow Muse hadn't actually failed at the Advanced Television Test Center, it had come awfully close. The test results were laid out in a fat binder that lay on every desk. So he tackled the problem head-on and seized on the easy excuse. "We consider the test results to be a good representation of our hardware—as delivered," he said. "But we had a couple of *implementation problems* with the hardware as delivered to the test center that affected our results. All those improvements are in the hardware. It is already fixed." There are several good reason to pick NHK, Kubota added in closing. "And one of them is that this will improve relations between American and Japanese broadcasters." This may have been the greatest leap of all, if NHK's dealing with John Abel and the others at the NAB was an example.

Bob Rast of General Instrument was next, and he spoke with confident, California salesmanship: "We've got a can-do team. We pushed the envelope. We made digital a reality. The issue now is, the U.S. is once again the leader in television technology." As for the test results, "we believe that in every case where we were not the best, we have worked hard to improve it. We have modified our system to include packetization. Now, is there a runaway leader here? No. But is there a clear leader? Yes, and we say it is DigiCipher."

Wayne Luplow wasn't interested in offering lofty thoughts like that. What Zenith liked most to do was count pennies, as Arun Netravali had once observed. And when Luplow stood up, he told everyone that Zenith-AT&T was the only contestant "that has a practical tuning system," adding, "The others used tuners that will cost $3,000." Then he complained about testing inequities. "We thought late arrival at the test center would result in disqualification, so we showed up on time. That was one of our problems." (Glenn Reitmeier of Sarnoff, sitting next to Luplow, offered no reaction.) "In summation," Luplow said, "the improvements we have made in picture quality are demonstrable and, I would say, *great*. We are ready to manufacture the equipment if we win."

As Reitmeier spoke for Sarnoff moments later, he rolled his hand over the trackball on his Macintosh Powerbook computer, scrolling his script up the screen. He seemed to be trying to suppress his usual boastful tone, but without complete success—he sounded as if he were reading a press release. "Our system wins or ties for each of ten selection criteria," he said. "No other system can match that." As for Luplow's remarks, Reitmeier shot back, "We are here to pick the best system, not the best tuner." (The Sarnoff consortium's tuner had not performed particularly well.) "We have committed the resources to create an HDTV industry in America." Then Reitmeier closed with the RCA spin: "We have a record of creating industry-wide revolutions, including NTSC television, color television" and the rest.

Jae Lim was last and offered up his new, unofficial test results. "We have no competition," he asserted. "And the purpose here is to establish the best system for our nation. Our actual performance at the test center was much below our hopes," Lim admitted. But naturally, he added, "we had *implementation errors.*" Still, Lim went on, "we were the only ones to finish testing ahead of schedule. We were the only ones to submit a six-channel sound system." He droned on for several minutes, but by now not everyone was listening. A representative from the State Department was dozing, and a liaison officer from the Mexican Embassy was nodding off, jerking his head back up every couple of minutes.

Technical discussions followed, and soon Robert Hopkins got his first opportunity to follow through on his warning, when Wayne Luplow, looking at the test-data book in front of him, asked in a whiny, combative tone, "Why aren't the improvements we offered in this book?" Hopkins slapped him down, swift and hard. "There's no way for us to have any definitive understanding among us of what the improvements are," he said, and quickly moved on to something else. Luplow's mouth dropped open in astonishment; Hopkins was suggesting that all the lab work and lobbying of the last few months would count for nothing.

A few minutes later, Luplow began quibbling about some of the data in the book, and Hopkins, teeth clenched, cut him off in midsentence. "This meeting will *not* be taken over by proponents!" he shouted. "I suggest you discuss that over lunch." Luplow looked as if he might choke.

Though the contestants grew angrier and angrier as Hopkins and the others at the head table cut them off over and over again, the discussion and debate seemed hollow; nearly everyone there suspected that the main decisions had already been made. Seldom if ever in his public life had Dick Wiley scheduled a meeting without arranging all the important decisions in advance, and the Special Panel was no different. As Wiley saw it, his job was to select the best HDTV system for America. All the contestants had come to the test center with one problem or another, but every one of them had made improvements in the months since. Al Sikes had left the FCC; he resigned the day Bill Clinton took office a few weeks earlier. Wiley was more or less on his own now; he could do what he wanted. So Wiley had decided to retest all the contestants. All but one of them, that is.

A few days before the Special Panel meeting, Paul Misener, Dick Wiley's assistant, called Keiichi Kubota at NHK's office in New York. Misener felt as if a decade of history had settled on his shoulders. "After all," he explained a few days later, "I came from the agency that set off all the alarms about HDTV" back in 1986. Before taking a job in Wiley's law firm, Misener had worked at the Commerce Department in the very office where Al Sikes had pushed the FCC to pay attention to high-definition television. NHK was still the problem back then; the world seemed close to accepting the Japanese technical parameters as a global television standard. And Misener had helped persuade Sikes and his successor at Commerce to change the official American position of support for NHK. Seven years later, the wheel had made another turn. Misener's tone was polite but firm when he told Kubota, "We've looked over the test data, and we believe the Special Panel will find that Narrow Muse is not comparable with the others. Mr. Wiley does not believe it will be recommended for retesting."

"I understand," Kubota said. To Misener it sounded as if Kubota was not surprised. He wasn't. "As soon as I actually saw the General Instrument system," he explained a few days later, "at that time I knew I was going to lose." When the Special Panel meeting opened, he offered the best case he could for his system—just to save face, it seemed. But then, toward the end of the fourth day of the meeting, Kubota asked to be recognized. "We are withdrawing from the competition," he announced to a quiet, respectful crowd. "I have already reported this result to our management in Tokyo, and they are comfortable with the decision. On behalf of NHK, I want to thank Chairman Wiley for your guidance. And I thank my colleagues on the Special Panel. We believe that NHK was treated fairly in this process. But I think a digital system is the best for the United States."

"We owe NHK a great debt of gratitude," Wiley intoned. With that, all of the Americans rose from their chairs and offered more than a minute of warm applause to the vanquished. Yellow Peril jingoism had begun the process that brought them here. Now the Americans were munificent in victory, full of warmth and praise as they tried to disguise the satisfaction swelling in their bellies.

All the data had been examined and discussed in numbing detail. The four-day meeting was almost over, except for one last bit of business. The Special Panel had to carry out Wiley's mandate and order a new round of tests for the four remaining contestants. It fell to Joe Flaherty, the CBS executive, to present the justification. Sitting behind the table on the amphitheater's stage, Flaherty spoke in a cold and matter-of-fact tone. "It's impossible to pick a superior system," he said, "since each of them has such serious faults that it could not serve as a national standard for an HDTV system."

"I concur," another Special Panel member piped up.

The contestants were stunned. They'd been pushed, pulled, and bullied all week long, and now the meeting was closing with another good, hard swat. They were fed up. Bob Rast spoke first, and even people who didn't know him could see that he was angry. "It would be very nice if this group would not back off of making determinations of things," he said, his voice steely. He was looking right at Flaherty. "You're just saying everything's inadequate. Well, I think what *you're* doing is inadequate!"

"There's no data to show inadequacy relating to any of these systems," Luplow chimed in.

Flaherty wasn't backing down. Far from it. "I do not believe we're doing an inadequate job," he said with a chilly, supercilious air. "It's not our job to give advice to proponents. And I for one am getting tired of these commercials!"

With that the Special Panel ended. As the contestants left the Sheraton hotel, every one of them was steaming mad. "We were treated like traveling salesmen," Rast complained. "That was my Negev low point."

"I was muzzled," complained Robert Graves of AT&T. He had not been permitted to offer the long, elaborate lobbyist's explanations of which he was so fond.

"We were bound and gagged," complained Eric Petajan of Bell Labs. And a few days later, Joe Donahue of Thomson summed up the feeling that had begun to jell over the preceding days: "A lot of us believe the process died last week."

All of them were angry, but one player was fed up. Donald Rumsfeld, GI's president, wasn't sure he wanted to go on with the race. Wiley had scheduled a full Advisory Committee meeting to take place two weeks after the Special Panel, to ratify the decisions made there. And like the Special Panel, the Advisory Committee's decisions seemed preordained. So Wiley also began making arrangements to start the retesting, and it quickly became clear that everyone could look forward to a long, expensive, and tortured process. Rumsfeld wanted none of that.

"I want this to end at some point," he complained. "General Instrument is the leader now, and the United States is the leader. But if they keep moving the finish line back, the contestants are going to look more and more alike, and the less advantage there is to being the leader. This is U.S. technology, and I'd like to see us use the leadership we have to enter the market around the world." In other words, Rumsfeld wanted the Advisory Committee to make some decisions—*now.*

GI's president was accustomed to being in charge. He and Dick Wiley had been in government at the same time, two decades earlier. While Wiley had been the chairman of the FCC, one of the government's cat-and-dog agencies, Rumsfeld had been the White House Chief of Staff, Secretary of Defense. Though he never said so, it doubtless felt odd to him to be Wiley's subordinate now, and Rumsfeld set out to change the process.

The way he saw it, retesting was going to cost each of the contestants several million dollars, given the testing expenses, staff time, and everything else. That was a lot of money for General Instrument. This would also delay the process for a year or more—another year in which nobody made any money from HDTV. As it was, he said, "when I talk to securities analysts about this, about HDTV, we have no revenue stream from this. Just costs." What was worse, once the contestants returned to the test center, each of them would have made improvements, fixed problems, added features that hadn't been thought of before. Most likely, the systems would have grown more and more alike. And if all the systems became nearly identical, how on earth would Wiley and the others make decisions a year from now? Maybe after all this time and expense General Instrument would actually lose—and then after spending an additional $2 million or more, to no useful end. What would GI's board of directors say to that? What about the quarterly results, the stock price? What if analysts saw GI's entry in the HDTV race as an expensive folly in the end? All of this was dispiriting at best. Rumsfeld wanted to turn things around.

Maybe we should join forces, he thought. End this competition and combine our individual entries into one supersystem. We'd certainly reduce our costs and assure profits for *everyone.* With an alliance, the contestants would also undercut the Advisory Committee's power. How could Wiley's group do anything but certify a winner when there was only one contestant?

Rumsfeld called Jerry Pearlman at Zenith first. Let's form a partnership and split the profits three ways: one-third for GI-MIT, one third for Zenith-AT&T, and one-third for the Sarnoff group. Pearlman was enthusiastic; Zenith couldn't easily afford the retesting. And if Zenith actually lost the race after all this . . . well, that could be the final blow. AT&T liked the idea, too; the company had already spent a reported $20 million on the high-definition program. What's more, the abuse everyone had taken from the Special Panel "gave the proponents a feeling of unity," as Petajan put it.

Jae Lim offered qualified enthusiasm, noting that "if there's an alliance system, I hope it won't be a cow's head on a horse body." Beyond that, Lim was keeping his thoughts to himself. The others would hear from him soon enough.

When the Sarnoff group got the call a few days later, they expressed some interest. But they also seemed—well, suspicious. Joe Donahue noted that given all the Euro-bashing of the last few weeks, "frankly, we were surprised to be included." Still, they said they were willing to talk.

Word of this got to Wiley, and he loved the idea of an alliance. But both he and Rumsfeld knew that forming a business agreement between institutions with such diverse and conflicting backgrounds, interests, philosophies, motivations, and goals was a long reach at best. So Rumsfeld came up with a backup proposal. Suppose we can't form an alliance. Maybe the Advisory Committee should make a further cut among the contestants now. Let's "minimize delay," he said in a letter to Wiley and all the members of the Advisory Committee (the day before their meeting). "We understand the arguments for additional testing, but our preference would be for the Committee to select if not a 'winner,' then at least a 'prime system,' and a 'backup system,' so that the process can go forward without the burden of testing all four systems." Wiley immediately dismissed these ideas. The decision had been made, and he wasn't about to change it.

Toward the end of the full Advisory Committee meeting at FCC headquarters the next day, Rumsfeld offered the same suggestions in person. Wiley rejected them with little discussion. "I know you are concerned about paying for retesting of two systems," he said, looking down at Rumsfeld from the dais. "But with two systems you have two chances to win. You can lower your costs by leaving one system behind if you want." Rumsfeld pursed his lips at that. Wiley called for a vote of the full Advisory Committee, and it was unanimous. All the systems would be retested.

"And now," Wiley announced as he rose from his chair, "we're going to hold a lottery to decide the order of testing." By now, Rumsfeld was really angry. Wiley had simply ignored his suggestions. As everyone stood up, he stormed from the room and called the other General Instrument people to go with him.

"What about the lottery?" they asked him, standing in the corridor. "We're boycotting it," Rumsfeld declared. Then he went downstairs, climbed into a taxi, and headed for the airport, leaving Bob Rast and Quincy Rodgers standing there looking at each other.

Rast was unhappy. It was fine for Donald Rumsfeld to make his grand statement of principle as he headed off to Chicago. But Rast knew that Wiley would be furious, and he was the one who would have to clean up after his boss.

Wiley, meanwhile, had moved the contestants and a few observers back to the chairman's office, empty now that Al Sikes was gone. Wiley stood behind the conference table that had been his almost twenty years before. Pearlman wandered in, followed by Jae Lim and Jim Carnes. But no GI. Wiley sent someone out to look for Bob Rast—to no avail. Then he turned to Lim. "Maybe you could draw for GI?" he asked. Lim shook his head. No way. "I prefer not to do that," he said.

"All right," Wiley said, irritated but impatient to get on with it. "We'll draw, and whichever card is left is GI's. I have four cards here, with numbers one through four." He held them up, shuffled them, and then laid the cards face down on the table. "I suggest we just draw them, make it easy." But by now Wiley should have known that nothing about this competition was easy.

"If we get number one, does that mean we get a choice of going first or last?" Wayne Luplow asked.

"I think number one goes first," Wiley said, but another Advisory Committee member quipped, "Maybe we should put together a working group to study that."

"Are they tradable?" someone asked, and Wiley said, "I have no problem with that."

"We object to that," Carnes barked immediately.

"You'll have to put that objection in writing," Luplow snapped back.

"All right, all right," Wiley said, hands in the air as he admonished his children. "Let's get on with it. Jerry, why don't you draw first," he said, motioning to Pearlman. Zenith's president walked over, picked up a card, glanced at the number, and then, poker-faced, pressed it against his chest.

"Aren't you going to show us?" Wiley asked. No, Pearlman said, but then he couldn't contain himself any longer. Grinning, he held up his card for everyone to see—number four, the choicest spot.

Jae Lim drew next and got number three. Carnes drew number two. And the first test slot was left for the absent player, General Instrument—the worst possible outcome for GI.

"You run a good selection process," Pearlman said, smiling, as he left. But then, perhaps not wanting to seem too exuberant, he added, "I guess the testing order really doesn't make any difference."

"So you want to trade?" Wiley shot back.

"No," Pearlman insisted, still smiling. "I'm going to sell my number." Behind him, somebody muttered, "It's probably worth more than your company right now."

Within a few minutes, Bob Rast and Quincy Rodgers heard about the number they'd drawn by default. "It was a squeeze play," Rast complained. "And it screwed us. I'm about to go over and have it out with Wiley." In fact, Rumsfeld was sending him over with instructions. He'd called moments earlier, still angry. "Tell Wiley we're considering dropping out of this race," Rast said his boss declared. "It's not worth it for us now. We're not going to play around with this anymore."

Rast was unhappy, too, but telling Wiley they were dropping out was the last thing he wanted to do. He knew just what Wiley would say. And he was correct.

"Well, we'll miss you" was the chairman's only response.

Rumsfeld cooled down after a few days. Wiley may have helped change his mind. In a testy phone call with the chairman, Rumsfeld had complained that GI would have won if the Advisory Committee had made a decision.

"No," Wiley told him, "you would have been a close second." The Advisory Committee's engineers had seemed to agree that the test results gave the Sarnoff group a slight edge. Rumsfeld polled the senior management of his company and got long memos back. The consensus was that it didn't make sense to drop out—at least for the present. So Rumsfeld refocused his energies on moving the partnership negotiations forward. Wiley liked the idea of a partnership. In early March, he sent a letter to all the contestants, offering them a deal: go through with retesting or form what he called a "Grand Alliance." The name stuck, but that proved to be the easiest part of the task. GI put out the first general offer: We'll split all profits and patent royalties one-third for GI-MIT, one-third for Zenith-AT&T, and one-third for the Sarnoff group. As Rumsfeld saw it, "that was a big concession." Not everyone had as much to offer as GI—a front-runner with a real shot at the grand prize.

Under this proposed deal the partners would decide among themselves which components from the individual systems would be included in the new "Grand Alliance" prototype, and they would use the test center results as a general guide. Then they'd divide the construction among all the labs. Zenith and AT&T went along with all this. So everyone waited for the Sarnoff group's response.

And waited, and waited. . . .

Philips and Thomson, of course, had to consult with their leaders in Europe. When they finally answered, the consortium members said they were interested in joining. "The Grand Alliance, properly structured," they wrote in a memo to the others three weeks later, "is a highly desirable outcome." But the business proposal that followed this letter "hit the table with a thud and then fell right off," as Peter Fannon, the test center director, put it.

"The parties agree to support the joint system as the HDTV standard for the U.S., Canada, and Mexico," the consortium's proposal said. In other words, the two European companies might reap profits from the Grand Alliance in the West, then try to quash it with competition of their own in Europe and the rest of the world. That idea did not sit well with the others. "It's really off the wall," Luplow told Misener when they discussed it in a phone call. Some of them started talking about leaving the Sarnoff group out of the alliance. "They're down now," Rast said after the consortium's offer crash-landed. "Maybe we'll go without them. We will be reasonable, offer a fair deal. But if they fail to pick it up. . . ." He shrugged. "I have to admit, when we've got them on the floor, foot on their neck, it's hard to take it off."

Other potential partners offered a more conciliatory view. One advantage of a Grand Alliance was that nobody would be a loser when the competition ended. No one would be left to file a lawsuit over the results, possibly delaying introduction of a new product for years. For that reason, if no other, most of the corporate leaders wanted to find a way to bring the Sarnoff consortium in. They asked for a better offer and waited . . . and waited some more.

While they worried about the Sarnoff group, another problem popped up—a big one. Jae Lim had been watching and ruminating for weeks. He felt he was being ignored—and that, of all things, he would not accept. "All the discussion is splitting profits one-third, one-third, one-third," he said in March. "But GI didn't check with MIT on this issue. Our position," he added, using the royal "we," "is, it's one-quarter for each of us. We paid just as much for testing as everyone else." (Of course, Lim failed to mention that GI had paid that bill and most of the others.) "And we have broad intellectual property rights and patents in our system."

Word of Lim's objections quickly spread among the other companies, and some of the players wondered why GI didn't just drop him. Pearlman asked Rast straight up, "He's your partner. Do you have control of him?"

Yes, Rast told him. But within a short time, Pearlman says, "he came back and said, 'Well, I guess we don't have as much control as we thought.' "

The truth was that Lim found the magic glove that enabled him to grab the Grand Alliance negotiators by the throat—and squeeze.

For Jerry Heller two years earlier, it had been a throwaway concession to his xenophobic new partner. When Lim made his last request over dinner in Cambridge, Heller had assented without thought: Sure, we'll give you full right of approval over any agreement we make to cooperate with any foreign-owned entity. Lim, Heller knew, had a classic Korean preoccupation with Japan, and Heller didn't believe that GI would ever be called upon to form an HDTV partnership with the Japanese. So what did it matter?

Two years later, it mattered plenty. Thinking about it one day in March, Lim suddenly realized: Maybe Thomson and Philips are not the Japanese. But they certainly are *foreign.* So didn't that give Lim full right of approval over General Instrument's participation in the Grand

Alliance? The Grand Alliance could not exist without GI. Wasn't Jae Lim now invested with absolute veto power over the Grand Alliance? As he saw it, the answer was a clear and unequivocal yes. And by god, Rast, Pearlman, and the others were going to have to start listening to him now.

"There can be no agreement without MIT," he told Rast. When Rast looked into it, he saw that Lim was right.

"MIT wants one-quarter," Lim said. And Rast thought, Oh, shit! The others didn't know about Lim's contract clause on partnerships with foreign companies. They were telling Rast, You guys are going to have to get Jae Lim under control. But it wasn't long before Lim grew to be only one of several competing irritations, because at just about the same time the partners of the Sarnoff consortium dropped a stink bomb into the arena.

Joe Donahue wasn't sporting the old RCA hubris just now. No, as he saw it, the Sarnoff consortium was being discriminated against in the Grand Alliance negotiations, and it just wasn't fair. "Jerry Pearlman's running around town politicking," he complained. "And Rumsfeld knows Washington far better than we do; he's here and there and all over the place. They have to know we can't let them walk all over us." He and his former RCA colleagues decided it was time to fight back.

Actually, the seed of their new strategy had been planted months earlier, when Clark Herman, a Sarnoff consultant, had read some of Robert Reich's writings on economic theory and concluded that his philosophies were friendly. Reich was a well-regarded professor at Harvard University's Kennedy School of Government, and as Herman saw it, Reich liked the idea of foreign companies owning plants in the United States, employing American workers to manufacture their goods. Now Reich was the Secretary of Labor.

Then in February, the AFL-CIO held its annual meeting at the Sheraton hotel in Bal Harbour, Florida. The Sarnoff consortium had been courting labor for months, so they put on an HDTV demonstration at the hotel. Reich was there, and when the secretary came by to see the demonstration, Herman buttonholed him. HDTV will create lots of jobs, Herman explained, and we are the only competitor that promises to manufacture the sets here in America. Reich was listening.

Everyone here at this convention supports us, Herman went on. Do you think you could give us some support? Reich nodded. By his own admission later, Reich didn't know a great deal about the HDTV race. But as Herman explained it to him, one thing seemed clear: The new high-definition television industry could create a lot of jobs, and early in the Clinton administration nothing seemed more important than that. A few days later, Secretary Reich wrote a letter to James Quello, the acting FCC chairman. "Final selection" of a winning system, Reich wrote, "should turn in part on an assessment of which system would make the greatest contribution to domestic, high-wage employment." Job creation should be given "significant weight in the selection process."

Well, that played right into the Sarnoff consortium's hands, just as Herman had planned. For months, the Sarnoff group had been boasting that it would manufacture HDTVs in its American plants, employing American workers. So on March 3, when the *Wall Street Journal* reported on Reich's letter, the article concluded, "Mr. Reich appears to be favoring an entry consisting of the David Sarnoff Research Center, Dutch-owned Philips Electronics. . . ." The story was only about two hundred fifty words long, and it was buried deep inside the paper. But it stunned the Sarnoff consortium's would-be partners. What in the hell do these people think they are doing? they asked. Is this their answer to our request for a new partnership proposal? And was the Secretary of Labor really saying he wanted a *foreign* group to win the HDTV race?

"I don't understand it," Jerry Heller said a few days later, confused and angry as he sat in his office in San Diego. "This heavy-handed PR is really very clumsy, to my mind. The jobs thing is economic blackmail. And the important question is, do we want to let this technology get out of the U.S.?" GI immediately began preparing a counter-letter to Reich, and so did Zenith. Meanwhile Wiley was furious.

"It's so stupid!" he spat. "I asked them, 'Why are you doing this?' " Then, affecting a high-pitched, whiny tone, Wiley repeated their answer: " 'They keep calling us Europeans.' Well, they *are* European. *Totally* European. Sarnoff is just a *front!* This has been an apolitical process until now. Are they deliberately trying to undermine the Grand Alliance?"

Donahue offered a defense, but the veneer was thin. "We didn't have any choice," he said. "And was this a lie? Did we distort? All we said was that we plan to make the TVs in the United States." But was it a smart, helpful thing to do? "Probably not," he admitted, looking at his shoes.

Soon, the other letters began landing in Reich's office. You don't understand, Pearlman told Reich. The real issue is, Who will make the semiconductors and other electronic components for these new digital televisions? "These semiconductors are where the new, high-tech jobs will be created," Zenith's president wrote. The Sarnoff group, he pointed out, would make most of its semiconductors in Europe.

Donald Rumsfeld was more direct. "The inference of the *Wall Street Journal* article," he wrote, "is that the administration may favor the predominantly European competitor over American companies, and that the Department of Labor is intervening in a matter which is the subject of action by the FCC, an independent regulatory agency." We're trying to form a Grand Alliance, Rumsfeld added, and all the *American* competitors have concurred. "But the response from the consortium was not encouraging. Importantly, they declined to endorse a combined U.S. HDTV system for the rest of the world, including Europe."

The underlying consensus was: This is really stupid. Does anyone doubt that TV manufacturers, domestic and foreign, will simply build high-definition televisions in their existing plants, no matter where those plants are, no matter who wins the race? And in fact, the choice of a winner in the HDTV race would probably have no effect on the number of jobs created in America. If HDTV was a success, every manufacturer would make the sets. If HDTV failed, then it wouldn't matter. The jobs argument was specious, but some of the other competitors thought Sarnoff might have had another aim in mind. The jobs debate might well torpedo the Grand Alliance negotiations, and without an alliance General Instrument was threatening to drop out of the race. With GI gone, the Sarnoff group's chance of winning would be far greater.

In any case, Secretary Reich was stung by the backlash. Apparently he hadn't known he was stepping into a snake pit. "I wasn't trying to politicize the process," he said two days later. "I was merely pointing out that economic considerations were appropriate." In a letter to Rumsfeld, Reich said he was "dismayed that my comments to the FCC were misconstrued. There was absolutely no expression [in the letter to Quello] as to which companies should be favored."

It didn't matter. Reich had started a wagon rolling, and the ever responsive House of Representatives jumped right on board as it careened downhill. The House passed a nonbinding resolution: "It is the sense of the House of Representatives that the Federal Communications Commission should consider domestic employment as a factor in its selection of a standard for high-definition television. . . ."

The Sarnoff group had been penitent at first, but they seemed to regain their composure. "We wish to commend you for the initiative you launched to make high-wage employment a significant factor in the selection process," the group wrote to Reich over the signature of all the principals. Then the consortium's lobbying started to spread.

"The decision that will be made by the Federal Communications Commission will have tremendous impacts upon my state," Evan Bayh, the governor of Indiana, wrote to Quello on March 5. "Many people are not aware that Thomson Consumer Electronics and its RCA brand are major employers in Indiana and in the United States. Thomson employs 6,000 people in Indiana and in the United States."

Bayh never specifically asked the FCC to select the Sarnoff group's system. But Governor Robert Casey of Pennsylvania showed no such reticence. Late in March, he toured Thomson's picture-tube plant outside Scranton and offered these thoughts, faxed around the country in a press release: "HDTV is more than just the greatest leap forward in technology in 40 years. HDTV means jobs. Good-paying high-tech jobs for working men and women here in Pennsylvania and elsewhere in America; more than 10,000 American jobs altogether. . . . The Thomson consortium is the only proposal pending before the FCC that pledges to build HDTV tubes and receivers in the United States—and more importantly, in Pennsylvania."

"The Casey endorsement," Heller said, "that really floored us."

"It's really amazing, frustrating, and astounding all at the same time," said Paul Misener, Wiley's assistant. Quincy Rodgers, GI's Washington representative, summed up the real mystery. "We are trying to create an alliance," he told *Broadcasting & Cable* magazine.[1] "I guess

---

[1]*Broadcasting* magazine, recognizing the ever larger role cable television was playing in the industry, had changed its name to *Broadcasting & Cable.* By now the Cable Mongols had managed to wire 63 percent of America's homes, and the magazine had enlarged its coverage of the cable business.

what puzzles us is, if you are going for an alliance, why are you trying to politicize it?"

The Sarnoff group responded to the complaints by disparaging the other competitors. "Mr. Rumsfeld depicts his company as the 'inventor' of digital television, with the 'European consortium' being a follower," the group wrote in still another letter to Reich. "In fact, the basic principle of the compression approach used by *all* of the proponents was first published in 1974." A tiny-print footnote referred the Labor Secretary to an obscure technical monograph.

James Clingham, a vice president at the Sarnoff Research Center, wrote to James Florio, the governor of New Jersey, "Our consortium is the only FCC proponent which manufactures television sets in the U.S.... You are well aware that Sarnoff has helped foster the development of high-tech jobs in New Jersey as a result of our high-definition work."

That was too much for AT&T. Bell Labs, of course, employed tens of thousands of people in New Jersey, and the corporation's leaders forcefully reminded Florio of that.

"We stopped it," said Bill Radwill, an AT&T division manager.

Grand Alliance talks were continuing, in theory. But the atmosphere was poisonous. When the National Association of Broadcasters convention opened in Las Vegas on April 8, 1993, the contestants were barely speaking. For years, ever since the HDTV race started, the NAB convention had been the place where the competitors showed their stuff and promoted themselves to the broadcasting community. At least seventy thousand TV and radio people came to the show each year, and many of them filed through the demonstrations each of the contestants set up on the convention center floor, the cavernous hall where equipment manufacturers from around the world showed their wares in elaborately furnished booths.

This year, the HDTV contestants' booths were grouped together in a hard-to-find area toward the back of the great hall. The NAB, as the competitors knew only too well, was not among their greatest fans. Still, each of them—Zenith-AT&T, the Sarnoff group, and GI-MIT—had set up displays offering promotional signs, pictures, and lectures, and an enclosed viewing room, where they put on demonstrations one after another all week long.

Some visitors to the Sarnoff group's stall might not have been sure whether they were visiting a booth belonging to a technology consortium or a flag factory. "Jobs, Jobs, American Jobs," the banners proclaimed, and American flags hung left, right, center, and in between. Big ones, small ones. Rows and rows of flags. Anyone reading the promotional handouts had to look long and hard to find any reference to the companies' European owners. "Investing in America's future," the front cover of a Thomson brochure declared. A Philips booklet boasted, "American Technology, American Products, American Jobs," and another described the American affiliate of the Dutch company this way: "Philips Laboratories is a research division of North American Philips Corporation—an American-based company with annual sales of about $6 billion." To one side of this showroom sat an easel holding a large map of America bristling with dots that purported to show where consortium members had research or manufacturing facilities, warehouses, post office boxes—maybe even the homes of retired employees. The Zenith-AT&T people could see this map from their own booth, just across a carpeted aisle. Bill Radwill wondered if it might be a blueprint for the consortium's jobs campaign. "We really ought to get a copy of that map," he said. "Then we'd know which governors they are going to call." So Radwill walked across the mauve carpet to the Sarnoff booth, pointed casually at the map, and asked, "Do you guys have a handout with that?"

"Two people from Zenith already asked," a Sarnoff official snarled back. "No, we don't."

A few yards away, at the GI booth, Jerry Heller was discouraged. The Advisory Committee continued planning for the retests, and in just three weeks everyone would have to ante up $306,000 as a test center reservation fee, unless the alliance negotiations succeeded. The first system was due to be retested in just over a month. Rumsfeld had made it clear from the beginning: If we actually have to go through with the retests, we are dropping out of these alliance negotiations, and the threat lingered that GI might drop out of the race altogether. Still, the negotiations were going poorly. "Everyone seems motivated," Heller observed. "But it just drags on and on. Nothing has jelled. The U.S. has a chance to set a standard for the world. But the European Community and the Japanese are going to catch up if we don't get going."

4

Politics

15 *Weekend at the Grand*

At the National Association of Broadcasters convention just a year before, in 1993, European engineers had been openly scornful of America's digital television research. The United States "didn't necessarily get the best analog system by being first" back in the 1940s, Bernard Pauchon, a senior French broadcasting official, observed, adding, "There may be the same risk for digital." John Forrest, the head of a British broadcasting company, was even more dismissive. He was "amused," he said, by the American boasts about digital, high-definition television. It had been tried only in laboratories, he pointed out, so there was "still a lot of work to be done" to prove that it actually works.

Meanwhile, the Europeans seemed intent on developing HD-MAC, the European Community's government-subsidized HDTV system. That system *had* been tried outside the laboratory, just a few months earlier. The problem was that it didn't work.

The MAC in HD-MAC stood for "multiplexed *analog* components," meaning the system didn't make even a passing feint at the digital transmission technologies under development in the United States. The European system had been created with only one real competitor in mind—Japan. Back in 1986, when NHK was campaigning to have its high-definition television standard adopted worldwide, the EC had rushed to form a Europe-wide government-industry consortium that was to create its own HDTV system. Like Japan's Muse system, Europe's research was far along when Woo Paik made the digital breakthrough, and the engineers had no interest in starting over. In fact, when Jerry Heller took a copy of Woo Paik's machine to Europe, the engineers "told us, 'Come back in twenty years,' even though they were extremely impressed," Heller recalled. "We showed it to some smaller companies also involved in the HD-MAC project, and they were impressed, too. But they told us, 'We really like what you've done, but we can't say so publicly. It would cut off all our money.' Nobody was willing to jeopardize that gravy train."

The first major real-world test of HD-MAC was at the Winter Olympic Games in Albertville, France, in February 1992. Jose Tejerina, the head of an advanced television group in the European Broadcasting Union, framed the results as positively as he was able. "HD-MAC works well when the optimum conditions are met," he told a European reporter. "But they weren't often met in Albertville."

The problem was, the *governments* of Europe had put forward huge sums of money to finance this project, and they intended to adopt HD-MAC as Europe's high-definition television standard when the work was complete. For European manufacturers, what could be better? Government pays for a good part of the research and guarantees that consumers will have little choice but to buy the product. Maybe the Albertville tests didn't go so well, the engineers and politicians were saying, but we'll fix the problem; this system is still under development. As for digital, let the Americans put their *early* proposal forward. We'll wait until they've worked out all the bugs, and then we'll put out a *better* system. For now, HD-MAC will be just fine.

By February of 1993, government and industry had together spent six years and, according to most reports, about $2 billion on the development of HD-MAC—far more money than the American contestants combined. But the truth was, the system *still* didn't work. Finally, in late February, the Europeans faced up to reality and confessed: HD-MAC was a dead letter. "The system of the future is digital," admitted Martin Bangemann, the EC's commissioner of industry and technology. So Europe formally abandoned HD-MAC. The engineering labs went dark, and the stillborn prototype was sent off to storage.

Washington greeted the Bangemann announcement as a declaration of victory. The American free-enterprise system had triumphed. The United States led the world. Republicans crowed, This proves it—"picking winners and losers" doesn't work. George Bush was right! "Congratulations, guys!" one of Wiley's friends wrote atop a fax that included a copy of a newspaper story recounting the European decision.

Maybe Europe was vanquished, but America's HDTV wars were not letting up. Every day brought a new partisan battle, and even as the contestants tried to negotiate a partnership none of them was willing to give an inch. Maybe they would form a Grand Alliance—"I'd say the odds are fifty-fifty," Rast kept saying—but they had to be ready in case the negotiations failed, so all of them were preparing for the retests that were to begin on May 24.

General Instrument decided that Jae Lim's system was ready for retesting but Woo Paik's needed some more work. That presented a problem: in Wiley's lottery, Woo Paik was up first and Jae Lim was last, many months later. So Jerry Heller gave Lim a call. "He offered me a financial incentive to switch places," Lim revealed a few weeks later, though he wouldn't say exactly how much—only that GI had offered to buy a piece of equipment for his lab that "cost more than $10,000 and less than $100,000. I said, 'This is not the way companies should work' "—Lim's voice was saturated with sanctimony—"but I did agree to switch with them, for no money." So GI wrote Wiley to inform him of the probable switch.

The Sarnoff group complained right away. "Allowing proponents to trade their slots if a willing trade is found, in essence, creates a quasi-private market for these retesting slots, something which should be discouraged," the consortium said in a letter to Wiley signed by Carnes of Sarnoff, Bingham of Philips, Sherlock of NBC, and Donahue of Thomson—all the principals of the Advanced Television Research Consortium. "Such trading is patently unfair and could taint the process," they asserted. Instead of allowing trades, why not just postpone the retesting of *all* the systems until September? That way everyone will be ready for testing, no matter who is up first.

Wiley read the consortium's letter with incredulity bordering on disgust. *You* of *all people,* he thought. He fired back, "If I do receive a request to trade test slots, my present inclination is to grant it. I do not find any of your arguments against this position to be persuasive. Moreover, and speaking very candidly, I can think of no proponent which has less standing to complain about an alleviation in the testing schedule than the [Advanced Television Research Consortium]. As you well know, I went to great lengths to allow flexibility in ATRC's original test slot in order to keep your system in the competition. I thought it was in the public interest to do so." And then, after everyone had made an exception for the Sarnoff group, the consortium had tried to back out of its agreement and bully the Advisory Committee into carrying out *all* of its tests, including those that had been dropped because the ATRC had showed up at the test center two weeks late.

As for the idea of postponing the retests until September, "I wonder if your suggestion is really a serious proposal, or simply a ploy," Wiley wrote. When the consortium leaders read that, they dropped the objection. This bit of RCA bluster had backfired.

When Jerry Pearlman first thought up the idea of supplemental testing, in the summer of 1992, he'd envisioned a few days at the test center—just enough time for each contestant to show that its improvements were real. With no hesitation, Peter Fannon and the other testing authorities rejected that idea. If we test just one part of each system, they said, the contestants could easily trick us by making sacrifices in other areas of performance just to make the tested areas shine. No, if we test anything, we have to test everything.

With that philosophy in play, the retests would last a lot longer than just a few days. By March 1, the projected time period for retesting was nearly seven weeks—and climbing. *Everyone* started campaigning to have his pet priority tested. So Mike Liebhold, the Apple Computer employee, was pushing hard to include what he and the MIT guerrillas called "interoperability testing"—evaluations to see how well each system worked with computers. "We find that the Advisory Committee process is configured inappropriately for digital interoperability testing," he wrote to Mark Richer, the PBS executive who oversaw testing. This had to be corrected "before any advanced television system architecture, including any proposed 'Grand Alliance' system, can be embraced with confidence." Richer wrote a polite letter back saying that a few of Liebhold's suggestions were being adopted but many more were not because they were "either not regarded to be critical to the system selection process, and/or impractical to implement." Maybe we can find another way to study these questions later, Richer said. Liebhold took that as a rank dismissal. They're ignoring us again, he thought. In Cambridge, Liebhold's disciples at the Apple-funded Program on Digital, Open High-Resolution Systems were publishing papers, trying to incite trouble of their own over the retest plan. As all this came bearing down on Wiley, he hardened his view of Liebhold and the others. They were devil figures now.

Wiley generally accepted almost any invitation that came his way to speak at industry meetings or conventions. Exposure is good for business, he believed. But he and his staff began closely examining conference agendas to see if Apple or MIT figures were invited. One advanced-television conference in Montreal that spring included "a Nicholas Negroponte disciple" on the speakers' list, a staff memo noted. And Wiley had absolutely no use for the Grand Vizier. Another speaker, from Apple Computer, was unfamiliar to Wiley's staff, "but we can guess at his viewpoints." Wiley didn't go. Liebhold and the MIT guerrillas were enemies now—unrelenting and wholly unreasonable, as Wiley saw it. Bring them in, and they'll do nothing but cause trouble. But he failed to anticipate how much fomentation they might brew from the outside.

In early April, Wiley began trying to encourage the Grand Alliance negotiations. He had set May 24 as the first day of retesting, though privately he acknowledged that it was "an artificial deadline." He had watched the jobs dispute with disgust, and then he began noticing that the contestants were hiring Washington lobbyists. Oh god, he thought. We're moving from a technical discussion to a classic Washington political fight. If they're all hiring lobbyists, we're going to be in a hell of a mess.

May 24 was approaching fast, and Donald Rumsfeld had said from the beginning that he wasn't interested in continuing the alliance negotiations once the retests began. On April 20, Peter Fannon sent each contestant a long memo by Federal Express setting out the financial commitments for retesting. That turned the screws even tighter. The new tests were going to cost each of the contestants $612,000, and half of that was due on May 3. The contestants groaned, but Wiley grinned. This was just the right incentive, he thought. If you don't want to pay, then form a Grand Alliance.

By the first of May, however, Bob Rast was "pretty sure it isn't going to happen." The contestants had met several times—at the National Association of Broadcasters convention in Las Vegas, in Washington, in a conference room at the Airport Hilton in Chicago, in Wiley's office—but the differences and disputes seemed to be growing ever more intractable. Rumsfeld and Pearlman weren't getting along; Pearlman thought GI's president had deliberately snubbed him. The others thought the Sarnoff consortium was maneuvering to win advantage for its European parents. And Jae Lim—well, "He's impossible," Wiley complained in the middle of it all. "A complete pain in the butt," snarled Peter Bingham of Philips. Lim still wanted his one-quarter share. That was bad enough. Worse, he was insisting that a Grand Alliance system use only progressive-scan displays, the type the computer industry favored. Liebhold and the MIT guerrillas were arguing for that, too. The television manufacturers vociferously disagreed, and each side was arguing its position with the intractable ardor of holy warriors.

To outsiders, the vigor of this debate on an arcane technical point remained perplexing, even absurd. Computer industry engineers saw interlacing as nothing more than a primitive abomination, one more sign that broadcast engineering was trapped in the Dark Ages. But for a host of reasons, broadcasters could not even conceive of abandoning interlace, the picture format they'd been using since the 1940s. They constantly touted the technical advantages, principally that interlacing enabled them to send twice as much signal information over the air.

But behind this lay several important competitive and financial motivations that broadcast-industry leaders seldom mentioned. Most television manufacturers also made TV cameras and related production equipment—a huge, profit-making part of their industry. Some of these companies also held income-producing patents on various interlace-related technologies that were key parts of their products. That income stream would dry right up if the television industry started broadcasting only progressive-scan pictures. Suddenly, the executives feared, they'd have to start *paying* patent royalties to the computer firms that had made progressive-scan monitors and related equipment. Besides, they said over and over again, nobody has even begun designing a practical progressive-scan television camera. It'll be *years* before anyone is ready to build marketable, affordable TV cameras for studio and field work.

Underlying all that, the broadcasters just didn't want the big computer companies mucking around in their business. Industry leaders knew that if they continued broadcasting interlaced pictures, they'd effectively block Microsoft, Apple, and Intel from taking any major role in the television world. None of them ever talked about that, however. Publicly they continually pointed out that interlaced pictures were *better!*

Everyone agreed that, in a general sense, a progressive-scan picture was purer, cleaner. But progressive-scan transmissions were twice as large as interlaced transmissions. As a result, the two progressive-scan systems in the HDTV race—Jae Lim's and the Zenith-AT&T entry—actually offered lower resolution than the others. HDTV had always been defined as a picture providing twice the clarity of conventional television, and back in the 1970s NHK had found that the best way to double the clarity was to double the number of lines on the TV screen, to 1,125. The two interlaced systems remaining in the race, by Woo Paik and the Sarnoff consortium, broadcast 1,050-line pictures, while the progressive-scan systems were able to offer only 787.5 lines. Zenith argued that the *perceived* resolution was actually just as good, if not better, because viewers were not bothered by the flaws inherent with interlacing. And the Zenith-AT&T system did look awfully good. Besides, Zenith and the other progressive-scan advocates argued, they were limited to 787.5 lines only for the moment. As compression technologies improved they would be able to increase that to 1,000 lines or more.

For now, though, the test results had showed that the two interlaced systems offered a better picture, giving real power to those advocates. This debate had grown only hotter over the previous months. And on May 11, Lim wrote a letter to the other contestants stating his position on the outstanding issues, including the debate over transmission formats. "The main issue here is whether or not we include interlaced scanning format as one of the possible transmission formats," he wrote. We should not, Lim added. "I fully recognize that some broadcasters will be unhappy about this choice. By jumping into the cold water and beginning to swim, we will achieve the ultimate goal of progressive scanning much earlier." Over the following days, Lim would brook no counter-arguments, and his insistence pulled the negotiations down into still another pit of muck. How on earth could all these people with such disparate interests ever move forward?

Early in the week of May 17, General Instrument moved Lim's system into equipment bay No. 1 at the Advanced Television Test Center. The competitors were no closer to forming a partnership than they'd been a month before, and most of them were ready to bow to the seeming inevitability of their failure. All of them had mailed their advance payments to Fannon—$306,000 each. But even as GI and MIT engineers were hooking Lim's cables to the test center arterial lines, all of the companies' principals decided they had better make one last effort.

They convened at the Grand Hotel, the first-class establishment on M Street near Georgetown where GI had staged its most recent demonstration just three months earlier. On Wednesday morning, May 19, 1993, about two dozen people from AT&T, GI, MIT, Philips, Thomson, Sarnoff, and Zenith checked into their rooms and then divided into three groups—engineers, lawyers, and corporate executives. Each group gathered in one of three conference rooms and settled around large walnut tables to make one last try. If they couldn't settle by Friday, they would run out of time. The retests would have begun.

The issues were straightforward. Jae Lim still wanted his own share of the profits, though he had gradually reduced his demands. Whatever he got would go to MIT, not to him. But the university had a profit-sharing arrangement with its faculty, so even a tiny share promised to make Lim quite rich. In his May 11 letter, he had offered to take 16 percent, leaving the three other groups with 28 percent each. By May 17, the active proposal under discussion would give Lim a 3.33 percent share. As everyone gathered around the Grand Hotel conference tables, they had agreed on that. So Lim turned his considerable disruptive energies to other matters. He was still stuck on progressive scan. On this he would not budge, and AT&T's engineers agreed.

Though that was the largest remaining disagreement—"the gigantic issue," Joe Donahue called it—it was hardly the only one. The Sarnoff group had backed down from its insistence that it would not support the Grand Alliance system outside the Western Hemisphere, but several potential partners still had problems with the patent-sharing plan. The active proposal was that everybody's patent-royalty payments would be pooled and then split; these were the profits that Jae Lim was worrying about. But the Europeans—Philips and Thomson—didn't think members of the Grand Alliance should have to pay royalties to each other, even if those payments went into a pool. They figured they would sell more HDTVs than anyone else and end up putting more money into the pool than they would get back. They also wanted to know whether European patents were to be included. Beyond that, each of the contestants wanted one or another of his own technical innovations, real or imagined, included in the final system. Everybody also had various additional pet requests.

These were the issues they had been debating for weeks, most recently during a marathon session at the O'Hare Airport Hilton the week before. There they had seemed to make some progress, particularly on the vexing question of progressive scan versus interlace. In Chicago they'd begun discussing a compromise—a system that could handle *two* display formats and switch from one to the other automatically, depending on what signal information came in. The set would display progressive *or* interlaced pictures, whichever was being broadcast. If we build the transmitters and the TV receivers so they can switch automatically, they said, we can make *everybody* happy.

Arriving at the Grand Hotel, the would-be partners hoped they could agree on that. Nonetheless, this proposed solution quickly fell apart. At the conference table, AT&T, supported by Jae Lim, suddenly delivered an ultimatum: We will not agree to any proposed system design that includes any interlace at all. But Donahue thundered back, "We will not accept zero interlace. We just won't accept it!" The other side countered with well-worn arguments: These TVs have to be compatible with computers. We have a chance here to kill interlacing once and for all. We have to take it.

They quarreled through Wednesday. On Thursday morning, Lim left the Grand and went to the test center to get his system ready for testing. By then a partnership agreement seemed a long shot at best. In fact, by Thursday afternoon, the progressive-scan ultimatum still sat on the table, and the negotiators had fallen back into loud arguments, angry name-calling and finger-pointing. One after another they settled back into their chairs, arms folded, sullen expressions on their faces, as the realization dawned: This is impossible. Finally, Donahue declared a deadlock.

"Thank you very much, but we're leaving," he suddenly announced, leaning forward in his chair and placing his hands flat on the tabletop. "It's all over. We have to leave.

"But first," Donahue added after a dramatic pause, "I want to get Dick Wiley over here. I want you to present your ultimatum to Dick Wiley so he won't hear ten different versions from ten different people later."

Robert Graves of AT&T, who saw himself as the most astute politician of the group, had tried to keep Wiley involved from the earliest days of the Grand Alliance discussions. And Wiley very much wanted to be a part of these final talks. Graves had promised to call if anything was agreed—or if failure seemed imminent. Wiley was working in his office a few blocks away but taking few appointments—"staying loose," he said, "expecting to go over to the Grand and shake hands, congratulate and bless them." By Thursday afternoon, he'd heard nothing, not one word, and he was "on pins and needles all day." Finally at about four o'clock, Wiley's secretary told him that Graves was on the line. Wiley was still expecting to be summoned "to bless it." But when he picked up the phone, Graves's manner was somber.

"We're about to break up," he said. "I suggest you might want to come over."

Wiley immediately called his assistant, Paul Misener, who was at the test center with Lim trying to work out final arrangements for retesting Lim's system, which was supposed to begin the following Monday. Wiley told Misener to get right over to the Grand. Then he hung up, grabbed some papers, and headed for the elevator.

When Misener told Lim what was happening, Lim fell into a near panic. "They're going to make a deal without me!" he stuttered. He dashed out the door and ran down the street waving his arms wildly to hail a cab. Misener followed, bemused.

Wiley, Misener, and Lim all pulled into the Grand's circular driveway at about the same time, and when they got to the fourth floor the would-be partners were still seated at the conference table, jaws set, faces locked in expressions mirroring the bitter intractability of their debate. At the table were Pearlman; Bingham; Carnes; Richard Friedland, a vice president of GI; Curtis Crawford, a vice president of AT&T's microelectronics division; and John Preston, a senior officer from MIT. Lim sat down next to Preston. Others sat in chairs along the walls. The men at the table—and as usual there was not one woman in the room—shuffled to make room for Wiley, who pulled up his chair and asked, "OK, where are we?" Right away the others started arguing again—"jabbering," as Wiley put it. "Screaming and yelling at each other. Pearlman started getting all mad. They sounded like *children.* This was going nowhere."

Wiley put up his hands. "Wait a minute," he interrupted. "Why don't we see if there's anything we *do* agree on." With that, everyone shut up, and Wiley jumped in. The few minutes of arguing and finger-pointing had shown him that the biggest hangup was still this damned progressive-interlace business, so he asked, "Can we all agree that our *ultimate* goal is a thousand-line progressive system? We can't get there now, but isn't that where we want to be in a few years? You know, the handwriting is on the wall for interlace. Progressive is the future. Can't we agree on that?"

That was a Mom-and-apple-pie offer. Sure, why not? It would be great if we could have progressive scanning *and* true thousand-line HDTV. But we don't know how to do that now. "Can we say we plan to *migrate* to that when we are able?" Wiley asked. OK, everyone said, including Lim and AT&T. We can agree to that. "Well, that's something," Wiley said, leaning back for a moment.

But what about now? the others were asking. We need to decide what the Grand Alliance will build right now. A few minutes earlier, while they were waiting for Wiley to come over, the AT&T team had left the room to discuss the progressive-scan ultimatum. Apparently they hadn't wanted to be painted as the villains behind the collapse of the Grand Alliance, because when they came back a dual scanning format proposal was suddenly on the table. They appeared ready to accept—begrudgingly, in the others' view—the earlier plan to build a machine that could handle both progressive *and* interlaced displays, switching from one to the other automatically.

We're making progress here, Wiley said after about an hour. We have the outlines of an agreement. Let's take a break. You can caucus and talk things over.

Everyone got up and left the suite. Wiley sat there with Misener for a few minutes, then got up to check in with the groups to see if any of them needed encouragement. There were three "caucuses," situated in a way that made the politics of the moment clear. The General Instrument, Zenith, and AT&T officers were together in one suite, and to Wiley everybody in there seemed fine. The members of the Sarnoff consortium were together in another suite; though they were not giving much away, Wiley saw no big problems among those people now either. The third "caucus," Jae Lim, was pacing the hallway. Wiley walked up to him and asked, Are you OK with this, Jae?

Wiley was a tall man, maybe an inch over six feet, and Lim was at least half a foot shorter. So he was looking straight up at the chairman as the sanctimony poured forth. "I can't support anything that isn't 100 percent progressive scan," Lim said, in a tone that suggested, I am the only righteous person here. "I think that is best for America."

Wiley groaned and rolled his eyes. "Give me a break, Jae," he said. Wiley was convinced: This is the moment. Now or never. If we don't come to agreement right now, it'll never happen. This whole thing will fall apart. America may *never* get HDTV. And now Jae is driving me nuts!

"Jae," he said with obvious exasperation, "we have a *migration* to progressive scan."

Lim just looked up at him. "We're going to lose this whole thing any minute," Wiley argued, and he was wagging his finger in Lim's face now. "Come on, Jae, this will be *good* for MIT. You'll be a part of this whole thing. You can't expect all these people just to *capitulate* to you. You can't! Jae, you've got to compromise."

Lim hadn't been in the room for all the discussion concerning the progressive-scan ultimatum earlier in the day. He hadn't seen how the compromise had come about. But it probably wouldn't have made any difference if he had. He wasn't budging. Bob Graves had anticipated this problem. Even before the meeting, "Lim was adopting his religious attitude, holier than the rest of us," he said. Graves had already decided that Lim needed what Graves liked to call "support" from somebody above him at MIT. That's why Preston was here. Preston took Lim into a room, and they called the leadership of MIT in Cambridge. This agreement potentially represented a lot of money for the university, millions of dollars, and eventually Lim was persuaded to agree to the compromise. That's what he told everybody when they reconvened in the conference room. One more big problem solved. Or so it seemed.

The conferees deferred some decisions, particularly the questions about which parts of which systems they would choose for the new Grand Alliance machine. Zenith and GI both insisted that they had the best transmission systems—a striking claim in GI's case, considering that Jerry Heller and the others at the VideoCipher Division had been certain that Paik would be unable to build any transmission system at all.

Lim insisted that his audio system was the best; he'd been an audio engineer before he took over Bill Schreiber's advanced television program. But the Europeans wanted theirs chosen. Musicam was set to be the audio standard in Europe. The others favored picking Dolby AC-3; it was vying to become an industry standard of sorts in the United States.

We'll hold our own competitions to select the best of these systems, they decided. Put the contenders up, run them through their paces, and then pick the best one. No politics, no arguments. We'll make these decisions on technical merits alone. Neat and clean. With that, they resolved the last of the major arguments that had vexed them for months. So Wiley suggested they move on to the final step: Let's put out a press release.

Just before rushing over from his office, he'd picked up a draft of a press release that Paul Misener had already written announcing that the Grand Alliance had been formed. "For immediate release: May 20, 1993," it said at the top. (Actually Misener had written it on Wednesday the 19th.) "The FCC's Advisory Committee on Advanced Television will review a single HDTV system proposed today by a 'Grand Alliance' of entities that, until now, had sponsored four competing systems. These entities today reached a business and technical agreement and submitted to the Committee a merged technical proposal." Misener had even filled in canned quotes from Wiley showing how satisfied the chairman was with the agreement:

"I believe the Grand Alliance proposal, subject to Advisory Committee review and approval, will lead to the best conclusion of a process that has fostered the development of highly advanced digital HDTV broadcasting and cable technology. The members of the Alliance should be commended for their accomplishments."

Wiley made copies of this and passed them around the table for comment and editing, then he placed a yellow legal pad in front of him and began drafting an insert to account for the new agreements. While the others looked over the draft, Wiley wrote, "The proponents unanimously endorse the objective of moving the standard to thousand-line plus progressive scan as soon as feasible and will work to accomplish a definitive migration path." When he read that aloud, suggestions flew at him.

"Say 'migrating,' not 'moving to' thousand-line progressive,'" somebody said, so Wiley wrote in the correction. "Not thousand-line *plus,* just thousand-line." Wiley scratched out the word "plus." "Get rid of this 'definitive migration path' business." And with a little discussion they agreed on a substitute sentence that fuzzed things up a bit: "All parties agree to eliminate the interlaced scanning format from the transmission path in the future."

In just a few more minutes, they worked out most of the other wrinkles. The deal was done, except for one thing: the Sarnoff group couldn't give a definitive yes until Joe Donahue approved, and the

Thomson representative still had problems with the patent-sharing arrangements, though he offered no specifics. They conflicted, he said, with previous agreements between General Electric and Thomson. But Donahue promised to get the questions settled quickly. Great, Wiley said. We'll have our press conference at ten o'clock tomorrow morning.

He wasn't about to tell them, but he saw this press conference as the critical last step to "lock them in," as he put it. Once this got into the newspapers, nobody would be able to change his mind. The deal seemed complete, or nearly so, and everybody left.

Wiley was back at his office, and the call from a member of the Sarnoff group came a short time later: Donahue still couldn't sign. He still hadn't worked out the patent questions.

Wiley sighed and called Donahue. "Joe, come on," he said. "We've got everything settled. You can't hold this up."

It's not going to be a problem, Donahue assured him. But by the end of the day Donahue still wasn't ready to sign—and now the patent issue was riling some of the others. Just what were these patents, some of them wanted to know. As Thursday drew to a close, Wiley worried: this thing could still fail.

Another problem flared at 7:30 Friday morning. Friedland, the GI vice president, called Wiley at home. "Jae wants to put out his own press release," he announced. He wants to say he doesn't agree that an interlaced format ought to be a part of the standard at all.

Wiley sighed aloud. "We can't put out *two* press releases," he said. Friedland agreed. Then Wiley got an idea: "What if we give Jae *a footnote*. Let him say what he wants there." I'll ask him, Friedland said.

A short time later, Donahue called. I still don't have everything settled, he said. Wiley knew in his bones that nothing would happen today. By now the workday was nearing its end in Paris, where Thomson's leaders lived. And it was Friday, too. The whole thing seemed to be unraveling. Wiley called all the principals and said, Let's reschedule our press conference for Monday morning. Get everything settled by then. He had already called his good friend Jim Quello, the acting FCC chairman, and asked whether they could stage their press conference in the FCC's eighth-floor meeting room. Sure, Quello said, and he'd agreed to attend. But now Wiley wasn't so sure the event would ever get staged at all.

By early Monday morning, Lim had agreed to the footnote idea, and Paul Misener added this new language to the draft. It appeared at the bottom of the second page, in a smaller typeface: "MIT believes that a digital video broadcast standard that exclusively utilizes progressive scanning, from the beginning, is in the best interests of the United States." Silly but harmless, Misener and Wiley agreed. The new draft was faxed to all the players, and at 10:00 A.M. Wiley got all of them on a conference call.

"Thomson is prepared to sign," Joe Donahue declared, to Wiley's great relief. But by now some of the others weren't so sure. Donahue had spent four long days working out his intellectual-property contracts—four days for the others to fester, maybe even imagine that Donahue was screwing them somehow. "A lot of stuff had been happening behind the scenes," Jim Carnes recalls, "a lot of stuff nobody understood." Among the other would-be partners, the tentative agreement had begun to unravel, and on the phone one after another of them spoke up to say, I've got two or three changes I want to make.

Donahue had guessed this would happen. Everybody's trying to get something, he thought. Just as he'd planned, Donahue broke in and said, "Oh, you want to make some changes, do you? Well, if you want to reopen negotiations, I've got ten or twelve changes I'd like to make. We can be here all week." Or, Donahue said again: "Thomson is ready to sign *this* agreement, right now."

Wiley groaned. More nonsense. He didn't understand all of it, wasn't really privy to the details of the *business* negotiations, but by god, he was not going to let this fall apart now. "I've got all day," he said. "I'll stay right here on the phone. Let's get it worked out."

Hour after hour, Wiley sat there at his desk talking into his speakerphone, prodding and encouraging while some of the others dropped on and off the call, working out of earshot. All the while, reporters were calling Wiley's office. Periodically his secretary carried the phone messages into the room. The *New York Times* had run a small story on Saturday noting that the contestants seemed to be close to an agreement, and this morning it seemed as if the entire national press corps was waiting for word. Trying to keep the pressure up, Wiley told the others that the press was growing anxious.

Finally, after lunchtime, they seemed to be ready. At 1:10 P.M., more than three hours after the conference call began, Wiley called for a vote.

"Sarnoff," Wiley said.

"Yes," came back from the speakerphone.

"General Instrument."

Yes.

"MIT."

Yes.

"Philips."

Yes.

"Zenith."

Yes.

"AT&T."

"No," Curtis Crawford answered.

Wiley rolled his eyes. "Come on now, Curtis," he said. "We've got to move. I can't tell you what to do, but get this settled. We've got to move."

Crawford got off the line, and Wiley kept all the others on the phone. He didn't know what was going on in the background, though fifteen minutes later it appeared that some sort of compromise had been reached. So at 1:25 Wiley called for another vote. This time AT&T offered a conditional approval, subject to clarification from Donahue on a business question that wasn't openly stated.

Go do it, Wiley said, and he still kept everybody else on the line. Reporters continued calling, meanwhile. Their deadlines were rushing at them, and this would be a complicated story to write. Finally, at 2:35 P.M. everyone came back on. More than four and one-half hours after the phone call began, Wiley called the third vote.

"Sarnoff."

Yes.

"General Instrument."

Yes.

"MIT."

Yes.

"Philips."

Yes.

"Zenith."

Yes.

"AT&T."

Yes.

Misener and Wiley smiled at each other across the desk. It was done. The new partners sighed, and with nervous laughter some of them asked,

"What have we done? I hope this isn't a big mistake." But Wiley was still pushing. "Come on now, we've got to get over to the commission for the press conference now. It's now re-scheduled for 4:00."

The FCC meeting room was packed. Television network cameras were lined up across the front, and copies of Misener's press release lay stacked on a table to one side. Wiley introduced the new partners, and Chairman Quello blessed the agreement. This partnership will actually *speed* the arrival of HDTV for America, he and others said, because it will reduce the possibility of lawsuits from losers.

The next morning, Ed Andrews's article about the new Grand Alliance was the lead story in the *New York Times,* and most other papers gave the story front-page treatment. For the previous year or so, HDTV had been ignored by the mainstream press. Now here was a bath of warm attention, reminding the nation that America led the world and that high-definition television was almost here.

16 The brass rat

Wiley and other Advisory Committee leaders held a dinner for the new partners a few days later. About twenty people sat at three round tables in a private dining room at the Mayflower Hotel, two blocks up the street from Wiley's office. The mood was good, though some bitterness lingered in the air. Jae Lim was nobody's favorite. In fact, Bob Rast had already adopted a tart retort for Lim's frequent complaints and objections: "Would you settle for a footnote?" Everyone would laugh but Lim.

Over dessert, Wiley stood up to commend the group once again, and he told one of his favorite jokes. An office worker in downtown Washington is rushing to get home, but he has to buy his wife a birthday present. He stops at a gift shop and looks quickly at one thing after another, but everything is too expensive. He has just $40. Finally the shopkeeper tells him, "I have this," and holds up a little sculpture—a brass rat. "It's $40."

"I'll take it," the man says, hurrying to leave.

"There's something I need to tell you about this," the shopkeeper says, but the man cuts him short. "Just wrap it," he says. The shopkeeper shrugs, wraps the rat, and hands it over.

Walking down the street toward his car, the brass rat under his arm, the man hears a noise. He turns around and sees a big rat running down the sidewalk just behind him. The man turns a corner and looks back again. The rat is still there, and it has been joined by another one. He gets into his car and starts driving down 17th Street, but when he looks in his rearview mirror, six rats are running along behind his car. He steps on the gas and looks back, and now there are ten rats, all of them running even faster.

The man pulls onto the Capital Beltway, thinking, Now I'll really leave these rats behind! He speeds up to seventy miles per hour, then eighty. But more and more rats are following the car and they're gaining on him. Finally he looks down at the package on the seat beside him and thinks, It's this damned brass rat! I've got to get rid of it. And when he crosses the Potomac River bridge he chucks the brass rat out the window, over the side of the bridge. In his rearview mirror he sees hundreds of rats jumping off the bridge after the brass rat and falling into the river five hundred feet below.

Damn! the man says to himself. I've got to find out about this brass rat! He turns around and drives back to the gift shop, bursting in just as the owner is preparing to close. The owner looks up with a knowing smile and asks, "So, now do you want me to tell you about the brass rat?"

"No, no, I don't care about that!" the man says in a rush. "What I want to know is, do you have any brass lawyers?"

Wiley told the joke with relish, and the room erupted in laughter. Jae Lim smiled for a moment, then offered his own comment: "I think he should have bought a receiver with progressive scan."

Loud groans filled the room, and Rast piped up, "Hey Dick, do you have any brass professors?" The laughter for that one was even louder.

For several happy, comfortable days, the new partners basked in the glow of their genuine achievement, even as press accounts wondered how these former bitter combatants could possibly get along after their shotgun wedding. Nobody was surprised when, within a week or two, the newlyweds began falling back into the same arguments that had plagued them all along.

The incentives of a multibillion-dollar competition had governed the behavior of everyone involved. At first, all of the contestants had been zealous advocates of their entries, willing in some cases to lie, cheat, and steal to gain advantage for the prototypes taking shape in their labs. But now the competition was over, and by Tuesday, May 25 it seemed almost as if a genetic sequence in each man's head had kicked in to prompt a change in both motivation and behavior. Now, all of a sudden, they were *company men,* dedicated to guiding the development of the Grand Alliance HDTV system so that it best fit their corporation's broad needs.

General Instrument served cable television customers. AT&T was an integrated-circuit manufacturer and wanted to be a player in the television world. Zenith wanted to survive. Thomson and Philips were eager not just to sell more TVs in America but to produce a product that could coexist with European business interests. Sarnoff wanted public recognition, so that it could win more contracts. Jae Lim was crusading to save America from invidious foreign forces. And all of them saw their work on HDTV as a potentially invaluable step toward these goals and other business opportunities. Under their agreement, each corporate entity would make the same amount of money no matter whose components were chosen for the joint system. Still, all of them knew that an audio or transmission system, a digital compression algorithm, and any other component chosen for the Grand Alliance machine might be marketed in other arenas, too. So everybody jockeyed for corporate advantage.

The partners planned a meeting in Chicago in July to divide up the major areas of responsibility. They had already decided that the sound system and the transmission system would be subject to internal "bake-offs," as Peter Fannon had begun calling them. They also set up a two-tiered system of governance. A technical-oversight group would direct planning and construction of the prototype. People like Bob Rast, Glenn Reitmeier, Wayne Luplow, and other supervisory engineers would staff that. A more senior steering committee would make the larger business decisions: Pearlman, Carnes, Bingham, and others of their rank sat on that. In both groups, each company would have one vote and simple majorities would rule. So all the rules of representative democracy—vote buying, influence peddling, and back-channel dealing—immediately came into play.

The largest remaining area of dispute was over who would build the encoder, the heart of the system. The encoder compressed the digital signal and readied it for transmission. Both GI and AT&T wanted to build it, and Eric Petajan had been arguing that General Instrument should not be chosen because GI was "an interlace company." Rast worried that Petajan's objections would rouse Lim, and sure enough, Lim began voicing the same worry. GI had already decided to make a business of selling HDTV encoders to television stations when everything was settled, but "nobody knows we want to do that," Rast said. "There's no reason to tell people yet." So General Instrument *had* to be involved in building the prototype.

The companies jockeyed and argued. Neither side gave up. Finally the partners agreed that GI and AT&T would build the encoder together—not necessarily the happiest solution. They would have to divide up the box circuit board by circuit board, a messy prospect. Besides, Rast didn't really trust AT&T. "We're nervous about working with them," he said. "The other systems that were built jointly"—the Zenith-AT&T machine and the one built by the Sarnoff group—"all had big problems." Rast also remembered all too well Zenith's complaints about "the sliding schedule" when the Bell Labs boys were building that earlier encoder. So, "while AT&T is building its half," Rast said, "we intend to build a whole encoder, even the parts that are their responsibility, and have it ready in case AT&T falls on its keister."

The Grand Alliance decided to let Sarnoff build the "transport," the part that arranges and organizes the digital bitstream. Philips was to build the decoder, the device inside each TV set that translates the digital signal so that it can be shown on the screen. The remaining elements—the sound system and the transmission—would be assigned later.

The Advisory Committee changed its structure to be more responsive to the new arrangement. With only one "entry" left in the competition, Wiley realized he risked being bulldozed. The Grand Alliance was now in a position to ask, If not us, who? What choice do you have but to recommend us?

Wiley wasn't ready to lie down for these people. His answer was implicit but unmistakable: We don't have to recommend anybody. So he appointed a Technical Subgroup of engineers to work with the Grand Alliance and provide "staged approval" of the new machine's various components as they were chosen and designed. Wiley's engineers could approve or disapprove of any part of the new system at any point along the way. The chairman of this new group was Joe Flaherty, the CBS executive who'd been so dismissive of everyone during the Special Panel meeting. He was not about to be bullied here, either.

Quickly, the Technical Subgroup began pushing the Grand Alliance in directions the members definitely did not want to go. Wiley, Flaherty, and some others decided on their own that the new digital signal structure should be a virtual mirror of the MPEG digital video standard taking shape in Europe, with significant American influence. MPEG (the acronym stood for Moving Pictures Expert Group) was intended to be a new world standard, and the Sarnoff group had used it for their entry.

Since then, the MPEG standard had been gathering support, and the Advisory Committee's leaders saw no reason that the United States should not sign on. But the Grand Alliance chafed at that idea. "People who talk about MPEG compatibility have never built an HD receiver," Wayne Luplow scoffed during one Technical Subgroup meeting. "Picture quality should be the most important feature."

Over the summer, Wiley's people pushed and bullied. "The Grand Alliance *will* be MPEG," declared Robert Hopkins, the former Special Panel chairman, who had sworn once before that the contestants were not going to push him around. Flaherty talked tough, too, and said: "Sometimes you're hired to be a bastard, so you just have to do it."

The new partners were in revolt over the new demands. "Remember, we are the ones who invented this," Rast declared. "*We're* the innovators. And MPEG isn't even a full system. It isn't finished yet!" But after weeks of technical discussion, the Grand Alliance folded to this and nearly all the Advisory Committee's demands. The partners had disagreed among themselves on some of the committee's requests, and were therefore unable to resist as a unified body. So at one meeting, Flaherty bubbled, "It's very impressive that the Grand Alliance has done virtually everything we asked for. It's almost a little scary sitting here on the other side."

Nonetheless, the partners still had left themselves two important decisions: Which audio and transmission systems would be used? With their planned internal competitions, they believed they would have no trouble deciding.

They were wrong.

The audio question came up first, and no one anticipated any problems. Digital audio was not a new technology, after all. Compact discs were digital, and engineers and consumers had been evaluating the relative merits of competing stereo systems for decades. During the first round of testing, hardly anybody had even cared about the sound systems. And as the Grand Alliance prepared to pick a sound system now, Rast quipped, "This is the easy one."

Philips was offering its own system, called Musicam, which was already nearing production in Europe. This was a multichannel digital surround-sound system, set to become the European standard for TV audio. But before the Grand Alliance was formed, Rast said Bingham told him he "didn't really care about audio."

Jae Lim wanted his own system selected, and his view was predictable. "Musicam is 100 percent owned by the Europeans," he preached. "Are we going to ask 300 million Americans to pay royalties to the European Community?"

Meanwhile, GI and Zenith favored Dolby's AC-3 sound system, which they'd used in their own prototypes. This was known as a "5.1-channel" digital surround-sound system. Like the other entrants, Dolby's system would give homeowners the chance to create a first-class, theaterlike digital sound system in their basements and dens using the audio signal coming out of the TV set. Three channels of Dolby's system would power left, right, and center-channel speakers. Two more were for speakers in the rear of the room, to provide the surround-sound effect. And the partial channel was for a subwoofer, the speaker that provides very low bass. Dolby already manufactured surround-sound systems for movie theaters and an analog surround-sound system for home theaters called Pro Logic. But with this new digital system, discrete signals could be sent to each of the six speakers, making AC-3 much more flexible and realistic sounding than its predecessors. (Of course, viewers could also choose to get their sound only from the TV's built-in speakers.)

All three sound systems were put up for test at Lucas film labs in California at the end of July. Using a time-honored system of audio testing, a panel of Golden Ears listened to selections of recorded music played first directly from the tape and then run through each of the three systems. This is called "A-B testing"; when the technicians switched from A to B, the Golden Ears did not know which was the original signal and which was being run through one of the systems. They simply noted the one that sounded better. It was sensitive work; differences were subtle. In fact, during one part of the test, the technicians mistakenly used the taped source for both the A and B signals, and some Golden Ears nevertheless judged one to be better than the other, prompting test director Jim Gaspar to note with a chuckle, "It shows that you can detect quite a bit of difference if you move your head just a little."

When the tests were over and the scores were tallied, the Dolby system had clearly won. Musicam, the Philips system, stumbled on certain passages of music, and Jae Lim's entry also scored below Dolby, based largely on estimates of its relative cost. The three contestants were quietly notified of the results. And if Peter Bingham had indeed not cared much about the audio system decision before, he sure changed his tune once he learned that Dolby had won. Along with almost every other member of the former Sarnoff group, Bingham was shocked, wounded, and wholly unwilling to accept defeat. He and the others cried out in unison, trying to evade their fate by offering the only excuse they knew: We had an *implementation error!*

"Our Eindhoven cousins put it together," Bingham explained, referring to the Philips engineers in Holland who had built the prototype sound system, "and they made a mistake in the hardware. It's fixed now." Philips still fully intended to introduce Musicam across Europe despite these unfortunate test results in California, so Bingham told Wiley, "If the rest of the world looks at Musicam after we've fixed it, and we've already thrown it out over here because of a screwup in hardware here, how will that look? We're not going to give up on this issue."

It's simple, Bingham said. We want a retest. That had worked before. Why not now? But some of the others cried foul. Does this process have so little integrity that a clear loser can demand a retest? they asked. "There has never been a losing Olympic runner who didn't have an ingrown toenail," Joe Flaherty quipped. If the *audio* issue couldn't be settled without a fight, how would this Grand Alliance ever build a complete machine? "We feel that the testing conducted at Skywalker Ranch in July was fair and valid and that any retesting is of questionable value and impractical," Jim Gaspar, the audio testing director, wrote to Wiley. "The FCC Audio Experts Group recommends that the Grand Alliance make a decision based upon the findings of the Audio Specialist Group tests."

The Grand Alliance set out to nip this budding problem. They planned to take a vote. Majority rules. Simple as that. The partners met in Chicago on August 31 with several issues on the agenda, and by now everyone had seen the audio test results. The outcome of the vote seemed clear: GI would support Dolby, and so would Zenith, since both companies had been Dolby supporters all along. Lim would probably support Dolby, too; he couldn't possibly vote for himself, given how poorly MIT's system had fared. And the Sarnoff group would cast what amounted to a protest vote for Musicam. Case closed.

Well, GI did vote for Dolby. But Jae Lim voted for his *own* system, even though there was not one chance in hell it could win. The Sarnoff group voted for Musicam, and Zenith voted for . . . MIT.

Rast looked across the table at Luplow. . . . Why on earth?

Jae Lim, the last-place finisher, got the most votes, though he was one short of a majority. They were stuck. Lim got up from the table and announced that with two votes he would start working to get a third, and the group scheduled another vote for the next morning. As soon as he could, Rast asked Luplow, "What in the hell are you doing? If you weren't sure, the least you could do was abstain."

I didn't know I could do that, Luplow told him. Overnight, Philips and Lim lobbied the Zenith people hard. The next day there was another vote, and this time Zenith abstained. Still nobody had a majority. They were deadlocked.

Very quickly, word of this odd outcome spread through the community, and at first no one understood it, though everybody knew that a private agenda lay hidden here somewhere. "Zenith is playing games," Rast speculated. "Maybe they're trying to be a player, so that someone will offer them something. Maybe this is part of the fire-sale mentality. I don't know."

"Everybody's gaming," Wiley said. "It's a barrel of snakes. I don't know why Zenith's doing this, but I'm against further tests. We had a test. We had results. There's a thousand ways this whole thing can slip. Audio is just the first challenge. If we allow this to happen, then everything will be challenged." Besides, he went on, warming to the topic as he leaned back in his desk chair, "the way the test turned out, there was a clear winner—Dolby. And it's an *American* system. That's great. It takes care of the boys on the Hill. I didn't set out for it to be that way, but there's synergy here. This is the way it came out, and it works."

Already that nationalist point had stirred up the boys (and girls) on Capitol Hill. "Clearly U.S. industry must be afforded the opportunity to have its audio technology considered on a level playing field," Senators Dianne Feinstein and Barbara Boxer and Representative Nancy Pelosi had written to acting FCC Chairman Quello. All three were from California, Dolby's home, and they were offering indignation in advance. "As the Grand Alliance and the FCC Advisory Committee progress toward a final recommendation," they wrote, "arguments will arise to adopt a European-developed audio system in the interest of 'international standards.' This decision by the FCC has the potential to strengthen the U.S. lead in this high-technology field. We must seize the opportunity." Quello passed the letter to Wiley without comment.

In the following weeks, there was even more bleating from Congress. Wiley thought the stalemate was stupid, but it was becoming a problem. He called Bingham, hoping to get the Philips executive to back down. "You guys are always telling us how American you are," he barked into the phone. "Now there was a competition between an American and a European system, and the American won. Are you going to fight this now? People on the Hill love it that Dolby won. You fight this, and it'll be a real political football." As Wiley recounted it later, Bingham grumbled something inconclusive; nothing was settled.

At about the same time, Jerry Pearlman, Zenith's president, called Wiley and urged him to support retesting of all the audio systems. Now this is really odd, Wiley thought. *Philips* hasn't called to ask for retesting. Neither has Thomson or Sarnoff. Why on earth is *Pearlman* fronting this? So (as he recollected the conversation later) Wiley asked Pearlman straight out, "How come Zenith is so wild about this thing?"

"I've got to stick with my Grand Alliance partners," Pearlman replied.

Suddenly Wiley realized what was *really* going on. "Come on, Jerry," he said. "Is there possibly something else you want, too? Maybe you hope one particular Grand Alliance partner will stick with Zenith when the *transmission*-system vote comes up in a few months?"

Wiley said Pearlman mumbled something back: "Well, yeah. Oh, you know." There it was. The answer.

Word of this conversation "zipped around the circuit in days," Rast said a week later. In Rast's retelling, the explanation was even more direct. He recounted the story this way: "Wiley asks Pearlman, 'Why do you want this? Why are you the ones asking for retesting?' And Pearlman tells him, 'We want to do this for Philips so they'll help us when the transmission decision comes up for a vote later.' Then Wiley tells Pearlman, 'Surely you don't expect me to help you with this.' "

According to Rast, Carlo Basile, the senior Philips engineer, was "waiting for Pearlman to call, asking Philips to return the favor."

Leaning back in the desk chair in his own little office, Basile summed it up this way: "For Zenith, it is clear they have only one piece of technology to offer. AT&T built everything else. That's different from Philips, Thomson, or GI. All of us have all of the fundamental technologies within our companies. But the only thing Zenith has is a transmission system. Zenith was a Dolby advocate before. Maybe they're trying to telegraph a message to us: Be objective when the transmission tests come."

Back in San Diego, Rast smiled. "The intrigue with all this is incredible," he said.

Nobody knew what to do. The paralysis continued into mid-September, even though the Grand Alliance had promised to choose an audio system by September 15. Until this was done, they couldn't settle questions about other parts of the machine. Lim finally consented to vote for Dolby; the company was *American,* at least. That made the tally two for Dolby, one for Musicam, and one abstention. No one had a majority.

As the days advanced, the partners begrudgingly concluded that the only way out of this impasse was a retest. The final decision would have to be delayed a month to accommodate that. But Jim Gaspar was adamant. He didn't want to do it. If the systems *are* retested, he warned, we're going to run them through a far more elaborate testing regimen "to further differentiate" them. And that can't be done in a month. The Advisory Committee's other leaders agreed. "The Grand Alliance can do whatever they want," Joe Flaherty said, "but we are *not* going to retest." Finally, Dolby's top management let it be known that they would not participate in a retest in any case. "A retest is clearly not in our interest," said a company official.

At month's end, the Grand Alliance was frozen solid. Philips officials were pushing to move ahead with retesting, but they were stymied by Dolby's refusal to cooperate. Still, the former members of the Sarnoff consortium were standing firm. "Look, someone screwed up in the hardware," Joe Donahue argued. "If it hadn't been fixed, why would we be asking for a retest? We've got to be compatible with Europe."

As the days dragged on, the leaders of the old Sarnoff consortium were growing angrier and angrier. They began to see a conspiracy. Finally, in early October, Peter Bingham started talking about dropping out of the alliance. "How can we continue in the Grand Alliance if people are stabbing each other in the back?" he asked. And the others knew that Bingham was just volatile enough to do it.

Worried about that, on October 8 Wiley invited everyone to his office for a talk. All sides argued their positions, but nothing changed. The deadlock remained. "We're heading into a total breakdown," Rast warned. And as the Grand Alliance slid toward dissolution, the partners scheduled one last meeting for October 14 at their favorite haunt, the O'Hare Airport Hilton in Chicago, chosen for its central location.

The out-of-towners arrived the night before, and when everyone convened in the hotel conference room in the morning most of them seemed to be in a foul mood. Rast had been calling this "the penultimate meeting," and the threat of divorce hung in the air. They'd barely settled into their chairs, cups of coffee still steaming in front of them, when Joe Donahue stood up and announced that he had something to say. Several of the others glowered up at him; the last thing we need, some of them thought, is another pitch for Musicam—still another complaint about "implementation errors" and "international standards." But Donahue surprised them. He was the elder statesman in the room, nearing retirement with a long list of stellar achievements to his name. And he spoke in a tone that was both wise and weary, occasionally glancing at a pile of photocopied papers that sat on the table in front of him. At the top, people sitting near him saw, was a sheet headed "Proposal for Grand Alliance HDTV System."

"I'm going to make a proposal," he said, pausing briefly to clear his throat. "But first of all, I think it's time we think about our customers and stop looking at our own navels. There's too many of us to do that. Our customers are the Advisory Committee process, the FCC, and so forth. They are number one. The people who are going to deliver programs to the consumers, they are number two. I consider them vital. And number three, eventually, the consumer.

"It's the hour to make decisions. We've fooled around with this thing now for six years. We're there. If we screw around with it anymore, there may never be high definition. That's my real view. So many other things are happening. Let's make decisions. And I'm going to make some that are very painful for my company. And I'm going to make suggestions that are painful for you. But it's the hour we should make decisions. Don't nitpick if you like one piece and not the other. I'm giving you a total proposal. If you don't take it all, then we're back to negotiating every point, all right?"

Donahue had everyone's attention now, and he began to describe his plan. Dolby will be the Grand Alliance audio system, he said. They won, and that's that. But the Grand Alliance will retest Musicam, to verify that Philips has corrected the problems. Should the Dolby system experience problems, Musicam will be the backup. Donahue outlined some proposed compromise solutions to a few other outstanding issues, none of them as divisive as that one. Then he said, "Musicam is a great blow to me. It's a *great* blow. We've got Musicam in early production. It's *in production.* But fighting forever for everything is useless, all right? Let's get it over with. There has to be an end to it."

And then he passed out his one-page list of proposals. Around the room, the Dolby advocates were smiling, while the others "looked shocked," as Donahue later put it. Finally, though, the partners embraced this initiative as a lifeline, since all of them knew they were sinking fast. Within a minute the mood had changed. Most of the Grand Alliance members were optimistic, upbeat—though Bingham continued his angry mumbling: "Others can be calm about this, but not me." The first great fight was over, and the risk of dissolution had been put aside—at least for the time being.

17 The most insidious enemy

The insurgency started so quietly that Dick Wiley barely noticed. Stalking him was a nontraditional predator, and the chairman's problem may have been that he remained focused on the accustomed enemy, John Abel and the National Association of Broadcasters. It's no wonder. While congratulations flooded into Wiley's office from around the world the day after the Grand Alliance announced agreement, the NAB broadcast a snippy little press release, as if to say, Don't forget us. "The Grand Alliance will shorten the time needed to achieve an HDTV standard by avoiding possible costly and lengthy legal challenges," it said. "But the agreement inspired by Dick Wiley is vague in technical details that are vital to broadcasters, and we have several concerns." It went on to complain that some technical parameters had not been set and the means of setting them were "yet to be defined."

As Wiley saw it, the Advisory Committee had a strong hand just then. Other broadcaster groups were full of praise, and in any case, by the summer of 1993, John Abel and the others at the NAB weren't doing or saying much of anything that anyone could see. "They're tranquil right now because their ox isn't being gored," Wiley said. "But if I were to gore it, I think they'd come running out of the woodwork."

A few other complaints filtered in, but none of them seemed particularly serious. So it was little wonder that Wiley wasn't overly concerned about the note from Mike Liebhold, the Apple employee who was the point man for the guerrilla faction up at MIT. It landed on Wiley's desk just two weeks after the convocation at the Grand Hotel. "Please accept my sincere congratulations for leading the negotiations resulting in the Grand Alliance agreement," Liebhold began. "There is much to celebrate. There is, however much remaining work to be completed. . . . I would like to accept your kind offer to provide an active role for my technical contributions. You may recall that during our last conversation, you assured me that you would enroll my *active* contributions in the final technical advisory process. Accordingly, please include my name among the active, voting members of the Technical Subgroup."

Wiley wasted no time answering that one. The chairman wasn't about to appoint this rabble-rouser to his most important subcommittee. He responded the next morning: "Thank you for your letter of June 8. I appreciate your interest in the work of the Advisory Committee. Prior to receiving your letter, the appointments to the Technical Subgroup were finalized. While you were not among the appointees to this limited body, your continued interest in its proceedings would be most welcome. I encourage you to be active. Please accept my best regards."

For Liebhold, that had been Wiley's last chance. Now *he* was angry.

"The major stakeholders in the National Information Infrastructure are not being included in the Advisory Committee process!" he started shouting all over Washington. "I'm talking about the people from the education, scientific, and computing industries." With help from Cambridge, Liebhold was now on a crusade, and his principal goal was to kill the interlace dragon once and for all. "They are saying that for all these economic reasons they have to include interlace," he argued. "The arguments are nonsense. The entire TV industry has plenty of time to gear up, to manufacture a simple device that can display an electronic textbook. They are going to populate the country with millions of TV sets that are not capable of displaying the simplest benefits of the National Information Infrastructure."

Liebhold laid out his complaint for members of Congress, technology officers in the Clinton White House, anyone who would listen. "The computer industry tried to use interlace scan years ago," he told the House Committee on Science, Space, and Technology in June, "but found that the display flicker produced on text and graphics rendered it unusable. . . . In an apparent attempt to compromise, the Grand Alliance has announced a preliminary intent to support both interlaced and progressive-scan transmission. In its current form, this compromise could result in a de facto interlaced standard." Equipment manufacturers would continue making interlaced sets, because they were cheaper, he argued. Millions of Americans would buy them. "Progressive will never be given a chance to flower. And the Advisory Committee is so dominated by equipment manufacturers that it can do nothing about this!"

Liebhold pumped out this view week after week, and soon Wiley began seeing the results. Members of Congress were sending him letters: Why aren't you being more accommodating to the needs of the computer industry? In a meeting at the White House, a senior technology officer asked, Why don't you just bite the bullet and make it all progressive? Wiley was getting angrier and angrier. Few things bothered him more than people working against him behind his back. "We *explained* that we have this migration path to progressive," he fumed, "and Liebhold seemed to accept it. I thought we had it worked out. But then he turns around and does all this. He's over at the White House complaining. And with Apple's ties to this administration. . . ." Wiley shook his head. "He's telling everyone this doesn't fit on the information superhighway, and then he's up there on Capitol Hill jangling Markey's chain. In twenty-five years in this business, I've never met a man as difficult to deal with as him. I told *Broadcasting* magazine that he was unsatisfiable. He is totally unreconstructable!"

Liebhold countered, "They're completely faking this migration strategy to progressive. They *have* no migration strategy. There is no migration plan. Important people in Congress know what's going on. Ask Markey!"

Sure enough, in early July a letter from Representative Markey landed on Quello's desk. The chairman of the telecommunications subcommittee—the man who had put on the very first HDTV hearing back in 1987—wanted answers:

Has the Grand Alliance fulfilled its commitment to consult with the computer industry and others involved in HDTV applications to ensure that their views are heard and their concerns are integrated? Please outline how, if at all, the Grand Alliance has fulfilled or is planning to fulfill its commitment to consult these companies or institutions.

Has the Advisory Committee on Advanced Television Service included representatives of these industries in its review process?

Wiley could barely contain himself, but he knew he had to do something. "The one thing I could see happening," he said, "is this White House—which is the computer-nerd White House—getting together and saying, 'Gee, this Advisory Committee recommendation was based too much on the interests of broadcasting.' " He frowned for a moment.

"There's going to have to be something in there for the computer people. Or else I don't think that our recommendation will be believed or accepted. It will be turned around. That's my view."

His relationship with Mike Liebhold was not improving. Far from it. The two of them fell into an angry shouting match on the phone. Wiley blew up at him, and the next morning a letter from Liebhold arrived by fax:

I am appalled that the chairman of such an important process would lack the fundamentals of civil communications! You appear to have no interest in dialog. You neither allowed me to begin or complete a sentence, nor gave me an opportunity to reply to your continuous stream of hostile assertions. If browbeating and intimidation of legitimate dissent is an acceptable behavior for the chairman of an advisory process, then the process is flawed. (And I am not intimidated.)

A short time later, Wiley called his advisers and allies to ask: Should we have a summit meeting with the computer people? Bob Rast told him he didn't think that would accomplish much. Well, Wiley next suggested, what about appointing "an interoperability subgroup." Good idea, everybody said, so Wiley asked Robert Sanderson, the Kodak executive, to head it. He was already working on one of Wiley's subgroups. As he saw it, Sanderson was a "reasonable" computer advocate. Not like Liebhold.

Wiley was much relieved when Sanderson agreed to serve. "I can work with him," he said. Sanderson made a few requests, and Wiley accepted them. Given the prominence of this issue, Sanderson and Wiley agreed that the interoperability panel should be a freestanding subcommittee—its name would be the Joint Experts Group on Interoperability—and Sanderson wanted to appoint the members. OK, OK, Wiley assured him, though he and Joe Flaherty insisted that some of their people be on the panel, too.

Well, when Sanderson's proposed membership list came in, one name jumped right off the page: Michael Liebhold, senior scientist, Apple Computer. Wiley swallowed hard, but there was nothing he could do. He had agreed to let Sanderson pick the members; he couldn't change his mind now. Another Technical Subgroup meeting was coming up in August, and now Liebhold would probably be there. Not sitting at the big table, mind you. Off to the side somewhere, but in attendance nonetheless—with official standing.

In late July, the Advisory Committee settled on a date for the Technical Subgroup meeting and sent out notices by fax. Among the replies, one came from Sanderson on August 4. "I will not be able to attend the August 11 meeting of the ACATS Technical Subgroup," he wrote. He planned to be away. But he had appointed a vice chairman, the man who would be acting chairman of the Joint Experts Group on Interoperability at the next meeting. The man who would sit in his place at the head table, just a few seats down from Wiley. The vice chairman was . . . Michael Liebhold.

Sanderson chuckled when asked later if he'd known what he was doing. He was well aware that Wiley had refused Liebhold's request for a seat on the Technical Subgroup, but he thought "it was important to bring Mike into the process," as a means of "reconciling and converging the views, instead of taking potshots at each other from the outside." But the way Bob Rast saw it, "The fox is in the coop."

The Technical Subgroup met in a conference room at the National Association of Broadcasters, next door to the smaller meeting room where, in 1986, the idea had raced through John Abel's head: HDTV . . . maybe that's it! This morning, however, nobody was thinking much about the broadcasters. All eyes fell on Liebhold, who sat at the table with the other Advisory Committee luminaries, about five seats away from Wiley. He wore jeans and a sport coat. His little rat-tail haircut fell over the open collar of his shirt.

Most of the decisions to be made here were technical, and the meeting slogged through a morning of tedium. Liebhold kept his remarks brief; he seemed to be trying to avoid confrontation. For his part, Wiley was working very, very hard to be a gentleman. Nonetheless, comments, looks, and asides made it clear that most of the people at the table just couldn't stand the man from Apple.

The Report of the Joint Experts Group on Interoperability was scheduled for 11:45, the agenda said. When the time came, another Advisory Committee member read Robert Sanderson's report out loud. It announced that Sanderson intended to hold a three-day meeting and then a second two-day event, both in September, "to evaluate the Grand Alliance proposal from the point of view of interoperability."

Rast exploded when he heard that. "What was just discussed is new to the Grand Alliance," he declared. "We've got to balance things. Let's not get carried away with this. We want to do the utmost toward interoperability, but I'm a little bit nervous here."

Liebhold tried to speak as gently as he was able. "In a strange way," he said, "we are trying to do the same thing."

"It's just a proposal for a schedule," someone else said.

But everyone knew what was really going on here. Five days of meetings to evaluate the interoperability potential of the Grand Alliance proposal—events run by computer people? The Advisory Committee and Grand Alliance were being set up for an interlace-bashing. So some people at the Technical Subgroup meeting tried to stanch the damage in advance. "You know that the cost of these receivers is extremely important to HDTV rollout and success," someone pointed out, noting that progressive-scan televisions would be more expensive. Liebhold's civil veneer cracked. "You're going to allow low-cost, interlaced receivers to dominate the market," he whined. "There will be no migration, so you're *always* going to have interlaced receivers in most homes!"

Rast spoke as if to a child. "This isn't our job," he said in a slow, even voice. "That's a receiver-manufacturer issue."

Liebhold threw up his hands in mock despair, while turning his head and rolling his eyes. Wiley stepped in to make peace. He had no use for Liebhold, but he wasn't going to allow *anyone* to turn one of his public meetings into a circus. "There may be an FCC decision at some point to discontinue interlace," he said. Case closed. Let's move on.

Liebhold left the meeting convinced that he had been ignored. The entire event had been "a frontal attack on progressive." He focused his energies instead on the two meetings he and Sanderson had planned. They did in fact turn into circuslike interlace bashes. Many of the forty-ish computer people in attendance sounded as if they might have been Vietnam War protesters in their youth. "Vested economic interests are insisting on the continuation of interlace," one of them told the gathering. "It's a corporate conspiracy," someone else warned. "Interlace promotes illiteracy!" said one of the buttons speakers wore on their lapels. "Interlace causes AIDS," said another. A poster offered this thought: "WARNING: The Electrician General has determined that interlace may be hazardous to your health."

When these meetings were over, Liebhold had compiled a fat binder full of letters and testimonials attacking the Grand Alliance and promoting progressive-scan displays. He wanted them added to the record for the next Technical Subgroup meeting, on October 21, where the final decisions on the progressive-interlace debate were to be made. But that was not going to happen. As with every Wiley meeting, all the decisions were made in advance. Liebhold learned that for all his work little had changed. The Grand Alliance system was going to be approved for construction using both interlaced and progressive scanning formats.

When the meeting opened—Thursday morning, October 21—Sanderson was there but Liebhold wasn't. "Guess who boycotted—Liebhold!" Wiley leaned over and whispered to an audience member, trying without success to hide his glee. "I feel bad. I brought him into the process," the chairman went on, a bit disingenuously. "I thought that if I put Liebhold into the mix, peerage would calm him down. That's how an advisory committee should work. I was wrong. But he had his own two-day meeting. He *ran* it!" What more could this guy want? Wiley was asking as he strode off toward the chairman's chair.

What Liebhold wanted, at a minimum, was to have his fat binder of complaints entered into the official record. But when that idea was proffered, the broadcast engineers on the Technical Subgroup shot it down. "Many of the letters are full of errors or misunderstanding," one of them said with a dismissive snort. "There are a few valid arguments, but mostly a lot of specious arguments," scoffed another. So the Technical Subgroup refused even to enter Liebhold's compendium of complaint into the record. In fact, the only direct official reference to his view was one in Sanderson's presentation and on an overhead chart that said, "Both progressive and interlaced scan have fervent supporters and certain technical advantages."

The next day, Liebhold was beside himself. "The ACATS process is so corrupt it is incredible!" he wailed. "They're going ahead despite all of this testimony!" He called Wiley to complain. Paul Misener came to the phone instead. Wiley had a new office rule: he would not take phone calls from Liebhold. Liebhold told Misener he would no longer participate in the process.

"I'm sorry to hear that, Mike," Misener said, in his most saccharine tone.

"I have no standing with the Advisory Committee," Liebhold whined. "I'm going to have to take this to alternative political processes."

"In fairness, Mike," Misener replied, "you've been doing that for some time."

In Cambridge a little while later, Liebhold's allies at the Program on Digital, Open High-Resolution Systems began trying to turn up the heat from the outside once again. The MIT guerrillas pulled together dozens of letters of complaint. Some of them were from Liebhold's collection and still more had been solicited from other members of the industry who didn't like the Grand Alliance system because it offered interlaced scanning. If the Advisory Committee was going to ignore these voices, the guerrillas were going to send these packets out by Express Mail to every member of government who might conceivably have a voice in this debate.

Wiley got a tip that the lobbying packages were about to go out, so he decided to stir up pressure of his own. At the next meeting of the Advanced Television Test Center board, he told all the broadcaster representatives, You'd better get off your duffs here and make yourself heard. You'd better stand up to these computer people, or they're going to run right over you.

A few days later a press release went out that said:

The Broadcasters' Caucus of the United States today expressed great satisfaction at the decisions made to date by the Grand Alliance in specifying its scanning parameters. Michael Sherlock, Executive Vice President of NBC and Chairman of the Broadcasters Caucus, said: "The parameters specified by the Grand Alliance include five progressive sets and one interlaced set. We fully support the inclusion of both progressive and interlace scanning parameters."

The MIT guerrillas duly sent out their Express Mail packages of complaint. But "our letters and the broadcasters' letter probably cancel each other out," Suzanne Neil, one of the guerrillas, conceded. "There certainly has been no stir that reached my ears." Their effort failed. A couple of weeks later, Liebhold withdrew from the war.

"A lot of people are really incensed about what happened," he said. "People are blown away. But I don't think it is productive to have me be the sole spokesperson and have them attack me personally. I am really looking to see other people pick up the banner here."

Wiley won this battle, another in a long string of threats to the race he had been trying to shepherd to completion for six long years. Created to be the vehicle for the FCC's cynical, delaying strategy, Wiley's Advisory Committee had seemed as if it might be smothered at birth by the profound distrust between the Broadcast Barons and the Cable Mongols. Then a year or so later, the old RCA publicity machine had tried to blow the committee away. Soon after, the budding broadcaster rebellion worked to thwart the whole endeavor. Technical failures and contestant ploys to evade responsibility threatened the race in 1991 and 1992. The European consortium's back-channel lobbying on the jobs issue threw a wrench into the works at a critical moment, and then Jae Lim nearly aborted the formation of a Grand Alliance. In the last few months, the Grand Alliance partnership had nearly dissolved over the audio-system dispute, and the computer industry had just fired a torpedo that could have sunk the ship. But now all of the malign forces at play in the race—greed, conceit, and selfish disregard—had been thrown into the contest. "This is a *saga*," Rast remarked in the fall, his voice filled with wonder. "This thing just won't quit." Through it all, however, Wiley and the builders of the new machine had prevailed. But nothing that came before could compare with what lay ahead. As 1993 drew to a close, they confronted the most insidious enemy of them all: indifference.

President Clinton appointed Reed Hundt as the new chairman of the FCC in the spring of 1993, but Hundt's confirmation was held up for months while the Senate dealt with other issues. Acting Chairman Quello sat in the chair ably enough, but he was in no position to do anything more than custodial work. So Wiley continued without direction through the spring, summer, and fall, waiting for Hundt to take office and let everybody know what he wanted. In the meantime, Wiley was apprehensive.

Wiley was a Republican, after all, and quite a prominent one at that. A large green stone elephant sat on his office coffee table, and over the years he'd served as an officer of the party time and time again. All along, the FCC had been run by fellow Republicans—Mark Fowler, Dennis Patrick, Al Sikes. What would a Democrat think of the Advisory Committee's work? Nobody was saying. At first, Wiley and some of the others thought the apparent indifference might be a good thing. The Clinton administration was fighting so many desperate battles—gays in the military, nominees' nanny problems, Haiti, Somalia, health care. With all their other troubles, "Why would they mess with something that's working?" Jerry Pearlman asked rhetorically. By the end of 1993, however, Wiley was still being ignored, even though Hundt had taken office. Nobody was telling Wiley a thing.

Reed Hundt was a corporate lawyer who had handled several communications issues before taking the FCC job. HDTV was new to him; nobody knew what he thought about it, and Hundt wasn't saying. Wiley had given him an HDTV briefing a few weeks before he took office, and Hundt had been supportive and pleasant enough—but noncommittal. He did say one thing that raised the hairs on the back of Wiley's neck. The new chairman said he'd been talking to Nicholas Negroponte about HDTV. Once again, the Grand Vizier's perceived wisdom threatened to muddy the waters, and his views had not changed. "This HDTV program is absolutely absurd," he told a gathering of MIT alumni in November. "If there's anyone here from the Grand Alliance, don't expect to leave happy. That's a really stupid way to end a very stupid process. Nobody wants HDTV, and if it does finally come out, there will be no receivers, no material, and no demand." He'd certainly said something similar to Hundt, Wiley knew.

Clinton administration officials were pleasant enough to Wiley, but they, too, were noncommittal. In November he paid a visit to the White House, "and somebody over there was very complimentary about the whole thing. But they were still asking questions about the progressive-interlace thing." Hundt had asked him those questions, too. Still, aside from these and a few other queries posed in a desultory way, Wiley was working in a vacuum, where rumors flourished.

Before he left office, Al Sikes had set up an elaborate regulatory strategy for HDTV. Every television station would get a second channel, and they would use it to broadcast HDTV. After fifteen years, time enough for every American to buy an HDTV set, the stations would give their original channels back. Under that plan, HDTV was *guaranteed* a market; that's what had kept the contestants going all these years. What if the new, Democratic FCC chairman threw all that out? The regulations had been published, but nothing said they couldn't be changed. "I'm just afraid the Clinton folks might abort this, us being Republicans and all," Wiley said one day. "Maybe say they want to reexamine everything. And you know, a six-month delay would kill us."

Vice President Al Gore had been one of the great HDTV screamers during the congressional frenzy a few years before, and Reed Hundt had got the FCC job because he was one of Gore's best friends. They had been classmates at St. Albans, one of the best prep schools in

Washington. Now one of Gore's staff members was heard to grumble that Sikes, Wiley, and Donald Rumsfeld had been running a Republican conspiracy of sorts. One day Wiley and Rumsfeld were discussing that, and Rumsfeld had volunteered, "Well, if that's the case, I should resign."

"No," Wiley said, "I'm the one who should resign."

"No, me."

"No, me."

"No, really," Rumsfeld insisted, "I'm the one who should resign." Rast and Quincy Rodgers were watching, and finally their laughter broke it up. But clearly they were all worried by the inattention from the people who were going to get the Advisory Committee's recommendation in a year's time.

"Nobody's saying a thing," said Rodgers, "and it leaves all of us spending money" without knowing if there'd ever be any return.

"You know, it could all come crashing down after we make our recommendation," Wiley said, his tone weary.

The lack of certainty was unsettling, but the inattention gave rise to a greater problem. Throughout Wiley's tenure as Advisory Committee chairman, the FCC chairman had stood squarely behind him; everyone knew that Wiley had access to the chairman anytime he wanted. "With Sikes there, I could rattle that skeleton in the closet," Wiley said. "I don't have that anymore." In fact, he was shut out of the FCC. Dick Wiley and the entire HDTV superstructure beneath him were all alone, and that gave power to his enemies.

John Abel and the NAB had lain low during the computer wars, but as Christmas approached Abel felt the change in the political winds and realized that the momentum was now with him. Nothing had changed among his members; most of them still didn't want to spend the money required to broadcast HDTV. "The truth is, we came through two administrations that were focused on HDTV," Abel said, relaxing at NAB headquarters in December, arms stretched out on the back of his sofa. "But this one is focused on the National Information Infrastructure. You can't ignore the fact that HDTV has not been mentioned by Reed Hundt. It hasn't been mentioned by Al Gore. The politics have changed. Wiley can't even get in to *see* Reed Hundt on this. This administration has said almost nothing about HDTV."

Abel saw that the time was right for a run to the goal. He seemed extraordinarily confident of scoring. "I don't think there's going to be anything happening in HDTV," he went on to say, whispering conspiratorially. "The process may be abandoned, done away with, put on hold, I'm not sure what. There's no interest in it now. I frankly think the HDTV process, it's dying."

18 The dreaded "H" word

From his earliest days in Congress, Al Gore had been an enthusiastic advocate of advanced technology. Now that he was in the White House, the vice president gloried in inviting reporters over to show that he could carry on E-mail conversations at his desk, and as everyone noted, this was in sharp contrast to the technology-averse Bush administration. So it seemed only natural that after a year in office, it was Gore who stepped up to announce the Clinton administration's new technology policy—an ambitious plan to wire the nation so that everyone could enjoy the benefits of the digital age.

In a carefully choreographed speech in Los Angeles, presaged with fulsome advance publicity, Gore reminded the nation that the federal government had built the interstate highway system—the envy of the world. And "today," he went on to say, "we have a different dream for a different kind of superhighway—an information superhighway that can save lives, create jobs, and give every American, young and old, the chance for the best education available to anyone, anywhere." Gore challenged telephone and cable TV companies "to connect all of our classrooms, all of our libraries, and all of our hospitals and clinics by the year 2000," so that every American would have access to "the National Information Infrastructure." The vice president's tone was lofty, and his speech was filled with metaphor as he outlined a legislative proposal intended to promote these goals.

Most telecommunications industry leaders welcomed the new initiative—although, as always, they wanted the government to put up more money. Over at the National Association of Broadcasters, however, the leadership had an entirely different reaction. They cried out in unison: What about *us?* Gore had spoken at length about new rights and powers for telephone companies, the computer industry, even cable television firms. But he had nary a word to say about television stations, and the broadcasters fell victim to the tired analogies that had long plagued these discussions: "The vice president of the United States left them on the shoulder of the information superhighway," an article in the following week's *Broadcasting & Cable* magazine said.

In truth, the NAB did feel wounded. After all, Abel argued, if you want to bring the National Information Infrastructure into every building in America, you don't have to lay wires, pass laws, launch satellites, or purchase computers. "We *are* the National Information Infrastructure," he began saying, one finger raised above his head as he preached the broadcasters' new gospel. "Nobody has the penetration we have. We're it. We're like the air. We're everywhere! No wire is ever going to achieve the universal service we already have."

Abel was right. Almost 99 percent of the homes in America had at least one television set—more homes than had telephones or flush toilets. Most hospitals, clinics, and schools had at least one TV. So, Abel said, "we are going to try to insert ourselves into this debate." He and others quickly began to see that the vice presidential neglect might be turned into a tactical advantage. And at that very moment the high-definition television debate took a decisive turn—one that threatened to push HDTV toward its demise.

For years, everyone had known that digital television would allow broadcasters to provide other digital services along with high-definition pictures. Nonetheless, neither the NAB nor anyone else had carefully considered all the implications of this. It took Al Gore's speech to focus the thinking.

Now Abel and the others began to concentrate on one clear fact that few people had thought much about before: A digital bitstream doesn't have to carry high-definition images. It is infinitely flexible. In fact, a television station could in theory forgo *high-definition* television altogether and use digital compression to broadcast four or five conventional TV signals over the same space on the airwaves. Broadcasters could also sell paging services, video cell-phone networks, pay-per-view movies, on-screen E-mail. They could enhance TV advertising—provide detailed information on request about the price or features of a new house, to complement a real estate advertisement. Or they could simply broadcast several conventional programs—"multichannel broadcasting," this came to be called. The possibilities suddenly seemed limitless. And as a positive *bonus,* some of them realized, the new political reality might finally have given them the arguments they needed to evade *high-definition* television once and for all.

Instead of squealing about the high cost of installing digital equipment, the broadcasters started saying: We'll spend what it takes to enter the digital age. But give us the flexibility to use this new equipment to earn back our investment. Give us that second channel Al Sikes promised us. Don't tie our hands, and we'll bring the National Information Infrastructure into every home. This new tack carried an implicit message: Don't make us broadcast high-definition TV.

Woo Paik and Jerry Heller had realized from the beginning that the digital technologies they were pioneering could be used for other purposes. In fact, years earlier Paik had created a digital compression system for conventional cable television that would allow cable operators to transmit three or four channels in the space previously occupied by one. But the rest of the industry had been slow to understand the implications of this. Now, however, "there has been a shift, indeed there has," noted Jim McKinney, the former FCC official. "It was brought on by digital broadcasting, which really doesn't have anything to do with HDTV, except that HDTV is what got it invented. All of a sudden, they know now that they can do a lot more with digital than they ever could with analog TV."

Michael Sherlock, the NBC vice president, agreed. He also happened to be the president of the Broadcasters' Caucus, the powerful political organization that represented all the major television networks and their trade groups. A few months earlier, the caucus had written that letter to counter the MIT guerrilla attack. "The NII [National Information Infrastructure], that's the break-off point here, and I think it's great," Sherlock said in February 1994, just a few days after a caucus meeting that had been called to discuss this. "Deep down there isn't any disagreement among us: All of the people around the table want the ability to offer digital services" on that second channel the FCC was supposed to give out.

As for high-definition television, the digital service for which that second channel was intended, "I'm sure everybody would say, 'Fine,' if there is a requirement that the second channel also be used at some time for HDTV—some minimum, to the extent that the market demands it. In the first years, maybe an hour a night of HDTV. Or an hour a week."

Then Sherlock confided, "We can't say that. If we came out and said we want to do only other services, then it would be turned around to say we don't want to do HDTV. That's the fear. We'll be misunderstood. Broadcasters will be painted as if they don't want any HDTV."

Which, of course, was exactly the truth.

Sherlock knew that open discussion of the second-channel issue would lead to news stories and widespread accusations that the broadcasting industry was engaged in a flagrant "spectrum grab." That second channel had been offered for one purpose only: to ease the transition to *high-definition* television. If the broadcasters came out foursquare with the truth—We don't want HDTV at all—they feared they might not get their extra channels at all. John Lane, a Washington attorney representing Land Mobile, was complaining that the HDTV race had now delayed Land Mobile's efforts to get those extra channels for *seven years.* Meanwhile, police and fire departments "definitely, desperately need" those vacant channels that had been set aside for HDTV.

So the broadcasters' strategy was set: Ride the Clinton administration's National Information Infrastructure initiative. Get the second channel, and use it for moneymaking advanced digital services. Describe them as central to the White House's new pet initiative, the NII. As for high-definition television—be vague.

Through all this, the Grand Alliance partners were getting ready for their transmission system bake-off, the last decision as they defined their new system. At the end of December 1993, Zenith and GI were to send competing systems to the Advanced Television Test Center, where both would be run through a series of tests. This was the Grand Alliance's own test; the partners would get the results and choose the winner. But Peter Fannon and his staff were to administer it and watch for any irregularities.

Even as they got ready for this, their final technical decision, some of the partners began to worry about the broadcasters once again. As Bob Rast put it, "We decided we ought to go talk to Abel, because we have some concerns about how the NAB positions itself with respect to the Grand Alliance." So he and Joe Donahue, the Thomson executive, asked Abel if he would like to have breakfast. Abel agreed and said he would bring along Michael Rau, the NAB's vice president for technology.

Abel, as usual, was more direct: "Basically their issue is to make sure we don't trash them." As it was, Abel's "trashing" was getting him into trouble with some of his own people. The Broadcasters' Caucus and allied organizations were convinced that constructive ambiguity was the only viable approach to the high-definition television debate. But ambiguity was not among Abel's virtues. "I received probably ten phone calls from Broadcasters' Caucus people telling me basically, to cool it, shut up," Abel said. "Get on board and don't give out any more left-handed comments. People from the trade press aren't calling me. Maybe they were told not to call me, I don't know. The point is, the Grand Alliance people summoned us to the meeting to make sure we are on board with them."

So four men, four minds filled with deep convictions and well-defined goals, sat around the linen-covered table in the Mayflower Hotel's decorous dining room. A huge vase of fresh-cut flowers sat on a marble pedestal in the center of the room. The breakfast started slowly; they exchanged pleasantries and trivialities as Abel and Rast, particularly, sized each other up.

Abel had "a lot of respect for Rast," he acknowledged. "I think he's sharp. I think he knows what he's talking about." And Rast realized, "I like John Abel. He may not be frank, but he is honest. If he is going to talk about something, he's always straight up." They'd never had a meeting like this before, and the mood was tense, until finally Rast explained why he and Donahue were there: We want to see if there is any room for increased cooperation between the NAB and the Grand Alliance, Rast said.

Abel knew full well that Rast didn't represent just the Grand Alliance. Rast worked for General Instrument, a company that was firmly, wholly, inalterably aligned with the *Cable Mongols.* "We have lots of suspicions about GI," Abel said, looking right at Rast. "You're a cable company, driven by the cable industry. Your products are consumed by the cable industry, and we are very skeptical that you would have any interest in making anything that could benefit the broadcast industry. You may not have a focus in intentionally disadvantaging us, but you don't have any interest in really advantaging us, cooperating with us."

Rast reared back in his seat. "Well, gee, guys," he said, his voice laced with sarcasm, "I'm really glad *you* called this breakfast this morning."

Rau, watching this exchange, "was judging it on who was winning what points," Abel said. Score a point for Rast.

"If we're the enemy," Rast went on to say, "hey guys, I want to tell you something that you need to realize: Your future lies in your enemy's hands. You better go find out what your enemy is doing. If that's GI, then it's GI."

But Abel wasn't going to lie down. After a few more minutes of acrimony, he showed his hand. "You know we stopped an FCC proceeding once before," he said, his voice low, cool. He was, of course, referring to the Land Mobile decision back in 1987—the one that got the HDTV race going in the first place. "We started this proceeding," Abel said. "We started it, and we can stop it. We are not afraid to take on Dick Wiley and the FCC."

This was a death threat, and Rast knew that "they do have some power." Score one for Abel.

"If you guys think we are just going to sit here meek and mild, you have another thought coming," Abel continued. "We have to do what we have to do." We may introduce legislation in Congress to further our goals, he announced, and the warning was clear. Congress was always vulnerable to the broadcasters' entreaties. Every member of the Senate and the House wanted to be on TV at home as often as possible.

The breakfast ended without any "resolve for friendliness in the future," Rast would later say, with significant understatement. As Abel got up from the table, he told Rast, "I'm glad we got together, and I think I see what you were trying to do here." That's kind of a cool thing to say, Rast thought.

When he and Donahue talked about it later, they could see what Abel was trying to do, too. "They don't have a strategic direction," Rast concluded, "but they don't mind a little chaos, because they are going to mine something out of the chaos. They feel they are good political players in the infighting. And they don't believe the FCC can prevail against broadcasters." Then there was Congress. Rast was more than a little worried about whatever legislation Abel had in mind.

Zenith's leaders were pinning their hopes on this transmission system bake-off. They had nothing else. If the company lost this competition, not one significant component of the Grand Alliance system would be theirs. Conversely, if the Grand Alliance chose their transmission system, they could sell it for other uses as well. Telecommunications markets were exploding everywhere, and the Grand Alliance transmission system was almost certain to become a de facto industry standard.

The company's financial position was dire as ever; Zenith showed a $97 million loss for 1993. As 1994 opened, *Business Week* reported that the company's directors were just about fed up with five straight years of losses that now totaled $332 million. Jerry Pearlman, one director said, "is on a short leash." Pearlman told *Business Week,* "Our strategy is to make money in color televisions and consumer electronics and not hang on by our fingernails until high-definition or digital technology bails us out." Brave words. But as he and everyone else knew, the company had been pursuing that strategy for several years with scant results. So Wayne Luplow could not even *conceive* of losing the transmission bake-off. "We're going to win this; we've got to," he said just a few days after both systems had been delivered to the test center. "Just wait and see. I am 100 percent confident—unless politics interferes." And Luplow was already working to avert any possibility of that.

In the Grand Alliance, as on Capitol Hill, politics came down to just one thing: votes. No matter how the bake-off turned out, the Grand Alliance members would vote to select a transmission system, and Luplow knew that many factors other than technology would play into that balloting. "GI's out there trying to set the de facto standard for the cable world," he sneered, leaning back in the fat, high-back desk chair in his Glenview Bunker office. "GI doesn't care one iota about the Grand Alliance. They have a lot of investment in this digicrypter crap—DigiCipher, whatever they call it. They think they have big, big volume sales all sewed up." Win or lose the bake-off, Luplow knew, GI would cast a vote for themselves. Zenith would do the same, so the decision would come down to the other two partners. The European group—well, Zenith's leaders figured they'd be objective. If Zenith's behavior in the audio debate had accomplished nothing else, at least Luplow and the others could hope for that. That left Jae Lim, GI's partner.

"Lim says he's independent," Luplow observed, "but he gets bought real easy." GI had done it once before, as Luplow saw it. The company had "bought" Lim's test slot in 1990. What if GI somehow "bought" Lim's vote this time? That would leave two votes for Zenith, two votes for GI. Another deadlock. And the Lord only knew how this one would turn out. "I don't think there ought to be any horse-trading on this," Luplow said. "But I'm just waiting for the other shoe to drop. I'm very wary." He wasn't standing by passively, however. "We've spent a number of days with Lim," he said with a grin. He and others at Zenith had been flattering and massaging Lim. "We call it the Educate Jae program," Luplow said. And if Lim could be "bought real easy," Luplow was dangling a carrot.

Zenith had been one of the earliest members of CATS, the MIT program that William Schreiber had started in the early 1980s. Like all the other members, Zenith had contributed $100,000 a year. But in 1990, Zenith's leaders decided they weren't getting any benefit. "Zenith was teaching the MIT people; they weren't teaching us," Luplow explained. "We weren't getting anything for our money. We were giving 90 percent and getting 10 percent back. I did get some benefit out of it by meeting these other people, but after a short while I'd done that, and that benefit wasn't there any longer." So on March 13, 1991, Luplow wrote to the chairman of the CATS board, letting him know that Zenith intended to resign from CATS:

"Regrettably, as a direct consequence of the financial difficulties the consumer electronics industry has been experiencing, there is a possibility that funding priorities later in this year may necessitate our resignation."

Ever since then, Lim had been urging Zenith to rejoin. ("Even when we were not getting along, I still asked them for support," Lim said. "I always do that.") But Luplow always demurred. Lim brought it up again in December 1993, as the two transmission systems were being moved into the test center. "And all of a sudden," as Lim put it with a cunning smile, "they got very interested."

Now, several weeks later, as Luplow discussed his company's sudden new interest in Lim's program, he was cool. A few moments earlier, he had disparaged CATS—just a bunch of college students. Still, he acknowledged, "as a practical matter I am busily talking to Jae about reentering CATS—and I emphasize 'talking.' We'd have eight or nine graduate students working for us. We'd have Lim working for us. We might gain something."

He swiveled around in his chair, reached into a stack of papers on the credenza behind his desk and pulled out the program prospectus that Lim had sent him. The date stamp showed that it had been received at the Bunker on December 24, a week before the transmission system bake-off was set to begin. "We've had a tremendously difficult time preparing for this bake-off, and I haven't had time to read it," he explained. "But I will." In the meantime, Zenith would leave its offer dangling.

Could Zenith come up with the $100,000 membership fee? The company hadn't been able to afford it three years before, and its financial picture certainly hadn't brightened. But Luplow swept an arm through the air. "Yeah, we could do that. If I brought that to Jerry, I don't think he'd have any trouble with it."

Apparently, though, he had never discussed the idea with Pearlman. Zenith's president very definitely did have trouble with it. In his own office a few minutes later, Pearlman said, "CATS cost us $100,000 a year, and we didn't have it. We told Jae we can't afford to be members. And he knows better than to come back to us as long as we are losing money."

No matter. Lim was saying his vote could not be bought: "I told them no money can buy MIT."

Zenith never did rejoin CATS, and as it turned out no vote buying was necessary. When the testing began, GI stumbled. The company didn't even have a complete system to offer. Both competing transmission systems had two main components: one part for over-the-air, "terrestrial" broadcasting and another for cable television. "Woo was working on his terrestrial system," Rast explained later, "and he didn't have time to do as much work as he could on the cable system." Paik said, "I didn't know I was going to have to do a separate cable system until it was too late to do it."

In the end, Rast said, "we had to reach the conclusion that we needed outside help." So at the last minute GI made a desperate decision. The company hired an outside contractor to provide the cable half of its transmission system. And when GI showed up at the test center with for-hire hardware, Zenith's people were stunned. "They bought some kind of brick outhouse from a military contractor!" Luplow shrieked. Even Rast had to admit that "to be fair, an unbiased party might slightly question what we're doing."

"Not the most desired approach," added Paik.

What was worse, General Instrument's system performed poorly. GI looked at the early test results and decided that they had no choice but to declare a system breakdown. That wasn't really permissible under the rules. The system wasn't really broken, it just didn't work well. But GI rationalized it this way: "The fact is, when we saw the results the system wasn't performing as it should," Rast said. "So it was broken."

The Zenith engineers started screaming so loud that it sounded as if each of them had been stabbed. But the testing stopped; Fannon had no choice. Paik and the other engineers poked around among their circuit boards and found, as Rast put it, "a vendor software error." In other words, Rast was saying that the contractor GI had hired to build half of the system had made a mistake. One member of the GI team (accounts differed on who exactly this was) decided to take the problem in hand. He removed the offending part and took it back to California over the weekend for repair. Then he returned to the test center on Monday and plugged it back in.

When Zenith's engineers heard about this, the red-faced yelling grew so loud it looked like the veins in their necks might burst. What if one of the original contestants had sneaked a part out of the test center for the weekend, taken it home, and then brought it back Monday morning? All the other testing controversies would have seemed pale in comparison, and even Rast had to say, "Nobody would argue that this is strictly according to the rules of Hoyle."

When the testing was over, Zenith's system had clearly performed better, and it seemed as if General Instrument's infractions would make the voting easier for everyone. Still, GI did not give up. A fierce debate erupted within the company. DigiCipher was central to the company's cable business strategy, and losing this race could be a huge setback. Should GI pull out of the Grand Alliance, sabotage the HDTV race to save DigiCipher? This makes the most business sense, some GI executives argued. General Instrument isn't even *in* the television business. Are we really going to make any money with HDTV? "There are influences here at GI that think we ought to go to war and trash the whole thing," Rast explained. "Fight for our own interests."

"General Instrument at first wanted to protest—tie things up," Lim said. "They wanted me to vote against it, make it a 2–2 tie, delay things for six months." Others noted that General Instrument would be setting off a public relations nightmare. The company might never recover from the black eye this ploy would give them, and that last argument prevailed—at least for the time being. "I don't think we're going to renege," Rast said. "It's called being professional. Take your lickings." In any case, Lim was saying he would not cooperate. "They should know that they cannot force MIT to do what we don't want to do," he declaimed from his soapbox. "I was more interested in promoting *America's* interests."

With little debate, the Grand Alliance chose Zenith, and in the Glenview Bunker the company's leaders could not have been happier. They were players once again. As Dennis Mutzabaugh, one of the engineers, said: "I can see Zenith getting out of the red with HDTV."

With the transmission vote, the last of the decisions had been made, so Wiley's Technical Subgroup gave the Grand Alliance full authorization to build the new system. Engineers from each of the companies retired to their labs. The completed machine was due at the test center in October 1994. But if the world's most advanced television system had faced an uncertain political future in the fall of 1993, by the time spring arrived, the picture looked so bleak that Wiley was near the point of despair. "The government mandated this thing," he said in March. "I didn't dream it up. But the way it's going, I'm worried that I may deliver something that's stillborn."

The White House remained largely indifferent, and the NAB was still plotting a killer blow. At the same time, though, an even more distressing enemy came into play: Reed Hundt, the new chairman of the FCC. During his first months in office, he'd said next to nothing about HDTV. If asked, he deferred, saying he hadn't studied the issue yet. Little by little, though, his views began to come out, and to most people Hundt seemed at best apathetic about HDTV. After all, he was one of Al Gore's best friends, and Gore was "Mr. NII." So it seemed likely that Hundt would be vulnerable to the broadcasters' new lobbying strategy: "*We* are the NII!"

Just as Abel had threatened, the NAB was in fact working on an amendment to a telecommunications bill taking shape in the House of Representatives, and the broadcasters were gaining strength from Hundt's perceived point of view. Even the Association for Maximum Service Television seemed to be joining the enemy camp. Wiley had always found this smaller group easier to get along with, but in the spring of 1994 they published a white paper on digital television, and as Wiley read it he grew angrier and angrier. It was filled with references to *digital* television and *advanced* television but said almost nothing about *high-definition* television. So he called Margita White, the association's president, and scolded her: "You've gone to great lengths to avoid using the dreaded 'H' word. The only way it is used is negatively."

Oh no, you misunderstand, Wiley says she told him. "She said they would revise it," he added, "but it didn't matter. They'd already sent it to Hundt."

Wiley, Joe Flaherty, and other Advisory Committee leaders decided the time had come to pitch Hundt themselves. Maybe he just doesn't understand. We have to show him, explain to him. If we can get the chairman to say something supportive, as Sikes used to do, that would solve our problems.

They decided to stage an event, give Hundt a forum to utter the magic words that would send their enemies reeling. Another Technical Subgroup meeting was scheduled for April, and they arranged to hold it in the FCC meeting room on the eighth floor, just down the hall from the chairman's office. Flaherty pulled together some footage from recent TV productions taped in high-definition, and Wiley invited Hundt to come in and sit with the Technical Subgroup for a few minutes, watch the tape, and then make a short supportive statement. Wiley figured he could wave it at his enemies later. He was chairman of an *FCC* advisory committee, after all—doing *volunteer* work for the commission! Shouldn't he expect the chairman to give him a little help?

The invitation was proffered, and Hundt accepted. Then Flaherty called a couple of Hundt's assistants and told them the FCC needed to get into gear. "Look, this is the most successful advisory committee process I've ever been involved with," he said. "We've got a thousand people involved, three hundred of them actively. You guys have an army out here marching to August 1992 orders. There's a cliff out there, and they are going to march right over it if you don't say something." Give us some support.

OK, OK, they said. Why don't you send over some talking points that the chairman can use at your meeting. He's not fully briefed on this issue yet. We'd appreciate your help.

Sure, Flaherty said. In fact, this was perfect. He was actually going to write the very remarks that he and Wiley would be using later to pull themselves out of their steep-sided pit. Flaherty drafted a "talking points" memo and shared it with Wiley. Its central message: That second channel the FCC is planning to loan to the broadcasters is for *high-definition television* and nothing else. Let's stay the course. The rules are set. Let's not change them now. Wiley suggested a few changes, and then they faxed the notes over to the FCC, printed in speech format so that Hundt could simply read them.

"HDTV, of course, has been and remains your goal and that of the FCC," the script said near the top. "Interoperability is particularly important, too, because HDTV receivers may represent the initial introduction of a digital bitstream into the American home.

"Indeed, the results of your work soon can benefit American consumers and broadcasters, and also create high-wage American jobs. As the International Brotherhood of Electrical Workers has said, 'The timing and extent of HDTV job creation relies in great measure on the speed with which the Advisory Committee recommendations are implemented.'

"To advance all these benefits, I ask that you continue on your present course with all deliberate speed."

Hundt's staff read the script and told Flaherty everything looked fine.

The meeting opened at 8:30 A.M., and Hundt was due for a brief visit at 9:00. The Technical Subgroup was arranged around a horseshoe-shaped table, and a high-definition television projector was set up in the middle, ready for a fifteen-minute demo. Wiley and Flaherty sat at the head of the table, the empty chair between them intended for Hundt. His office was just down the hall, and as the meeting stumbled awkwardly through its first half hour, it was painfully clear that Dick Wiley was just as anxious about *this* visit from the FCC chairman as John Abel had been when he staged the first HDTV demonstration in this very room seven years before. Nine o'clock came and went with no sign of Hundt, but nobody really expected him to be on time. Flaherty droned on about Grand Alliance technical specifications, but everyone present knew that the Advisory Committee's leaders were just treading water.

At 9:15 Wiley heard some commotion in the hall. His back was to the door, but he popped right out of his chair and stepped outside, where Hundt stood huddled with some of his aides. The two men shook hands and exchanged greetings for a moment. Both men, the former and present chairman, were tall—about the same height. But Wiley's hair was graying while Hundt's was all black. They chatted briefly, and then Hundt turned back to his aides. Wiley returned to his chair. Flaherty had stopped the meeting, but when he saw that Hundt wasn't going to sit down immediately, he continued the plodding technical discussion.

Finally Hundt bustled in. "Perfect timing!" Flaherty declared as the chairman sat down beside him, twenty minutes late.

"I'm late already, and already behind," Hundt muttered, his words rushed. "I apologize." He laid his talking points down in front of him, and Flaherty leaned over to look at them out of the corner of his eye. Perfect, he thought. Just what I sent over.

Then Hundt launched into his remarks, turning first to look at Wiley.

"I want to thank Dick publicly for the wonderful work he has done, for the public service he has offered—but not just him, lots of people. This process has brought together some of the best minds in America, some of them in this room. It's a fantastic industry effort. I also recognize that there were many people along the way who said you weren't moving fast enough, or you were doing the wrong thing. But I can say now that they were flat wrong."

Flaherty nodded. So far, so good.

"I don't know whether advanced television is going to be the new mass home-entertainment center or the digitized telecomputer of the 1990s," the chairman said. "My own view is it will be both. But not many people realize that advanced television—or high-definition television, whichever you call it—how critical it will be as the portal, the gateway to the National Information Infrastructure."

Uh-oh. Hundt was off the ranch. The chairman was skipping right over those warm, supportive sentences he was supposed to say. Instead, his remarks sounded as if John Abel had written them.

Hundt continued in that vein for a few minutes, ending with a promise that he would *positively* conclude the advanced-television proceedings during his term in office. When he wrapped up, he'd said not one thing the Advisory Committee's leaders could use to show that the FCC chairman was on their side. Wiley and Flaherty were disconsolate. But they went on. Time for the demo. Maybe *seeing* high-definition television might sway him.

"We have here some recent productions taped in high-definition, and we thought you might like to see them," Flaherty said. Hundt nodded with a smile. Somebody lowered the lights, and the tape rolled. The opening sequence was some NHK footage taped at the 1994 Winter Olympics, and as Hundt watched, Wiley and Flaherty worked him from both sides. "Look how clearly you can see the faces in the crowd," Flaherty whispered, leaning over to him.

"Uh-huh" was all Hundt said as the tape moved on to a scene of skater Nancy Kerrigan skating across an arena a few days after the revelations about her problems with Tonya Harding. "I've been real busy; who is that?" he joked. Wiley and Flaherty both laughed. The next sequence was a new television production, taped in high definition, about the last days of World War II. This scene showed Churchill and Stalin facing each other over Stalin's desk in the Kremlin. They were dividing up Europe. "The background's so clear," Wiley whispered.

"Yeah, it's amazing," Hundt said with an obligatory tone. A few minutes later, the tape ended, and Hundt was already rising from his chair. "That's great" was all he said. After a few handshakes and backslaps, he was gone. The next day, Flaherty called Wiley and rated the success of their effort "a three out of ten," Wiley said. "Hundt had the script. Flaherty looked over and saw it. But he didn't read it. It was disappointing."

Hundt was disappointed, too. The demonstration "didn't tell me anything about the telecomputer side of it," he said, sitting for a discussion in his office. Apparently the crystal clarity of the pictures had made no impact on him, and for the first time he stated his real view. "I think HDTVs will be sold for all sorts of uses, not just better pictures. But no one in this whole process has demonstrated what it can do. I've never seen a demonstration of medical uses. Have you? The Grand Alliance people, they keep saying, 'We'll put it out there and let the market decide.' Well, that's just not good enough. It's not enough to say this is just a pretty picture."

There it was. Hundt didn't seem to care about *high-definition* television at all. "Suppose a market study shows there's no interest in this—none! And we've given the broadcasters a second channel. Do we take it back? From a political and vision point of view, they need to be able to articulate the other uses for this: health care, library access, interlineation of documents. These things will fuel the intensity of the debate. Pretty pictures will not decide it. What material is out there? They haven't shown us. I started telling Dick this last fall."

Hundt also showed great sympathy for the broadcasters' arguments, no matter the history. "Preserving the long-term viability of the broadcasting industry, that's a very important goal, in my view." For the Grand Alliance he had this advice: "At the moment the public can't relate to them. They have to mobilize the public to a certain point of view."

As it happened, the partners were already talking about putting on high-definition demonstrations at AT&T's Washington headquarters, a couple of blocks away from the FCC. But Hundt scoffed at that. "I don't know who will go," he told his interviewer, "except maybe a few people in the process, and people like you."

The "people in the process" were stunned to learn of Hundt's remarks. "I think it's just great that he's able to do a better job for us than the market would," Rast said, his words heavy with scorn. "Maybe if we can convince him, maybe the market will get a chance to see HDTV. God forbid that we have to outlive this guy." And Flaherty said, "I don't know if he is overwhelmed by this job or just incompetent in this field—or if he's playing political games. My guess is he's playing political games."

Whatever the truth, without support from the FCC Dick Wiley was alone in the wilderness. What if the Grand Alliance finished its system, the Advisory Committee forwarded its recommendation to the FCC, and Hundt threw out all the rules that Al Sikes had ordered before him? What if Hundt gave in to the broadcasters, abandoned the notion of high-definition television and offered them a second channel with permission to use it for added digital services? No HDTV. The public might never get a chance to see it. The government's compact with industry would be broken, and seven or eight years of work would have gone for naught. The Grand Alliance system, if it was ever completed, would be packed up and sent to a warehouse somewhere.

What was worse, members of Wiley's own Advisory Committee were beginning to turn on him—especially George Vradenburg, the CBS executive who had been instrumental in convincing GI to enter the HDTV race back in 1990. Later he had become a senior vice president at the Fox television network, and at first in the new job he had remained among the broadcast industry's few boosters of high-definition television. "Fox will get in right away," he averred in 1993. "We'll have a full season of high-definition programming right away."

Since then, however, Rupert Murdoch, Fox's leader, had come to be an open critic of HDTV. In March of 1994, he told *Forbes* magazine, "The current proposal is that the FCC will give us that spectrum for high-definition television. But high definition is a luxury. Compared with a modern TV set it's not that different. Why shouldn't that extra spectrum be given to me or you or anyone to put on that extra number of channels?"

Now Vradenburg had grown a new opinion, and one afternoon Wiley related the story of his latest encounter with him, which occurred at a Technical Subgroup meeting. "Vradenburg buttonholes me and says, 'We've got to change all the references [in one Advisory Committee report] from HDTV to advanced television, or digital television.' I had to talk to him out in the hall, and I said, 'George, you know the program. It's going to be HDTV. I can't have you serving in these leadership roles if you're not with the program.' "

Wiley said Vradenburg shot back, "You're not going to fire me because I disagree with you, are you?"

"No, George, I'm not talking about firing you," Wiley replied. "But you know what the program is. I'm just going to tell you: I'm not going that route. I may be the last body alive, but I am going to stay the course."

Wiley clearly saw the danger behind that encounter. His Advisory Committee had always been a rubber stamp for him. But at its birth seven years before, Dennis Patrick, the FCC chairman then, had stacked it with broadcasters. Patrick's purpose was to calm them down, get them out of his office, stop all the lobbying and complaining about the Land Mobile rule. Now, however, these broadcasters were forming up behind a new view. They were turning on Wiley. "You know," the chairman said, "there's some uncertainty about whether I can get the votes in the committee to pass this recommendation."

The members of the Grand Alliance, meanwhile, weren't paying much attention now that their own wars over the audio and transmission systems were concluded. They were feeling pretty good. But in early spring Wiley called the Grand Alliance's leaders to a meeting in his office. When the partners assembled in chairs around Wiley's desk, Paul Misener was watching closely. "It was clear," he said a few days later, "that they were looking for a pat on the back." They didn't get it.

Wiley was convinced that the Grand Alliance had to start lobbying. He told them, You've been concentrating all your energies on getting the system built, and that's good. But you haven't been concentrating on the other big issues. That's education (otherwise known as lobbying). You've got to go to Capitol Hill, the White House, the FCC—tell them what you have, show them, or you may lose the whole thing, no matter how good your system turns out to be. The partners slunk out of Wiley's office "feeling really terrible," according to Rast. But for the next several weeks they strategized. Let's hold demonstrations, they said. Hire a lobbyist, court cable television as a counterbalance to the broadcasters, lobby the *other* FCC commissioners. But it was hard. This was unaccustomed terrain. They were floundering. "We have no presence in Washington at all," Pearlman complained. "We have no lobbying firm, nothing."

"Maybe it's time for a demo," Wiley offered. "I see some of the steam coming out of this." But the Grand Alliance had nothing to demonstrate. They were just starting work on their new machine.

The Grand Alliance had one indisputable ally: the world's television manufacturers. Wide-screen, high-definition television represented an exciting new product for all of them—the first since color TV came on the market almost forty years before. Over the previous decade or two, almost every other segment of the consumer electronics industry had moved through several generational changes that allowed manufacturers to convince people to abandon their existing gear and buy something new. Cassettes replaced reel-to-reel tape, CDs replaced LP records, and new products flourished: VCRs, personal computers, cellular phones, fax machines. Through all this, television had barely changed at all. There was little cachet in buying a new TV before the old one wore out. But HDTV promised to put an end to that in a big way.

Perfect television pictures, and a wide, wide screen—just like at the movies. Set makers saw a marketing bonanza, and they were already planning the public relations strategy to bring the nation along. The potential for unspecified digital services was a nice additional benefit, but they didn't expect this to be the feature that actually sold the sets. "Services of high complexity don't appeal to most people," Joe Donahue said during a presentation to an industry group. "They're intimidated." At the same conference another speaker repeated a current bit of industry lore: When AT&T ran a trial of its new interactive TV service, at the end "40 percent of the people never took the plastic wrap off the instruction book. *That* is our typical consumer."

Donahue added, "Surveys and test projects show that most people—more than half—want to use the device to watch TV, watch movies. In simple terms, we must capitalize on the huge interest in TV." To the broadcasters' assertion that nobody wanted HDTV, Donahue offered a chicken-and-egg truism: "There's never been a demand for a new product before it was introduced." And indeed, the public had not been calling for the ability to pluck music out of the air when radio was introduced in 1920. Nor was there a clamor for television before RCA introduced the first set in the 1940s. The same could be said for VCRs, compact discs, cellular phones, personal computers, fax machines, and a host of other successful consumer products. (Of course, consumer electronics manufacturers who cited this litany generally didn't choose to mention some of the less successful products for which there had been no advance demand, like picturephones, RCA's VideoDisc players, and personal digital assistants.)

The Grand Alliance system was due to be finished toward the end of 1994, which meant that the FCC should set a standard and formulate regulations sometime in 1995. If that held true, then the first HDTVs would arrive in the stores late in 1996. This was the year, the manufacturers decided, that they should stage some sort of huge public relations campaign to persuade the American public that everyone *had* to own an HDTV. A big event. Something everybody would be talking about. A show everybody would watch. Like the Olympics.

That's it, they decided. The 1996 Summer Olympic Games in Atlanta; *that's* where we'll debut HDTV.

The Electronic Industries Association, representing the world's consumer electronics manufacturers (many of them Japanese), promptly appointed an Olympics Steering Committee. By the end of 1993 it had a proposal: "The Steering Committee envisions Advanced Television display sites in eight major U.S. cities and selected cities in Canada and Mexico. The eight are: Washington D.C., Los Angeles, New York, Chicago, Atlanta, Denver, Miami, and San Francisco. There would be a minimum of one display site in each city at which ATV receivers from a number of manufacturers would be shown. We envision ATV signals being provided to the display sites primarily by local broadcast and cable systems."

NBC held exclusive broadcast rights to the '96 Olympics, and in early discussions the network seemed amenable to the idea. What could be more natural? RCA and NBC had practically invented broadcasting, and the companies had been the masters of big-event product promotion since the industry's earliest days. David Sarnoff had sold millions of Americans on radio by broadcasting the Dempsey-Carpentier fight in 1921. He'd promoted television with a demonstration at the 1939 World's Fair. The Sarnoff labs had shown off ACTV on the fiftieth anniversary of that event. Now, a high-definition Olympics broadcast on NBC in 1996? History seemed to smile on the idea.

Greg DePriest, the head of Toshiba's "listening post" in Princeton, was the chairman of the EIA's Olympics Steering Committee, and in October 1993 he wrote to an NBC division president formally asking for "cooperation to provide ATV coverage of the 1996 Summer Games." The network said it was considering the request, and enthusiastic planning continued. By the early spring of 1994, the proposal was rolling along. The Grand Alliance had seemed ambivalent at first; the partners hadn't built their new machine yet, and they wondered whether the summer of 1996 might be too early. "I personally don't see the Grand Alliance system broadcasting to 10 to 20 percent of American homes by then," Luplow said early in 1994. By spring, though, the partners realized that wasn't the point. Whether or not anyone actually owned an HDTV, stories about an Olympics demonstration would fill the newspapers. People would flock to the demonstration sites. Enthusiasm would build and build—at least, that was the dream. "Our best opportunity is the Olympics," Rast declared in May. "Lots of people want it. Manufacturers want it. PBS is going to announce that they will inaugurate high-definition service with the Olympics. This is it!"

Rast saw another benefit, too. The partners had returned to their labs to build the machine, but motivation was becoming a problem. Why should the companies push to build this expensive, complicated device if they couldn't be sure anybody would want it when they were finished? "There's a malaise inside the Grand Alliance, born of what's going on outside," Rast acknowledged. Some of the key players had been promoted or had moved on. At GI, Jerry Heller was in semiretirement and Woo Paik had been promoted into Heller's old management job. At

Zenith, Wayne Luplow was promoted to a more senior position, though he still held responsibility for HDTV. Arun Netravali of Bell Labs was promoted, too. Joe Donahue announced he would retire. And almost everyone who remained took on additional responsibilities. In all of the companies, HDTV just wasn't the priority it had been when they were locked in a race to the death.

Bob Rast remained virtually the only executive who devoted nearly all his time to the Grand Alliance—an odd development considering that his company had once pondered sabotaging the process. But he had been hired as General Instrument's point man for HDTV, so he stuck with it, gaining extraordinary influence over the process. Slowly, unofficially, he came to be the Grand Alliance's principal spokesman within government and the industry. (Public relations officers from each of the companies managed inquiries from the news media.)

As the spring advanced, he and the other partners quietly admitted that several of the labs were falling behind in their work. "So what I'm saying," Rast explained, "is that we need to try to create this compelling need to get high definition done in time for the Olympics launch." Jim Carnes had always argued that engineering projects needed visible goals—demonstration dates, anniversaries. So the Grand Alliance agreed: This is our strategy. Olympics, here we come.

The broadcasters had little to say about the Olympics strategy. Some of them probably thought it was irrelevant, since they were planning to kill high-definition long before the Olympics anyway. As the House of Representatives fell deep into discussion of a major telecommunications bill, the NAB and allied groups quietly proffered something they called the "broadcast spectrum flexibility amendment." It would make the following idea a part of law: If the FCC gives broadcasters a second channel for advanced television services, as proposed, "the Commission shall adopt regulations that allow licensees or permittees to offer such ancillary or supplementary services on designated frequencies as may be consistent with the public interest, convenience and necessity." The amendment would also "limit the broadcasting of ancillary or supplementary services on designated frequencies so as to avoid derogation of any advanced television services, including high-definition television broadcasts, that the Commission may require."

In other words, the broadcasters were trying to turn the process on its head. Al Sikes had offered to loan them a second channel for one purpose only: the transition to high-definition television. Now the broadcasters were saying, Give us the second channel and let us use it for moneymaking digital services, unless they conflict with required high-definition broadcasts. And everyone knew full well that Reed Hundt was not enthusiastic about HDTV. He was unlikely to require any high-definition broadcasts at all. If the amendment was enacted, high-definition television might very well die.

At the same time, just after the first of the year, the NAB held a summit meeting at La Costa Resort, in Carlsbad, California, to discuss strategies. There John Abel went public with the organization's newly devised view. "There are plenty of ways for broadcasters" to make money offering advanced services on their second channels, he told the gathering. And high-definition television wasn't one of them. "You don't want HDTV," he said. "And consumers say they don't want it, either." His comments found their way into *Communications Daily,* the widely read industry newsletter, and set off a storm. It made Wiley so angry that he wrote Abel a terse letter—something he rarely did, because, he admitted, "I don't like to be at cross-purposes with the NAB. It's not good for my business." Wiley's huge law firm represented numerous broadcasters. What's more, one of the firm's attorneys had recently begun helping the NAB with its legislative strategy. Even with that, Wiley thought Abel had gone too far.

PERSONAL, his letter said at the top:

I had an opportunity to read your recent California speech. I also have seen press accounts of what were perhaps ad-libs by you to the effect that neither broadcasters nor consumers want HDTV.

I believe that it is not wise for broadcasters, who seek an additional channel, to discard the concept of HDTV altogether, as you seem to counsel. I say this in part based on political strategy but also, in part, on my view that HDTV could well prove to be a popular video service. Speaking personally, I must also express some degree of disappointment that you would write off HDTV without even giving me the courtesy of some advance discussion on this point. As you know, I have invested a lot of personal time and effort over the last seven years, working on a project that I suspect you personally initiated. It seems to me that before you dismiss all of us and all of the efforts that have been made, you might wish to at least consider other views.

Abel reacted with unbridled scorn. "Wiley's saying, 'You guys changed your policy, and you didn't consult me,' " he said, affecting a whine. "He called Eddie Fritts, all mad. Well I didn't know I *had* to check with Dick Wiley. It never *occurred* to me to check with Dick Wiley. And remember, it was an attorney in his own firm who *wrote* the amendment for us that is now being debated on the hill—the very one he hates."

When Wiley called Fritts to complain, the NAB president offered a few soothing lobbyist's promises: Don't worry; we're with you. But a short time later, Fritts explained his organization's position this way: "Those who say we are trying to kill HDTV are entirely wrong," he said, with barrel-chested authority. Then Fritts offered a more precise view: "On the other hand, if we have to reinvent the entire system, and it turns out that nobody wants HDTV—well, it's a damn serious matter. We have no knowledge of any market tests that show demand for HDTV at this point. If demand for HDTV blossoms, you'll have broadcasters racing to it quickly. Where will the public see it first? They'll see it first on satellite and cable." In the meantime, "the government should not *require* HDTV. It should *allow* HDTV. That's our position. The process is evolving, and we are comfortable with the way it is evolving at this point."

And no wonder. The broadcasters' "flexibility" amendment was passing through Ed Markey's telecommunications subcommittee with relative ease, even though Chairman Markey didn't like it one bit. "The risk inherent in the amendment before us," he said in a March press release, "is that in the rush to provide a vast array of new services, broadcasters will forget the reason for assigning them this spectrum in the first place." HDTV is a success story for America, Markey added. "Let's not snatch defeat from the jaws of victory." Mike Synar, a Democratic representative from Oklahoma, was even more direct. "We're being played for Uncle Sucker," he complained. But the odd thing was, the amendment still seemed destined to pass.

Wiley and the Grand Alliance partners were panicky. As chairman of a government advisory committee, Wiley believed he couldn't lobby. But he did urge everyone he could think of to call on members of the House. None of it seemed to be having much effect, even though most of the representatives preparing to vote for the broadcasters' amendment were the very same people who'd been happy, drunken partygoers at the high-definition television bash just a few years earlier.

"Four years is a lifetime in politics," Markey explained with an apologetic air. Most congressmen hadn't thought much about HDTV since George Bush closed down Congress's HDTV jamboree in 1989, and now "this subject remains below the radar for most members. That allows special interests with the most energy and passion to make their cases, even if they are out of sync with the record. The broadcasters came in very late. We were working on this very big telecom bill; really it was twenty bills in one, and they came in at the eleventh hour with this little amendment. It was just four lines: Give us complete flexibility to use this second channel." So hardly anybody noticed—a classic Washington-lobbyist strategy.

"I said, 'Wait a minute, slow down, let's look at this,' " Markey went on. "But all of this was hard for most members to digest. I thought it was inappropriate. They were trying to take advantage of us. In the end, we improved the language. But I couldn't stop it."

The amendment was included in the telecommunications bill, and the House passed it. With that, the NAB proceeded to the other side of the Capitol, where the Senate was debating its own telecommunications bill, and Wiley once again pushed his friends to lobby. His office even circulated "talking points": "A flagrant spectrum grab is quietly under way on Capitol Hill," it began, and went on to articulate forceful arguments against the amendment. Still, it seemed destined to pass in the Senate, too.

At the FCC, Reed Hundt was watching all this with interest. He was still asking questions, though his basic view already seemed set. He didn't care about "pretty pictures," his aides said. In June he attended a lunch in New York celebrating the Museum of Television and Radio. Several TV network leaders were there, and so was Al Sikes, who now worked on electronic-media issues for the Hearst Corporation in New York. Sikes had been watching with alarm as all his plans for HDTV seemed to be falling apart. "I worry that HDTV is going to be delayed on an extended basis now near the end," he said. "I could act as something of a heat shield for the Advisory Committee when I was at the FCC. But now I'm beginning to worry. I can't intercede now." At the lunch, Sikes asked Hundt what he intended to do about high-definition television. "I'm still looking at that" was all the new chairman would say.

"He gave some very general response," Sikes reported afterward, scornful of his successor. "In fact, he said that in answer to every question offered that day." But, Sikes said, Hundt did tell him, "I haven't sensed anyone pushing for HDTV now, aside from the companies that are doing the work."

Hundt asked Larry Tisch, the CBS head, Do you think you can make any money with HDTV? Tisch told him, No, not in the early days. Hundt's line of questioning ended there, even though Tisch's view was actually more complex. Many of Tisch's broadcaster brethren were talking about using digital compression to broadcast three or four conventional television programs on a single channel, instead of high-definition programming. Tisch hated that idea; around CBS's executive offices he often said, "HDTV will cost me some money. But multi-channel will *kill* me." Where would CBS and the other networks get the programming and advertising to support all these new channels? They were having enough trouble keeping just one channel afloat. But that day Hundt didn't press his questioning far enough to understand this.

The chairman did ask his predecessor what he thought, and Sikes gave him an earful. "I feel that the process should be completed on a timely basis," he told Hundt. "It's important that we move to an advanced television standard, whether it's a digital standard or a high-definition standard or whatever it is. I think it's time we moved. It is important in no small part because the government set up a process and announced that there would be a reward at the end of the process, and companies have spent tens of millions of dollars to bring us a generation ahead of the rest of the world. If we somehow back out of the process now because we can't reconcile the differences between the computer and the broadcast industries, or because the broadcast industry wants to slow-roll it, doesn't want to make the investment—well, I think that would really be too bad."

Hundt nodded but didn't say much. Sikes left no less worried that his most important government initiative might have been for naught.

The news over the following weeks did nothing to cheer Sikes or the others. Wiley got word that NAB lobbyists were pressing the FCC on the same issues they were pushing before Congress, and he noted in an internal memo: "It's all-out war." Apparently the NAB felt the same way. Finally Wiley decided he had no choice but to ask the attorney in his firm who was working for the broadcasters to drop the NAB as a client. She did, and Wiley felt wounded.

The broadcasters also reached out to give the Grand Alliance partners a good, hard slap. The partners were proceeding enthusiastically with plans for their Olympics demonstration. Then in June letters from an NBC executive began arriving in the mailboxes of everyone involved.

"As you well know," said one of the letters, this one to Greg DePriest, the head of the EIA's Olympics Steering Committee, "National Broadcasting Company Inc. is the sole and exclusive licensee of all television rights to the 1996 Summer Olympic Games in Atlanta for the territory of the United States, its territories and possessions. High-definition television rights are part of the exclusive rights.

"You, through your participation on the Electronic Industries Association's Olympic subcommittee, have individually and collectively engaged in a misleading media campaign designed to encourage a belief in third parties that you have acquired rights or will acquire those high-definition television rights that have already been vested irrevocably and exclusively with NBC. Please take notice that any further attempts by you to secure through media or other intervention those rights NBC holds exclusively will be met by swift and appropriate action."

Similar "cease and desist" letters blanketed the community. Asked to explain, NBC officers said the network wouldn't have the staff or equipment to provide a proper high-definition demonstration. But DePriest and others suspected there was more to it. The broadcaster community was engaged in a broad public attempt to evade high-definition television. Why should one of the networks actually help *publicize* HDTV? "I think the effort is pretty much dead," DePriest said, "absent some intervention from some enormously powerful entity."

Rast was glum. "We're trying to create an alternative," he said, "and there is no alternative. The first four best things to do are all Olympics, so we are now trying to find out what number five is. I'm real concerned about this."

At about the same time, the Grand Alliance partners announced that they weren't going to meet their construction deadline. They said they'd be at least two months late. "I think part of the problem is that in all of our companies, it is not a high enough priority now to get this done," Rast explained. Why dedicate important resources to a project that might be stillborn? The Senate, meanwhile, accepted the broadcasters' "flexibility" amendment and added it to their own telecommunications bill, which seemed all but certain to pass. If Rast thought he had reached his "Negev low point" months earlier, now he and the others had to reach up to touch bottom.

Wiley tried to remain strong. "I'm going to stay the course," he said, in a tone at once stoic and defeated. "I may be the last man alive, but I'm just going to go on."

19 The Manhattan Project

While the American high-definition television program lurched toward its likely demise, Japan caught the world's attention in late February 1994 with a startling pronouncement. Akimasa Egawa, the director general of the broadcasting bureau in the Ministry of Posts and Telecommunications, stood up to announce that Japan was about to abandon Muse, the analog high-definition juggernaut of the 1980s that had been struck down by the invention of digital TV in the United States.

"The world trend is digital," Mr. Egawa acknowledged at a news conference he had called. Of course, everyone already knew that Japan's high-definition system was outmoded and largely irrelevant, even though NHK was broadcasting a full schedule of Muse programming to the few thousand people who owned the ridiculously expensive sets. But who ever expected the government of Japan to come forward and admit that one of the nation's proudest technological achievements had been a failure?

As soon as Egawa spoke, Japan's TV manufacturers howled in protest at this clear betrayal. They were still making Muse sets. How on earth could they ever sell them now? And Muse proponents pointed out that Egawa's remarks were as much the result of internal political struggles as of any genuine recognition of the system's technological irrelevance. The next day, Egawa was forced to retract his statement, in part. Japan wasn't exactly abandoning Muse, he said. But the damage was done, and the message was clear: Europe had already dropped out of the arena. Now Japan. Could there be any better demonstration that the United States had triumphed? Nonetheless, the news gave the American players little cheer.

A few weeks later, Joe Donahue announced that he was retiring after more than four decades in the industry, with RCA, GE, and now Thomson. The entire HDTV crowd was invited to a reception at the Grand Hotel in Washington, where the competitors had held their final Grand Alliance negotiating session almost a year before. The mood was bittersweet. Everyone liked Donahue, and many in the room still gratefully recalled his last-minute proposal that broke the audio system deadlock. The toasts and speeches were warm and appreciative. But Donahue was sad.

"I had a dream last night," he said, drawing out his words for emphasis when his turn at the dais came. "My dream is that we finished our system and decided we didn't need all this testing," adding as an aside, "Sorry, Peter." Fannon smiled.

"My dream was that we got our system to Dick," Donahue continued, with a nod toward Wiley. "Dick approved it, recommended the standard, and passed it on to Reed Hundt. Hundt published the standard, and we got our product to market."

Simple as that, and everybody laughed—for that scenario seemed as about as likely as a congratulatory telegram from the National Association of Broadcasters. In fact, Quincy Rodgers, General Instrument's lobbyist, was scurrying around the back of the room as Donahue spoke, urgently passing out copies of the NAB's "HDTV-killer" amendment to the Senate telecommunications bill, now in the final stages of debate. Like the House version already passed, this one asked the Senate to let broadcasters use their second channel however they pleased.

"This is bad," Rodgers was saying. "This is really bad." And sure enough, a few weeks later the Senate passed the bill with the broadcasters' amendment firmly in place. The Senate and House bills were similar but not identical, so they were sent to a conference committee, where senators and congressmen would work out an agreement. Whatever else happened there, however, the HDTV-killer amendment seemed certain to glide through, because the Senate and House versions were nearly identical.

Abel and his ilk seemed on the verge of winning, and they were getting cocky. As Eddie Fritts had put it, "The process is evolving, and we are comfortable with the way it is evolving at this point." But just because they were about to slay the HDTV dragon once and for all, there was no reason they couldn't use high-definition television as a lobbying tool one more time, just as they had in 1987. In the summer, the Department of Energy proposed a new rule that would apply certain electrical appliance energy-conservation rules to the television industry. No, no, we *can't* accept that, the NAB squealed. If you put this requirement on us, it'll delay the introduction of important new technologies, *like HDTV*. The department, obviously unaware of the hypocritical nature of this argument, dropped the proposal. At about the same time, the ABC television network was trying to reduce its tax burden by dropping broad hints that the network might move its headquarters out of New York City. ABC argued that it was about to face substantial new costs—maybe as much as $150 million dollars—to finance the start-up of HDTV services. ABC had good reason to believe the tactic would work. Using precisely the same argument, CBS had already worked out a similar deal. ABC got the tax cut, too.

The TV station in the far suburbs of Charlotte, North Carolina, had once been the offices of televangelists Jim and Tammy Bakker, though the only remaining indication of their tenancy was a label pasted to a shelf that said "Bakker/PTL tapes," next to an arrow pointing toward the floor. More recently the offices had housed the local NBC affiliate, but when that station grew more prosperous it moved into town. Now the flooring was pulled up in places, exposing bundles of cables half torn from their housings, as if the building's entrails were being ripped out. The building had been abandoned, which was why the Advisory Committee's Field Test Task Force had been able to win use of it for free. Amid all the clutter and decay, Advisory Committee engineers had installed state-of-the-art digital transmission equipment, and out back, atop a 1,330-foot TV tower, they'd also mounted a new antenna.

Early on, Wiley and the others had worried that all the measurements taken at the Advanced Television Test Center would need to be validated by real-world observations in the field. What if Peter Fannon's computer simulations weren't borne out when people started receiving digital, high-definition broadcasts in their homes? So they drafted a field test plan. After laboratory testing, the prototype equipment would be shipped here to Charlotte, a city with typically varied terrain. A digital, high-definition TV signal would be transmitted from this tower to a receiver mounted inside a panel truck. Under a carefully devised test regimen, the truck would drive to dozens of spots around town, each one a little more distant and difficult than the last, to measure signal reception.

All this had been planned years before, when the Advisory Committee's leaders had assumed they would choose a winner in 1993. Promising a quick, nine-month turnaround, in 1992 they had borrowed hundreds of thousands of dollars' worth of transmission and test equipment from two dozen manufacturers, who liked the idea that their goods would be used for this important test. Now, two years later, some of the companies were growing restive, and Ed Williams, the field test director, was fighting a daily battle to keep the facility intact while the Grand Alliance procrastinated in the decaying political environment. Almost every day, one company or another called to say they wanted some piece of equipment back. Today Williams was asking his men to pack up a repeater (a device that amplifies a weak signal, received far from the transmitter, and retransmits it to locations even more distant).

"The owners want it back, and they're coming today," Williams explained.

As his men loaded the repeater onto a pallet, the field-test center was setting up for the day's tests of the Grand Alliance transmission system. While the other partners dragged themselves through their work, Zenith had managed to build the transmission system rather quickly. The men in the Glenview Bunker were more motivated than the others. The company's financial situation had not notably improved, and they still saw this system as the key to Zenith's very survival. Once it was finished, Wiley and his lieutenants had decided they should test the transmission system alone, even though the rest of the new machine wasn't yet built. So a squadron of Zenith engineers had carried their new baby down here to Charlotte, and by now the tests were more than half done.

The field-test panel truck stopped by the side of a two-lane semirural road. When the passengers threw open the doors, they saw nothing outside but mud and trees. A locator box inside the truck showed that they were only 9.98 miles from base—a close-in spot. And as they raised the antenna boom to take another reading, the engineers were thinking, This one will be a piece of cake.

The field test plan called for readings from 202 sites around town, and this was site number 154. At each previous location, they had radioed back to base when everything was ready and then watched a small TV monitor on the front wall as the base broadcast a conventional

NTSC television signal. They'd studied the broadcast and noted the measurements of signal strength and accuracy. After that, their colleagues at the base had used Zenith's transmission system to broadcast a digital signal—just data, not an actual picture, since the rest of the Grand Alliance system wasn't built yet. Then the men in the truck had measured that signal's strength and accuracy, too. And in every one of the 153 locations so far, the Zenith system had performed beautifully—as well as or better than the conventional broadcast. The men from Glenview were mightily pleased.

Lief Otto, a Zenith engineer, was manning his company's equipment this day. He was a tall, rugged, good-looking young man, whose naturally cheerful outlook had been beaten back by several years of battle in the high-definition television wars. Now he was wary, fully vested in Zenith's bunker mentality. It was no wonder. A few months before, he'd been a member of the Zenith team at Fannon's test center in Alexandria during the transmission system bake-off. General Instrument's behavior there had stunned him. "I learned a lot in Virginia," he said with a sad shake of the head. "This isn't about technical merits. It's about politics."

The equipment was ready for this morning's first test, and Otto sat poised at his console. A couple of minutes before, he'd found a box turtle in the mud just outside the door. He had brought it inside and placed it atop his equipment. "A mascot," he said with a smile. Now the turtle sat there quietly, locked tight inside its shell.

Behind Otto was an Advisory Committee engineer, relaxed as he set the scopes and monitors. Standing next to him was Douglas Miller, an FCC officer, up from the commission's Atlanta offices. "My job," he explained as he stood to one side with his arms crossed, "is to keep these guys straight, make sure they write down what they say they are going to write down." But after 153 field test stops, Miller was ready to go home. He had no patience any longer for "the box lunches, peeing in the woods, and coming back with ticks and chiggers crawling on me."

The person who seemed to interest Otto most was Carl Scarpa, a short, unassuming fellow who stood quietly before a computer console with his back to the others, just to Otto's right. Otto kept glancing over at Scarpa—eyeing him warily. "I don't know why he's here," he'd complained on the drive out.

Scarpa worked for Hitachi's lab in New Jersey, one of those Japanese "listening posts" just up the road from the Sarnoff Shrine. And a few weeks before, his lab had proposed to let the Advisory Committee use a new piece of equipment they'd designed. The device was able to capture and record digital video broadcasts so they could be replayed and examined. Nobody else had anything like it. Hitachi had offered to record tapes of these field test broadcasts and give copies to the Advisory Committee—if Hitachi could keep copies for themselves. After considerable debate, the Advisory Committee had agreed. Now the Japanese manufacturer had moved its listening post right here into the field test truck, and Otto didn't like it, for more than one reason. Hitachi had designed the prototype device, but an American company had provided some of the hardware—the same company, Otto noted, that General Instrument had hired to build part of its own transmission system a few months earlier. Otto saw a possible conspiracy. "On the surface, they are gathering data for all of us," he said. "But are they working with GI, too?"

The first broadcast here at site 154 came in crystal clear. The conventional NTSC signal was given a rating of 4.5—almost perfect on the 1–5 rating scale. The parameters were noted on a ratings sheet and then Otto got ready for the digital broadcast. If the conventional signal was a 4.5, then the digital signal ought to be a 5. A small computer at Otto's left elbow displayed a horizontal line that crawled left-to-right across the screen as the signal came in, rising and falling with the signal strength. As the transmission began, the line entered the plot just where it should. Then, suddenly, it fell to the floor. "We're getting some pretty severe fading," Otto calmly noted. No problem, he and the others thought. Just a momentary glitch. Happens all the time. It'll clear up in a minute.

They ran three 20-second tests, and in two of them the signal looked OK. Then, on the third, a signal analyzer started chirping. "Wow," the FCC officer breathed, and Otto's eyebrows raised as he said, "This is the first time I've seen fades that bad." He was shifting in his seat now; this was a *close-in* site, and Otto wondered out loud, "Maybe a weather system's moving through?" The others shook their heads: The NTSC signal had come in just fine. Something's wrong with Zenith's transmission system. Otto suggested reorienting the antenna. That did no good. He ran a computer plot of the terrain, looking for unusual obstructions. There were none. His consternation grew. "This is the first time I've seen this," he said, "the first time ATV has performed worse than NTSC."

The turtle on top of his console poked its head out of the shell just then. The console was getting hot, so it tried to crawl away, only to find a sharp drop from all sides of its perch. Otto plucked it off the equipment and put it down on the floor. His mascot crawled into a corner head first to hide.

"The signal has dropped down to *zero!*" Otto exclaimed when he looked back at his equipment. The others in the truck were saying nothing, heads down as they busied themselves with their tasks. "I'm going to fax data sheets out to Glenview when I get back," Otto said. "They'll be real interested to see this. This is the worst I've ever seen." To Otto's right, Scarpa stood quietly, looking down at the controls. "I don't know of anything that would cause this," Otto was muttering, "short of sabotage."

There was no sabotage. In fact, when the field tests ended, Zenith's transmission system was judged to have performed superbly. As for site 154—well, the engineers concluded that it was an anomaly. Unexplainable problems like this often cropped up in complex engineering projects. But in the charged atmosphere of that moment in the field test truck, no one had been able to see this.

Later, Jerry Pearlman, Wayne Luplow, and Zenith's other leaders began waving the field test results like a magic talisman. Then in the third quarter of 1994 Zenith accomplished the impossible: the company actually showed a $9.4 million *profit,* its first in almost a decade. In the Bunker it felt as if the good old days were back.

Searching for good news himself, Wiley also began distributing copies of the field test results to congressmen, FCC officials, and others. As he tried to save his program, the chairman suddenly found himself in the odd position of actively promoting the Grand Alliance. When he paid a courtesy call on one of the new FCC commissioners who had just moved to Washington, his host was not entirely sure who he was, and asked, "You represent the Grand Alliance, don't you?" No, no, Wiley said with a smile. In truth, though, the new commissioner was not entirely wrong. If high-definition television was ever to be, Wiley knew, he had to talk up the Grand Alliance. It was the only game in town.

As the summer of 1994 drew toward a close, Wiley and the others found that the field test results were not the only favorable development. Several others fell into place. First, the Senate and House conferees lapsed into argument over disputed parts of the telecommunications bill, and to everyone's immense surprise the entire bill died. The conferees had barely noticed the HDTV amendment, but when they gave up on the bill the NAB's amendment was buried with it. The broadcasters were disconsolate.

At about the same time, the Clinton administration got a new idea to raise money. Why not auction off some of the electromagnetic spectrum? Until now, the FCC had simply given away lanes on the airwaves. The government took applications from the competing potential users—Land Mobile or the broadcasters, for example—decided which of them had the strongest public-interest proposal, and then issued a license for free. But as technology created more and more potential uses, and the number of applications grew, the Clinton administration took up an idea first raised theoretically during the Reagan years: Since the airwaves are a valuable, government-owned resource, why simply give them away? Why not sell them? The first auction was for a thin slice of spectrum to be used for advanced two-way paging services, and auction supervisors gasped when the first bids came in. The sale raised far more money than the government officials had hoped for even in their greediest dreams. When the first auction for the tiniest imaginable slice of spectrum ended, the government had raised more than $200 million.

With that, auction fever washed over Washington; auctions for far more valuable lanes on the airwaves were scheduled later in the year, and congressmen's heads were filled with anticipation of future wealth. Boy, they'd say, when we start auctioning larger chunks, we can raise billions of dollars—and at just the time we are going to have to show progress in balancing the budget. With all this auction money, maybe we can keep Medicaid, save Social Security, salvage farm price supports, along with a thousand other pet projects—and *still* balance the budget. None of this seemed to have much to do with high-definition television, until Bob Rast and several other members of the Grand Alliance began to see that they could use auction fever as a weapon.

Every autumn *Broadcasting & Cable* magazine staged a one-day conference in Washington called "Interface," filled with panel discussions on current trends and controversies in the industry. For Interface '94, high-definition television stood as a central problem. So the organizers scheduled a public debate between two of the leading antagonists, Bob Rast and John Abel.

"You're playing Russian roulette with the second channel!" Rast fired at Abel, as the two of them stood on a low stage in front of about two hundred people at the Shoreham Hotel. "By saying you want to use it for digital services, not HDTV, you may not get it. The government might decide to *auction* that spectrum, and that means we may not get it for HDTV at all!"

Abel shot back some of his favorite one-liners: "HDTV is just for rich people who can afford $3,000 or $4,000 TVs. Spectrum is scarce. Why use 12 megahertz of it to do the same thing we are already doing in 6 megahertz?" By that, he was referring to the FCC's simulcasting plan: each broadcaster would transmit regular programming on one 6-megahertz channel and again on a second channel, in high-definition.

Dick Wiley moderated this debate, and periodically he posed questions of his own. "Why deny the public better pictures?" he asked Abel toward the end.

"Why deny the public additional digital services?" Abel fired back. "The consumer right now isn't asking for higher-quality pictures, isn't knocking down the door saying, 'I want HDTV.' But they *are* embracing digital."

Even with that, under pressure Abel made a comment suggesting that new political realities were forcing him to modify his hard-line position ever so slightly. Wiley asked him if he would be willing to accept a rule requiring broadcasters to transmit HDTV for just a few hours a day as the price for getting free use of a second channel.

"I would accept a requirement of a couple of hours of HDTV a day," Abel responded, "if we are given permission to do other things, too. But there can be no standards. We have to be able to dial the resolution up or down." For a football game, "maybe dial it down during the player lineup, up during an action shot."

Digital television may have arrived as a result of Woo Paik's quest to create a high-definition system for satellite subscribers, but a digital TV signal doesn't have to be high definition. As some Advisory Committee engineers liked to say, "Digital television is so flexible that you can make the picture just as bad as you want." Now Abel was saying he wanted permission to reduce the picture resolution so broadcasters would have extra space to offer additional services.

A digital television signal is a data stream like any other in the computer world. With the Grand Alliance system, a television station would be able to transmit 20 megabits of data every second. A 'megabit' is just a measure of the amount of computer code being sent, and 20 megabits a second is a raging gusher of data, the equivalent of sending the entire text of *War and Peace* every few seconds. Typically, a full high-definition television signal would use 90 percent of that—18 megabits a second—leaving 2 megabits free for other uses. A TV station could use that extra capacity to transmit other services—stock quotes to clients who paid for them, as an example, or access to chat groups on the World Wide Web.

But the Grand Alliance system was being designed to allow even more flexibility than that. As a result, in control rooms across the country, engineers would have the power to turn a dial that would lower the resolution little by little. Instead of 18 megabits a second, the TV signal could use 16 or 14—all the way down to 4 or 5 megabits, yielding a TV signal that offered roughly the resolution of conventional TV. That would leave a tremendous amount of space for additional services. And there was nothing to prevent the engineers from turning the dial down even lower—which is why the Advisory Committee engineers liked to say that digital TV could produce a picture "just as bad as you want."

Playing on that, Abel answered Wiley's question by saying, "HDTV has to remain undefined. It can't be defined just as the Grand Alliance sees it. What happens if some stations dial the picture down even during an action shot" in a football game—a challenging sequence for digital TV. "Will the consumer even know? Will he care?"

Dialing down the bitstream would give broadcasters a huge, open digital highway to the home, and they wanted to use it however they chose. While broadcasting a *low-definition* TV signal, so much of the bitstream would remain unused that a TV station could transmit the entire national zip code directory in about four seconds. Nobody really knew what moneymaking opportunities might come from that, though Abel and many of his brethren were certain they existed. Some of them wanted the opportunity to run paging services, compete with the phone company. Others wanted to "dial down" the picture resolution to a point where they could fit four or five low-definition digital TV programs into the same space on the airwaves now occupied by one conventional, analog program.

By the time of the annual HDTV Update, staged by the Association for Maximum Service Television at another Washington hotel a few weeks later, all of the Grand Alliance partners and the leaders of the Advisory Committee had figured out the broadcasters' strategy and had settled on the best counterattack. At this conference, Abel found himself on a panel with Jim Carnes of the Sarnoff Shrine and Joe Flaherty of CBS, among others. Land Mobile, the demons who had pushed Abel to lobby for HDTV in the first place, had been complaining of late that if the broadcasters weren't interested in HDTV anymore, why shouldn't Land Mobile be given the promised extra channels? Flaherty picked up on that. "Land Mobile is suggesting that if you can compress four TV channels into one," he said, with a touch of glee, "they will be happy to take the other three."

Carnes came next, and he slammed the broadcasters hard, using the technojargon of the moment. "We've got to stop talking about 6-megahertz TV channels," he said. "It could be that if you want to do only standard-resolution programs, the FCC will give you only 4 megabits and auction the other 16 megabits." (That may have confused outsiders, since he alternately referred to both megahertz and megabits. A television signal occupies 6 *megahertz* on the airwaves, that figurative 6-inch pipe. A Grand Alliance digital television transmitter sends 20 *megabits* of data through that pipe every second.) "The beautiful thing about high-definition it is that it takes up the whole 20 megabits." If TV stations broadcast high-definition programs, he said, there would be no space left to auction off. "So I would think that the broadcasters would *insist* that high-definition is absolutely necessary for broadcasters."

Abel didn't have anything to say about that, and the audience of broadcasters remained silent.

Don West, editor of *Broadcasting & Cable* magazine, moderated this discussion, and he asked Abel straight out: Given your present strategy, "do you think you are at risk of losing your channel?" Abel answered something else first and then asked West to repeat his question. West said, "Well, I guess my question is, politically speaking, do you know what you are doing?" Around the room the broadcasters laughed and applauded. Abel grumbled, "I guess that remains to be seen."

By late fall of 1994, the momentum was shifting, and the high-definition advocates could feel it. "I think we bottomed out some time ago," Bob Rast said, "and now it's improving." And Wiley said, "This thing is working out great. I think we need to relax a bit."

Even if the political environment was improving, Wiley quickly found he had other problems. During the uncertain days of spring and summer, the Grand Alliance had slacked off. Rast had whispered about this a few months earlier: "The risk for us in Grand Alliance," he'd said, "is that we have a tenuous relationship among the partners, and the development work inside the Grand Alliance is at the whims of key people in six different organizations. If anyone sets inadequate priorities and puts another project that actually makes money ahead of this one, then we don't make schedule. That's my secret worry. And I don't have a way to deal with that."

The partners had told Wiley they would be a couple of months late delivering a finished prototype to the Advanced Television Test Center. The deadline, October 30, was pushed back to the end of December. Now this new deadline was just a few weeks away, and it began to look as if the partners wouldn't make it. Zenith, of course, had finished the transmission system, and Dolby Laboratories (not a member of the Grand Alliance) had completed the sound system. The Sarnoff labs, responsible for the transport, the hardware that arranged the digital signal into coherent sequences for transmission, was finished with this work by fall. Philips Labs had finished the decoder, the box that went inside every television set and translated the digital signal into picture and sound. But the most complex piece of hardware was the encoder, the sophisticated computer that held the compression algorithm and turned a video signal into a digital bitstream. GI and AT&T were building that together. At the beginning, Rast had boasted, "We'll build an entire encoder, even AT&T's part, so we'll be ready if they fall on their faces." In truth, GI had struggled to complete even its half. They finished only in November. That left AT&T to complete its part, and the Bell Labs boys had fallen far, far behind.

Rast now held the informal position that Robert Graves had filled during AT&T's tormented partnership with Zenith—that of arbiter between the partners. Little of the enmity that afflicted the earlier relationship had developed, but Rast was having a hard time figuring out what was really going on at Bell Labs. The deadline was close, and every time Rast called John Mailhot, AT&T's lead engineer for this project, Mailhot would tell him something like, "Two weeks. Just two more weeks and we'll be done." Two weeks would pass, and once again the Bell Labs boys would say, "Two weeks, Just two more weeks and we'll be done."

"We get these very optimistic reports," Rast was saying, "but they're not being honest." And engineers at Zenith were saying, See, See! It's the sliding schedule all over again.

But the stakes were higher now. This was the final system, and Wiley was getting anxious. After seven years he wanted the process to end. In addition, the partners could feel a light wind at their backs. Now was *not* the time to stumble. They would lose every inch of ground they had struggled so hard to gain. Nonetheless, the encoder couldn't even make pictures yet.

Late in the year, the partners gathered for a meeting to decide whether they could meet their new deadline. Talking it through, they quickly realized they could not, though nobody wanted to tell Wiley about still another delay. They decided to ask for just one more month; maybe Wiley would swallow that. The newest setback would push delivery to January 31, 1995. But Rast wondered if even a month would be enough. He took Mailhot out to the hall and asked him: John, how much time do you really need? Tell me now, and we'll get it. Mailhot said a month would be enough, and so it was settled. Few of them really believed they could finish that quickly, but they were afraid to ask for more.

A few days later, one of them told Wiley of the new delay. If that's what you need, Wiley said, well, okay then. The next day Wiley said: "I pretended to be magnanimous, but actually I was relieved. From what I'd been hearing, I thought the delay was going to be a lot longer." Given the many years he'd run this program, Wiley wasn't too concerned about another thirty days. He dutifully informed the FCC—even though almost every message he'd sent over there in the last year had seemed to fall into a bottomless hole. Evidently, Reed Hundt just didn't care. But this time Wiley found that something had changed.

"We heard secondhand that Hundt was really upset," Paul Misener reported a few days later. "We're being told he wants to get this moving. And I'll be honest, we don't understand."

A week later, Hundt gave an interview to *Electronic Media,* an industry weekly, and the interviewer asked him what his priorities were right now. One of them, he said, was advanced television. "I'd like to make sure we act as promptly as we possibly can on the advanced television or HDTV issues, whichever you prefer to call it," Hundt said. "I would like to push that process. I was a little bit concerned when I read in the newspapers lately that there's some kind of slowing down by the Grand Alliance. I think it's very important for the country and the economy that the process move very, very expeditiously, and I'm interested in doing whatever we can to hurry it along."

The partners were stunned—and thrilled. "We're delighted he's concerned," Carnes said with a broad smile. "It's the first time I got a warm feeling from finding that the chairman of the Federal Communications Commission was angry with us," noted Bob Graves. But nobody could explain why Hundt suddenly cared.

The truth was, Hundt hadn't been paying much attention until now. He tended to look at issues in sequence, and he hadn't taken up high-definition television until he had settled several other more pressing matters. A year earlier, he had sat on his office sofa, stretching his arms high above his head, and yawned as he said, "I can be very relaxed about this, because I don't have to make a decision for a year."

Now that year had passed, and when Hundt finally did turn his attention to HDTV, lo and behold, he found that the Grand Alliance partners really *were* creating the "telecomputer" he had been calling for all along. As soon as Hundt understood that, he started pushing for quick introduction of these new devices. As he saw it, they would connect every home to the information superhighway. This was, after all, the pet project of his good friend and chief political sponsor, Vice President Al Gore. At the same time, however, Hundt still couldn't care less about *high-definition* television.

"The term 'HDTV' was coined to describe the possibility of getting pretty pictures," he said, as he leaned back on his office sofa once again, one morning in December 1994. "But what's clear to everyone now is that advanced television is not about pretty pictures anymore. It's about the digitization of television and a huge range of new services—a true watershed moment for the broadcast world." He smiled. "The commission asked them to build us a better bicycle, and they came back with a Ferrari. I think it's great for the country. But now we have to decide what to *do* with this Ferrari."

Just as Wiley's long race was heading toward its end, Hundt was saying that the FCC really had no idea what to do with the winner. But whatever rules the FCC decided to set, Hundt made it clear that he wanted to move fast. Even with all the construction delays, he said he intended to start writing regulations very soon. That was the reason for his impatience with the latest delay. "It seemed to me," he said, "that the date of shipping was slipping and might continue to slip. It has been made crystal clear to me that American industry is leading the world, and we as a government have the obligation to give them the opportunity to exercise that leadership by bringing a product to market."

The change in view could not have been starker. Now Reed Hundt was an evangelistic convert to *digital* TV, though he remained underwhelmed by *high-definition.* HDTV, in his view, contributed nothing to the information superhighway. The problem was, the Grand Alliance partners believed that high-definition pictures would sell the sets—not some ill-defined notion of advanced services that no one could yet describe. If the FCC didn't require broadcasters to offer at least *some* HDTV, who would buy these expensive new TVs?

A few days later, Wiley escorted Hundt over to see an HDTV demonstration that the partners were putting on at AT&T's Washington office (the same demo that Hundt had scoffed at a few months before). The Grand Alliance was trying to capitalize on the changing political mood, even though their machine wasn't finished yet. So they were running a simulation.

Wiley knew perfectly well that the partners weren't likely to meet even their new deadline. When the next delay came, he wanted to make sure Hundt wasn't taken by surprise. As they walked up the street, he told Hundt that another postponement was probable. Hundt stopped in his tracks and turned to Wiley. "Are they looking at this like the Manhattan Project?" he asked. Are they managing this as a crash program —three shifts of workers going at it around the clock?

"No, I don't think so," Wiley said, surprised by the question. Right then Wiley realized that when the partners called to tell him of the next delay he couldn't be so magnanimous. Just after the first of the year Wiley got the call: We can't make the new deadline. All the system's parts were supposed to have been completed by now and delivered to the Field Lab at the Sarnoff Shrine for final assembly. But AT&T still wasn't finished, and nobody knew when John Mailhot and his team would be done. We need three more months, the partners told Wiley.

Wiley's response was cold, and the next morning he faxed a letter to all of them. He'd been working on the draft for weeks.

With considerable disappointment, I learned last evening from your representative that the Grand Alliance will not be able to complete a prototype

HDTV system by January 31, as planned under the schedule that you confirmed to me less than one month ago. Instead I am now advised that you intend to deliver the system for testing within one or two months after the end of January.

In response I must tell you that I am unwilling to agree to the new timetable in the absence of Advisory Committee oversight of the remaining system integration effort. Specifically, I intend to appoint an inspection team of Technical Subgroup experts, under the leadership of Joe Flaherty, to visit your facilities in Murray Hill and Princeton as soon as possible in order to determine the actual state of system progress. . . .

In closing, allow me to again express admiration for the pioneering efforts that the Grand Alliance has made. Nevertheless, given the importance of advanced television to the country, I believe that the time has come to complete this very lengthy effort.

A long "cc" list followed, and at its top was "FCC Chairman Reed Hundt." That made it easier for the partners to swallow. The letter is for Hundt's benefit, they agreed. Wiley has to show he's being tough. If he wants to send an inspection team—well, that won't be so bad. Obviously, though, they'll ask to see the unfinished encoder. With that, all eyes turned to Bell Labs.

"Everybody knows that the encoder is the long pull," Mailhot said, standing in his lab a few days later, raising his voice so that it carried over the hum of fans cooling the equipment racks behind him. "It's not a secret. But we are not far from having something we can integrate with the decoder." Behind him, half-a-dozen engineers worked at consoles, running data streams through computers, putting test probes to circuit boards pulled halfway out of the equipment racks. For all the expectations and concern that hung over their work, the atmosphere was relaxed. This was no Manhattan Project. And Mailhot, a normally even-tempered young man with a thick black beard, bristled at the suggestion that AT&T had anything to explain.

"Remember, the decoder is a product of AT&T and GI," he said, a sharp edge to his voice. "And to say that GI is finished and we are not is not entirely accurate."

Then his tone softened: "If we don't make some real progress in the next two weeks it could get ugly. But everyone wants this thing to work. When will it be ready to integrate, to go to Sarnoff? Sometime in the next several weeks."

Behind him, Michael Acer, one of his engineers, sat at a computer station, where he was "trying to work out some timing problems," he explained. Acer ran test signals through one of the encoder's boards and then examined the signal that came out the other end. "We think we have a minimum number of problems," Acer said. "But we keep finding new ones, maybe created by the other fixes we are making."

"We're going to make this work," Mailhot insisted, "And we're going to make it work in the next couple of weeks."

But no one really knew how long Mailhot and his crew would take. Actually, nobody had a full understanding of the work in all of the various labs. The partners were strictly equal; they had never appointed a project supervisor; each company proceeded at its own pace. Until now, they'd held conference calls every week to coordinate their efforts, but if one company was falling behind, no one else had the power to redirect the work. They had muddled along over the last year or so. But as they approached the point when they would have to put all the separate pieces together to build one grand machine, the lack of a project director stood as a glaring problem. Wiley and Joe Flaherty decided this was the real reason behind the continuing delays. So Flaherty ordered the partners to appoint a project manager—right away. Without a supervisor, Flaherty said, "this thing will *never* get done."

The partners had no real problem with the idea. Many of them had known it was the right thing to do, but Grand Alliance politics had made it difficult. Now that they were under orders to do it, they didn't debate long before choosing their man: Bob Rast.

Everybody liked him, and he was the only senior player who carried no other significant responsibilities within his company. Rast accepted the assignment happily. Long ago he had decided that the project was important. We're making *history* here, he believed. Already he had forgone other, more promising opportunities within General Instrument to stick with this program because he wanted to be able to look back one day and say, I did something that really mattered! Still, Rast was nervous. "There's an undercurrent from the Advisory Committee that they want somebody who can crack the whip," he said. "But there is no real power in the Grand Alliance." None of the others really had to do anything Rast suggested. "The only power anyone has," he said, "is power that is earned."

In February 1995, almost three months late, AT&T finally sent the encoder down the road to the Sarnoff Shrine, even though it wasn't really finished. Mailhot had argued that they could work faster if they kept it at Bell Labs, where he had the staff and the sophisticated equipment he needed. But Advisory Committee inspection teams were coming up to Sarnoff every few weeks, and Rast was insisting that the new deadline was real. "We're going to ship the system to the test center on March 31st even if it isn't totally finished," he said. That wouldn't work if part of it was still up at Bell Labs.

Through February, work continued at a busy pace. The prototype was being put together in the Field Lab, where Sarnoff had scrambled to finish its first entry back in 1991. This was the old VideoDisc press room, but now a red sign on the roadway just outside pointed to: "Field Lab, GA HDTV." Engineers from all the companies traveled in and out of Princeton every few days. Generally they worked in two eight- to ten-hour shifts. Rast set goals and deadlines timed to show progress for each new visit from the Advisory Committee inspection team.

The most important visit was set for March 10. This was just three weeks before the final deadline, and the Advisory Committee inspectors were supposed to decide if the Grand Alliance should be given permission to ship the prototype down to the test center.

On March 9, the Field Lab was a hive of frantic activity. The prototype had been assembled, its pieces hooked together in equipment racks that stood in two twenty-foot rows, back to back, ten feet apart. One row was the encoder, the other row the decoder. Each of the racks prominently displayed its origin: "Philips MPEG-2 decoder . . . David Sarnoff Research Center Grand Alliance Transport . . . Thomson Grand Alliance Transport Decoder Version 4.2 . . . Dolby AC-3 Audio Decoder . . ." Dozens of cables snaked from the back of one row, up to an aluminum rack hanging from the ceiling and then down to the racks on the opposite side of the room. Beside the equipment racks, engineers from the various companies sat on stools, minding their part of the machine. In front of them, smaller cables spilled off the circuit boards, and yellow paper-tag labels hung from many of them, showing the date of the last fix: "Updated to Rev. C, 2/21/95," said one.

Rast presided from a small, spartan office just outside this main room. The Advisory Committee inspection team was due at ten o'clock the next morning, and Rast had scheduled an internal status check for six o'clock this evening, so that everyone could see just how far along they really were. The machine produced pictures in the two principal modes, interlaced and progressive scan—the end result of the long interlace-progressive dispute. The HDTV sets would switch automatically from one to the other, as the incoming signal changed.

The interlaced mode worked pretty well. That wasn't surprising; engineers had decades of experience building interlaced TVs. "We're getting pretty good pictures in interlace," Rast explained. "We still have bugs"—small flaws and imperfections that popped onto the screen. "Overall, though, we're almost there." But the progressive mode was far behind.

In the main lab, a monitor next to one of the encoder racks was showing a test tape of tropical fish swimming in a tank, photographed close up, and every few moments a bright shimmer, like a Fourth of July sparkler, flashed across the screen. One of the engineers was poking at chips on a circuit board he had slid out of a rack, keeping one eye on the monitor. "We're still sparkling," he said. "We swapped boards with GI, and we're still sparkling." Then he grabbed a small part, a capacitor, between two fingers, and the sparkling stopped. He looked over at the guy next to him with a surprised smile. "We could include one of you with every set," his colleague quipped. Another engineer looked over and said, "Your fingers are probably adding some capacitance of their own. Why don't you just put a larger capacitor in there." He did, and the sparkle went away.

So it went, as they attacked the bugs one by one.

When a group of Grand Alliance leaders looked things over at 6:00 P.M., the interlaced mode was in pretty good shape. But the progressive mode—well, that was another story. "I feel we'll get there," Rast said. "But we're not there yet." Some of the time the progressive picture just looked bad; other times they couldn't get a "proscan" picture at all. Later in the evening, an engineer arrived from Philips Labs to work on the interlace decoder. He didn't leave until morning. By then the interlaced picture was virtually perfect. But any work on a device this complex carried a risk: Fix one part, and you might inadvertently mess up another part, since at one place or another all of them were interconnected. And at 9:30 Friday morning, half an hour before the inspection team was supposed to arrive, the progressive picture was *gone*. Turn it on, and all that came up was a blank screen.

Engineers scrambled, shouting panicked orders and suggestions as they tried one thing after another. "Proscan is not talking to the transport!" somebody barked. Rast walked into the lab, and six engineers were pulling at an equipment rack, jiggling wires and chips as they looked at a monitor to see if anything made a difference. "I've got signal," one of them declared, and the monitor clicked on, showing... nothing. The clock now read 9:45. "If we don't get proscan back in ten minutes...," one of them moaned, and Rast noted, "Well, that would definitely be a negative." At 9:53, the first of the Advisory Committee inspectors arrived; they'd come up from Washington on an early morning flight. Robert Hopkins, the former Special Panel chairman, and Paul Misener, Wiley's aide, walked in with a couple of others. They were shuttled into a lounge. The lab was fifty feet away, out of earshot. In the lab, Aldo Cugnini, a tall, thin young engineer from Philips, was shouting, "We're making pictures. We're making pictures! Where's our video? Come on, guys, give me a video feed!" Across the room someone started a tape player, and a test pattern flashed onto the screen. But it was jittery, unstable. "God, I don't believe this," Cugnini sighed.

In the lounge, more Advisory Committee inspectors were arriving now, including one who plunked a videotape box down on the table. The last time the inspectors had come up here, some of them had thought the Grand Alliance demonstration was using particularly easy material—mountain vistas and other scenes that included little motion. So they had prepared an especially challenging test tape. The Grand Alliance was supposed to use it this time.

"The way we see it," Robert Hopkins said as he waited in the lounge, "this is a critical moment. If they aren't ready now, there's a real question whether they can ship on the 31st." Given Reed Hundt's new stance, everybody—inspectors and Grand Alliance partners alike—knew that another delay would be disastrous.

At ten o'clock, Cugnini walked over toward the lounge, and by the time he had crossed the hall he had assumed a studied calm. "Hello, gentlemen," he said, nodding to the inspectors one by one. He sauntered among the inspectors for a minute, just long enough not to seem rushed. Then he turned around and hurried back to the lab, where he plunged back into the panic when he saw that the others had produced a proscan picture that was a mess—a black-and-white screen filled with incoherent images.

"We're back," Rast said with a laugh, as he looked askance at the monitor.

"We're missing color," Cugnini said, as if that were the only problem. "They've switched a million cables around. Some of the cabling must be wrong."

"But I looked through every single cable," another engineer whined.

Rast looked at his watch and realized he had just about run out of time. "There's a lot of confusing recabling going on in here," he said. "It'll either work in the next two to three minutes, or it won't." Then he walked over to the lounge, greeted everyone, and pulled Paul Misener aside and told him exactly what was happening. Misener simply nodded. After a few introductory greetings, the inspectors said they wanted to see the demonstrations. The whole lot of them walked back to the lab, where the frantic work continued. "It's working," an AT&T engineer confided to Rast as the group walked past.

"Is that good or bad?" Rast whispered back.

"I can't say it's good."

A moment later Rast saw just what he meant. The ten Advisory Committee inspectors walked through the lab to a viewing room on the other side, where a large-screen high-definition monitor sat on a tall metal stand. As they clustered around, someone in the lab flipped a switch, and the progressive-scan picture flopped onto the screen.

"God," somebody muttered.

The left side of the picture was in black and white, the right side in a sickly color. A black vertical bar split the screen into two mismatched halves that vibrated and shimmered. "This is called putting our worst foot forward," Rast said with a nervous laugh. The inspectors were quiet at first. Too quiet. Finally Hopkins said, "Last time we were here, there were problems. But this . . . I don't even know how to evaluate this. The last time it was almost *working*." The parody of a picture played on for several more excruciating minutes. "This would be great for wall-eyed viewers," one of the inspectors cracked. Just then a jet plane seemed to be taking off at their feet. The sound for a TV image they could not see roared through surround-sound speakers sitting on the floor.

For all that, the experienced engineers among the inspection team said they could see that the problem might really be just mismatched cabling, not anything more fundamental. And besides, it was in everyone's interest to say that the prototype would be ready in time, even if no one was really sure. No one wanted to call Wiley and tell him the Grand Alliance was in trouble. None of them wanted to hear that Reed Hundt was upset all over again. With three full weeks to go, maybe the Grand Alliance could fix this problem. So the inspectors decided to put progressive scan aside for this visit—they would look at it again next time—and turn their attention to the interlaced mode instead. Lab engineers put the new test tape into the player and routed the signal through the Grand Alliance system, then onto a monitor. The group clustered around it to watch. A picture popped onto the screen bright and clear, the first image in a six-minute montage of Helsinki street scenes. The picture was stunning—almost perfect.

"The audio and video are both significantly improved," one of the inspectors remarked with an antiseptic voice as the others nodded their approval. They wanted to seem neutral; still, after the progressive fiasco, several of them let out long breaths, clearly relieved. One of them hit a button that switched the display from an unaltered feed coming directly from the tape player to another feed that had been run through the Grand Alliance system. A-B testing, this was called, like the comparisons the judges had made during the audio system shoot-out. Not all of the inspectors could tell the difference. "Hey, are we on source or what?" one of them asked. "Let's face it," said another. "It looks pretty damned good." Misener stood in the back listening to the comments. If these guys are impressed, he decided, it must be pretty good. A few minutes later, all of them were smiling as they filed out of the viewing room. Rast pulled Misener aside and whispered, "You OK?"

"Yeah," he said. "There's still an upside." Proscan still wasn't ready. "But I'm happy."

As the inspectors walked back through the lab, the engineers looked up at Rast. He flashed a thumbs-up sign. Several of them smiled. Flaherty was flush with enthusiasm all of a sudden. He turned to a Sarnoff engineer and pointed to the equipment racks. "Let's get some slides of this so we can show them at the NAB," he said. The 1995 National Association of Broadcasters convention was just a few weeks away. "We can show this and say, 'Here it is, guys.' It may not be quite ready, but it's obviously on the way."

20 One hundred eighty degrees

The National Association of Broadcasters convention, the huge extravaganza held in Las Vegas each spring, attracted between 70,000 and 80,000 broadcasters. So it was always the event of choice for showing off new products, making grand statements, and gauging the industry's mood.

Here at "the NAB," as the convention was known in shorthand, Al Sikes had stood in the doorway of the vast exhibit hall nine years before, looked up at the gaudy neon signs, and remarked, "I don't see a sign for a single American company out there." A couple of years later, mellifluous announcers had stood on carpeted rostrums in the same great hall, importuning the passing crowds to "Come see the TV of tomorrow. Come see ACTV!" General Instrument offered its first public showing of DigiCipher here, and in the following years the other contestants in Wiley's race had put on dueling demonstrations of their competing systems. In fact, there was no better place than the NAB to judge the broadcasters' real view of advanced television.

In 1994, the Grand Alliance had no booth, no real presence here—partly because the partners were still building the machine but also because the environment had seemed so hostile. After that year's convention Wiley had said, "I almost didn't go to the NAB this year. I didn't feel like I was welcome." Every year he'd been invited to a certain formal dinner, but in 1994 "they withdrew the invitation. They tried to squeeze me." As NAB-95 approached, however, the Grand Alliance was preparing to unveil its new machine to the broadcasting world. Everyone expected an interesting week. None of them knew just how right they were.

The Grand Alliance engineers at the Sarnoff Shrine had careened through the final three weeks of March, crashing through all-nighters until finally they had got everything done by the morning of March 31. The progressive format was finally working well enough now, and meeting the deadline stood as a genuine triumph for Bob Rast, who had pushed and coaxed to the very end. That last morning, Paul Lyons, a Sarnoff engineer, watched from the window in Rast's office as workmen loaded the truck a little before noon.

"It's done," Lyons said. "It's really done."

A few days later, the prototype was in Alexandria, hooked up to the Advanced Television Test Center's cables, as Peter Fannon's engineers ran preliminary tests. So far, everything was working.

Anticipating just this moment, the Grand Alliance had built a backup system so they would have one to demonstrate at the NAB. After all the delays, arguments, and controversies, they wanted to show their stuff. They rented a large piece of floor space in the convention hall, and on Sunday, the day before the NAB formally opened, they were setting up a sophisticated group of demonstrations, large and small.

A few months before, Wiley had addressed an industry gathering and opened his comments by saying, "I have good news, and I have bad news. The good news is that the Advisory Committee process is almost over. And the bad news is that the Advisory Committee process is almost over." Since then, the situation hadn't seemed to change very much. The broadcasters had reintroduced their "spectrum flexibility" amendment for another telecommunications bill that was taking shape in the new, Republican Congress. Everyone expected the congressmen to accept this HDTV killer, just as they had before. Then in early March, Robert Wright, NBC's president, surprised everyone by offering a clear, frank opinion about high-definition television, even though most broadcasters had seemed to favor deliberate ambiguity before. Asked in a *Broadcasting & Cable* magazine interview, "How do you feel about HDTV?" Wright said:

I'm in the digital camp. The digital picture is a better picture than the picture that we, as broadcasters, can provide today. And the digitization of the spectrum gives us more flexibility to provide services, in addition to the picture, and that's a benefit.

The HDTV concept that utilizes all 6 megahertz to make a better appearing picture is less attractive than the benefits the consumer and broadcaster get from digitization. And pursuing HDTV as an objective by itself, requiring you to have entirely different production technology and cameras, I don't think that's working.

Wright could hardly have been more direct. He couldn't care less about HDTV; he was interested only in the new, moneymaking digital services. A month later, at a preliminary session in Las Vegas Sunday morning, the day before the NAB convention's formal opening, Wiley told five hundred broadcasters just how he felt about opinions like that one. "Now, as success beckons," he said, "discordant sounds are being heard across the land from government and industry. And after all the effort to get this done, I think it would be a tragedy if it didn't happen. If we'd had this kind of vision in the 1950s, we would never have had color TV." Joe Flaherty spoke a moment later and reminded the audience that General Sarnoff had faced similar skepticism from radio broadcasters fifty years before. Many of them hadn't been the tiniest bit interested in television.

When John Abel's turn came, he seemed to be enjoying himself. This was his convention, after all, and he was smiling his Cheshire cat grin as he said, "I think it is truly revolutionary that the Grand Alliance now spends more time talking about how flexible their system is than they do about high-definition. So I don't think I have to talk very much about high-definition. I feel totally vindicated. They have heard us." This session, the only one that NAB officers had bothered to schedule for a discussion of HDTV, lurched through a morning of all-too-familiar argument. Toward the end, though, Abel did catch people's attention with one quick aside. "We *want* to do HDTV," he said in response to a question, seeming to choke on the words like a schoolyard bully who'd been ordered to apologize. Several people in the audience wondered what that was all about. Very shortly they would find out.

Broadcasting & Cable, a weekly, came out every Monday, and the NAB was an extraordinarily important event for the magazine. With almost the entire industry gathered in one city, the editors wanted high visibility. This year, 83,408 broadcasters had come to Las Vegas, a record. And as conventioneers got out of bed Monday morning, they heard the sound of something plopping on the floor just outside their hotel door. It was this week's issue, hand delivered. The cover was jet black with a bright red number "2" burning from its center. Over that was the headline "The Fateful Battle for the Second Channel." Inside lay word of a sea change so remarkable that mouths dropped open in hotels from the MGM Grand to the Stardust all the way down to the Hilton.

There on page 28 was another interview with Bob Wright, NBC's president, who had disparaged HDTV on page 44 of the same magazine just five weeks earlier. But in this article, written by the magazine's editor, Wright had another message.

"Among broadcasters," the article began, "NBC President Robert Wright has a reputation for pushing the envelope. Not only has he kept that network in the forefront of the conventional television business, he has also been busy developing [alternative network services].

"But Wright is pushing another envelope now: HDTV. He thinks that the new medium has been pushed aside in public policy discussions by the enthusiasm over flexibility. The latter, he says, has been overplayed; it's not, in his view, the issue but a subissue.

"Yes, digital's promise of enhanced productivity during the transition to HDTV is exciting, Wright says, but it is not the goal. That remains a 100 percent rollout of HDTV, however long it takes."

Then the article quoted Wright: "There are many satellite programmers who will embrace it. There are manufacturers of hardware who will build toward the standard. Of all the potential programmers, the broadcaster is the last one who wants to have anything but the best picture in the home. I don't think we have a business if we end up with less than the best picture. It is inconceivable to me that broadcasters, of all people, would want to be in any way left out of complete parity in high-definition."

Neither the NBC executive nor the writer bothered to explain why Wright had so suddenly reversed his view. In truth, no explanation was really necessary, for a good part of the answer was in the same magazine a few pages back. "Congress Sees Gold in Them Thar Second Channels!" a headline declared on page 23. "Association Marshals Forces to Resist Movement in New Congress to Sell or Auction New Spectrum for Digital TV."

Throughout the previous fall, Grand Alliance leaders had been warning the broadcasters that, as Rast had put it, "you're playing Russian roulette with the second channel!" Auction fever had been sweeping Washington then, and if Congress's attention swung around to the second channel that was to be used for high-definition television, "we may not get HDTV at all," he had warned. Heedless, the broadcasters had resubmitted their "flexibility" amendment in Congress—the one that would let the broadcasters use the second channel however they chose—baldly announcing that they had no interest in HDTV. Meanwhile the FCC had staged even more spectrum auctions, and the new ones made the first look like a rehearsal. By now several auctions had raised almost *$8 billion* for the Treasury. And sure enough, Congress took the bait.

"A lot of us are asking questions that have never been asked before," said Representative Jack Fields, the new chairman of the House Telecommunications and Finance Subcommittee (the panel Ed Markey had led until the Republicans took over the House). "Is it really fair," he asked, for broadcasters to get a second channel to be used for subscription services, if that channel "was never subject to the auction process?" As for the old assumptions about the second channel, "That was then, and this is now," said House Budget Committee Chairman John Kasich, puffing up with the new power he now held. "It's a new day in Washington."

Ten-thirty Monday morning, the Association for Maximum Service Television, the smaller industry group whose leaders were often at odds with the NAB, staged its annual meeting in a convention center conference room. Year after year, the meeting took place in the same room here, and every year Jonathan Blake, the association's attorney, stood up and offered a briefing on developments in Washington. Most of the time, HDTV was a topic of central interest.

Blake had been the one, eight years before, who had finally convinced FCC Chairman Dennis Patrick to abandon the decision to give Land Mobile those vacant channels. HDTV was coming, no question about it, Blake had told Patrick, sitting in the chairman's office in the spring of 1987. All the other services—cable, satellite, and the rest—they'll be able to transmit it without asking permission from anybody. Only the broadcasters were constrained by FCC regulation. Could the industry survive if they alone were prevented from providing HDTV? Patrick had sighed with resignation and abandoned the Land Mobile decision a short time later.

Like the rest of the industry, Blake and the organization he represented had changed their minds. In the spring of 1994, MSTV, as the group was known, put out that "white paper on broadcaster flexibility" that had so strenuously avoided any explicit mention of high-definition television. Blake had stood up at the 1994 annual meeting and offered a hard line: They can't make us broadcast HDTV, he had said in sum. Simply upgrading to *digital* television is justification enough to get the additional channel.

Now, almost exactly one year later, Jonathan Blake stood in the same room, behind the same dais, addressing the same audience, and he didn't even try to hide his embarrassment. "In its early days," he said, "the *Saturday Evening Post* ran a series of cartoons in which one of the installments ended with the heroine drinking at night with her married boss in his home while his wife was away. The following installment began with them having breakfast the next morning in his home. Shocked readers complained. The *Post* responded, 'The *Post* cannot be responsible for what the characters in the serials do between installments.' There are times in these annual reports on Washington developments to MSTV's membership when, similarly, we'd like to disown responsibility for the happenings in the interim. This may be one of them."

Thus began a revisionist apologia every bit as stunning as Bob Wright's in that morning's issue of *Broadcasting & Cable*. Unlike Wright, however, Blake explained why his view had so suddenly reversed. Many years ago, he said, MSTV first began asking for additional spectrum "to allow broadcasters to participate in the new HDTV technologies then on the horizon." In the years since, Blake added, "we preserved existing television spectrum in which to implement HDTV," and now, at last, high-definition television was about to become a reality. "But all this could be undone by seismic changes in Washington. It started small. Broadcasters began to investigate, perhaps too publicly, how they might use the sometime excess capacity of the new advanced television channel for additional, ancillary services. But some in government and on the outside reasoned that if broadcasters used the ATV channel for paging services or other nonbroadcast uses, broadcasters should pay for that spectrum, just as other mobile-service providers have recently had to bid for their spectrum.

"Now, however, the tail threatens to wag the dog. Flush from raising $7.7 billion, government is eyeing the ATV channels for similar purposes. If broadcasters intend to use those channels for nonbroadcast purposes, why shouldn't others be eligible to apply for those channels, and thereby jack up auction revenues?"

That cannot be allowed to happen, Blake said, because "the public—all of the public, which only broadcasters serve in its totality—deserves the opportunity to participate in the benefits of digital TV, most centrally HDTV."

The audience, several hundred broadcasters, sat quietly through this, though near the back, Joe Flaherty could not keep a smile from spreading across his face. Blake and the association's other leaders had pressed this view on the organization's board of directors Sunday morning, and after a somewhat raucous meeting the MSTV board had decided it had better adopt this new position.

"It wasn't conceived in advance," Margita White, the association's president, explained a few days later. Then, with a thinly veiled slap at John Abel, she added, "There's been a lot of noise about flexibility, but the owners of the stations have not really been a part of it. When they sat down to talk about it, they decided flexibility is nice, but HDTV has to be the priority. There was complete unanimity on this." So her board formally adopted a new resolution by unanimous vote:

The MSTV board reaffirms its commitment to high-definition television and the use by broadcasters of their ATV channels substantially for HDTV. The goal of public policy and the local television station community should be to transition our nation's free and universal television broadcast system to the digital era. The totality of the American public, which only broadcasters serve, must have the opportunity to participate in the benefits of digital television, and, most centrally, HDTV.

Word of this about-face swept through the convention halls like a fresh breeze. "One hundred eighty degrees, one hundred eighty degrees!" trade press reporters and HDTV partisans chanted back and forth among themselves. "The broadcasters have turned around one hundred eighty degrees!" At the Grand Alliance booth, the partners fairly bubbled.

"They're saying they *want* HDTV," Carnes effervesced. "This is perfect, it's *perfect.*"

By midday Monday, network executives were stumbling over themselves to announce their plans to broadcast high-definition programming. "NBC, I can tell you, is set to do some programming in HDTV as soon as the standard is set," said Michael Sherlock of NBC. Thomas Murphy, the president of Capital Cities/ABC, said his network was "committed to offering high-definition television on the additional spectrum." And if Joe Flaherty spoke for CBS, that network would not be far behind.

Mark Richer, the PBS executive and Advisory Committee leader, could hardly contain his amusement as he tried to explain what was happening around him. "I think it's a lot of things," he said. "Yes, it's concern about the spectrum. But I think they also realize that HDTV will help them differentiate themselves from the competitors. And the success of DirecTV has also made a lot of people think."

DirecTV was a new satellite service that offered almost two hundred channels, picked up on a small gray dish easily installed on the roof or in the backyard. As advertisements for the service pointed out over and over again, the programming was transmitted in digital form. Engineers at the Sarnoff Shrine had designed the algorithm; it was a standard-definition version of the Grand Alliance HDTV signal. DirecTV had been planned for years. In fact, Stephen Petrucci, the Hughes executive, had had this in mind almost ten years before, when he asked Larry Dunham, the head of the VideoCipher Division, if Dunham's company could do high-definition.

Now that DirecTV was finally being marketed, consumers were buying the units far faster than anybody had ever dreamed. With installation, a DirecTV system cost nearly $1,000. And yet after just one year, more than a million units had been sold. No consumer electronics product in history had taken off faster. DirecTV's success was forcing lots of people to reexamine their assumptions about how eager Americans might be to buy a new HDTV. Even now, Jerry Pearlman, Zenith's president, was floating around the convention hall telling every interviewer who asked that his company would begin selling HDTVs "two years from right now," and at first they would cost about $1,500 more than an equivalent conventional TV. Soon after, prices would fall.

Would consumers buy them? That was the big open question. From the very beginning in 1987, both sides in the debate had drawn the conclusions that fit their needs, though there was not one solid piece of evidence to support either opinion. All through these last few years, John Abel and other broadcasters had been arguing that nobody was complaining about the quality of TV pictures—that there was no demand for HDTV. The Grand Alliance, obviously, had a different view.

For months, the partners had debated among themselves about how best to demonstrate their system's many features at the NAB, and the consensus was that they had better show off its digital flexibility. The broadcasters had been crying for that. So the Grand Alliance set up five or six freestanding televisions around the floor. Each showed one or another of digital television's many possible uses. One displayed a BMW advertisement and then showed how the TV had downloaded additional information while the ad played. If viewers were interested in the car, they could click a button and get a list of its features, then see a street map pointing out the closest BMW dealer. Another set showed that digital televisions were in many respects little different from desktop computers—meaning that as the Internet grew more and more central to American life, people wishing to explore the on-line world or simply pick up their E-mail would be able to do so on their television sets. Knowledge of desktop computers would no longer be necessary.

Those and other exhibits drew appreciative remarks. But nothing got more comment than the simplest demonstration of them all: a wide-screen HDTV next to a conventional set, in an alcove in front of a black leather sofa. A demonstration tape playing simultaneously on both sets showed a couple of minutes of a pro basketball game, the changing of the guard at Buckingham Palace, mountain vistas, and shots of tourists milling about in front of the Smithsonian Institution in Washington. The scenery was ordinary, but the comparison was stunning. "Wow," viewer after viewer gushed, so stunned they spoke without thinking. "Would you look at that!" "Boy, that was good. I've never seen a picture so good."

For all the grand promise of the digital revolution, nothing stirred the hearts of these jaded professionals more than the same basic demonstration that had started this long drama more than eight years before, when NHK showed Muse next to a conventional TV in the FCC meeting room. Now without realizing it, dozens of broadcasters were affirming that Bob Wright, Jon Blake, and the other industry leaders were probably correct.

"I don't think we have a business if we end up with less than the best picture," Wright had just said. Certainly that was a tactical statement, not a heartfelt belief. But one fact was clear: When digital televisions did finally come to market, the salespeople at Circuit City and Best Buy were not going to sidle up to customers and croon about multichannel capabilities, paging services, interlineation of documents, or Internet access to the Library of Congress catalog. No, in TV showrooms across the country, salesmen were going to do just what the Grand Alliance was doing here at the NAB. They would put a big new wide-screen HDTV up on a carpeted stand, right next to a conventional set, then run a tape of the latest *Star Trek* movie through both of them.

No contest.

Monday afternoon, Reed Hundt stopped by the Grand Alliance booth. All the partners were there, red carnations in lapels. Glenn Reitmeier, from the Sarnoff Shrine, was the designated tour leader, and as he took the chairman around, the other partners trailed behind. At each free-standing demo, Reitmeier explained and Hundt listened while offering noncommittal nods. Finally the tour wound its way over to the side-by-side show, and the chairman took his place on the black leather sofa. The Buckingham Palace guards' uniforms were sharp, clear, and a vivid red on the HDTV set, while on the conventional TV they seemed blurry and dull. Hundt pointed at that. "You guys have fuzzed that up a little bit, haven't you," he said, in a teasing tone.

"Oh no!" Reitmeier protested, recoiling from even this joking suggestion. "You're just not used to the comparison."

The tour ended with a stop in a small theater, where the Grand Alliance was presenting a fifteen-minute high-definition production taped at a Hollywood studio. It showed off a variety of studio special effects. Hundt sat in the front row next to Pearlman and watched without remark as Zenith's president pointed out one wondrous feature after another. When the show ended, Hundt leaned back toward the other Grand Alliance partners sitting behind him and asked straight out, "When the testing at the test center starts, are you going to be testing it on the dynamic scalability level?" In other words, would the test center measure the Grand Alliance system's ability to offer multiple channels of conventional television as well as HDTV?

That was just about the last question they wanted to hear; Hundt's question fell right into the Grand Alliance's fault line, and Rast spoke up first. "Yes, we are going to break the signal into four bitstreams and show that it can do that," he said. "But we are not actually testing for that per se."

The Grand Alliance had been trying hard to avoid any involvement in this issue. Even though their system was, in theory, perfectly capable of transmitting four channels of conventional TV in the same space as one high-definition program, most of the partners believed that *high definition* was the feature most likely to sell their sets. And General Instrument vehemently opposed promoting or even testing the multichannel feature. GI worked for the cable industry; it couldn't be involved with a project that would give broadcasters several additional channels and a new competitive advantage over the Cable Mongols. So, with Wiley's concurrence, the Grand Alliance had agreed to offer a paper showing that its system was capable of multichannel broadcasting. That was all.

But Hundt wasn't satisfied. "Lots of people have lots of different ideas, and it's important that we don't get hit by somebody with the question whether you are able to do this or that," he said. "They'll say, 'You assert you can do this, but you didn't even test for it.' Better to test for everything now."

Rast answered fast. "Well, my answer to them is, It was an open process, all these things were decided at public meetings—where were you?"

Hundt shook his head. "With great respect to you, Bob, that won't do it. You've got to look under the rug and anticipate everything. Otherwise it just means delay." The others were quiet.

Rast was seething as Hundt got up to leave. The Chairman shook each partner's hand one by one, and to Pearlman he said, "You ought to be very proud." But as soon as he had walked away, Rast exploded, angrily rebuking his partners for keeping quiet as Hundt pushed them in a direction they had already agreed they weren't going to go. "I'm going to have to work on this now," Rast told them through clenched teeth. And in fact the dispute did not die there.

Even with all of this, the next morning it became clear that Hundt had modified his view a bit. Over a breakfast with the broadcasters, he tackled the high-definition television issue head-on. "There has to be the possibility of delivering full HDTV over the air," he said. But then he added: "I am wary about the wisdom of the government mandating how you should take advantage of the opportunities that the digital revolution creates. I suspect you know better than the government what you should send."

In other words, if Hundt had his way the FCC would not *require* the industry to broadcast HDTV. A few months ago that would have seemed like bad news, but now it was not so clear. The broadcasters' competitors—the Cable Mongols—were lining up to promise high-definition television service. When a magazine asked Amos Hostetter, the chairman of Continental Cablevision, the nation's third largest cable company, if he planned to offer HDTV, his answer was unhesitating. "Absolutely," he said. "I think it is going to be a significant competitive disadvantage for anybody who doesn't get there. The picture quality is discernibly superior. I think you'd pay a big price if you didn't introduce it. In the new competitive world, if you are a late adoptee of new technology you're going to get passed by." Could the broadcasters really allow the Cable Mongols to offer a service they didn't have?

John Abel, of course, was the father of high-definition television in America. Everyone also knew that he had been working hard to abort his child from almost the day of conception. Even now, Abel barked while the rest of the industry moved on.

"This is the digital revolution," he grumbled, sitting alone with an interviewer in his small, windowless office on the second floor of the convention hall. "There are going to be all kinds of twists and turns ahead that we don't even know about yet." As for HDTV, "many things could forestall it still. This thing is not clear-cut at all. This could still be strung out for a very long time." Still, not even Abel could ignore the rumbling under his feet. "Production costs for HDTV have come down quite a lot," he admitted. "If the FCC issues a standard, I think at least three of the four networks will offer high-definition programming almost instantly."

Below Abel's office, on the convention floor, the broadcasters were proclaiming the new gospel with no hint of Abel's tight-throated tone. They were lining up to chant their new-old mantra. "Alone among our rivals," Jonathan Blake was saying once again, "we have to use government parceled-out spectrum to implement these new technologies. We preserved the spectrum in which to implement HDTV. But now all of this could be undone."

And from that, the rallying cry was reborn: If we are not allowed to offer high-definition television, that will bring *the death of local broadcasting as we know it!*

21 This crowns it all

Eight display-size color photos, matted and framed, hung as artwork on the walls of the Advanced Television Test Center's waiting room. And as Grand Alliance engineers left for lunch, or just to get a breath of fresh air during the three long months of testing, sometimes they stopped to have a look. The photos had been shot during the first round of tests, when NHK, GI, MIT, Zenith, AT&T, and the Sarnoff consortium were competing for the grand prize. Now, several years later, in the summer of 1995, the pictures evoked the sort of bemused nostalgia a middle-aged man might feel as he looked at a photo of himself caught in an adolescent conundrum.

The first picture showed Keiichi Kubota standing before three blue cases of Narrow Muse equipment as other NHK engineers behind him were hooking it up for the tests. Kubota was biting his lower lip, and the slight smile on his face suggested ambivalence, apprehension.

Jae Lim's face evinced no uncertainty. When the test center photographer asked him to pose, he stood with one hand draped possessively over the MIT-DigiCipher encoder, and his thin-lipped smile projected certitude.

When the photographer asked the Zenith engineers for a picture as they scrambled to get their system ready, none of them even responded. The lobby photo showed them from behind, hunched over with heads buried deep inside their equipment boxes.

Now, three years later, the Grand Alliance system was being poked, prodded, and examined just as its predecessors had. But this time few of the problems or controversies that had bedeviled the first round of testing were evident. The circumstances had changed. This wasn't a competition any longer. In fact, everyone wanted this entry to succeed. Certainly Peter Fannon and the test center's other leaders were intent on ensuring that the test regimen was precise and untainted. But the competitive tension was gone, and when the system faltered, both sides made efforts to fix it. "The test center is being reasonable," as Rast put it a month into testing. "They have made a substantial cultural shift relative to the first round and are trying to help—within the bounds of what is proper."

One problem came up quickly, in an area hardly anybody had thought much about before: audio interference. During the first round of tests, the test center hadn't even examined the audio systems. Nobody had worried much about them; audio engineering was an old field. But as it turned out, when a full Grand Alliance digital signal was broadcast next to a conventional signal—the HDTV signal on channel 13, for example, and a conventional signal on channel 12—the audio signal on the conventional channel was badly distorted. "Everybody is surprised by the audio problem," Fannon said a couple of weeks after it was discovered. The problem was in the transmission system, built by Zenith.

Fannon and his test center colleagues were greatly disturbed by this, but they were just about the only ones. Not surprisingly, Zenith's engineers insisted it was no big deal, and the Advisory Committee's leaders seemed to accept that. Everybody agreed that the system's picture quality was superb, especially in the interlaced mode, and to most people that was the important thing. By now no one really had anything to gain by highlighting any flaws. If there were problems—well, TV manufacturers could deal with them when they started manufacturing the actual sets.

"The great news is that the interlaced picture is clearly better than any of the earlier interlaced systems have been," Fannon said. The progressive picture was quite good, too, though not as good as the other. And when the testing ended in July, Dick Wiley looked over the preliminary results and decided that there were no serious flaws. "The end is in sight," he declared.

If the end truly was in sight, then lobbyists had little time left to affect the outcome. So in the summer of 1995 they turned out in force. Some broadcasters were still lobbying for the option to offer *standard*-definition digital television at least part of the time so they could broadcast several programs in the space normally filled by one. Early in the summer, Reed Hundt called Wiley and Paul Misener over to his office at the FCC and said he wanted them to put together a proposed *lower*-definition digital standard along with the high-definition standard they were preparing. I promised the broadcasters this, Hundt told Wiley. Please take care of it.

Wiley didn't like this idea any more than the Grand Alliance did, but he went along with the chairman's request. His aides began trying to form a consensus around a lower-resolution technical standard. It was only a paper proposal, but of course it set off all the old debates: should it be interlaced or progressive, wide-screen or square, interoperable with computers or not? In the end they settled on several proposed standards, just as the Grand Alliance had done. Everyone's interests were addressed in one or another of the proposals. Later, somebody else could decide how they were deployed.

Through the summer of 1995, complaints seemed to pop up from every quarter. Hollywood filmmakers argued that the shape of the new wide-screen TVs—the "aspect ratio"—was inconvenient for them, even though Hollywood figures had been heavily involved in the original decision almost ten years before.

Computer manufacturers, led by Apple, continued to complain about interlaced scanning. And the major broadcaster organizations began mewling all over again that the transition time was far too short. Under current proposals, every TV station would have to give up its second channel after a decade or so. But the broadcasters argued that millions of people would not have bought digital TVs by then. Hell, they said, some people still have black-and-white televisions *today*!

Transmitter manufacturers complained that if every TV station in the nation had to buy a new transmitter within a two- or three-year period, all those transmitters would have to be replaced at roughly the same time some years later. No one would buy transmitters in the middle years, leading to a boom-and-bust cycle for the industry.

Then the Cable Mongols noticed a problem. Looking at the new standard-definition proposals that would allow broadcasters to offer several channels of programming at once, cable operators wondered what *they* were going to be asked to transmit. Current law required cable companies to carry local TV channels, if the stations requested it. Would cable companies have to carry all this new programming, too—forty or fifty programs instead of nine or ten? What if a station decided to broadcast a single high-definition program at night and multiple standard-definition programs during the day? Would the cable companies be required to switch back and forth at their competitors' whim?

New dilemmas appeared every week. Even Hundt grew frustrated. "The path to the digital future has a lot of S-curves," he said with a sigh in July. But slowly, warily, his agency moved ahead anyway. On July 28, 1995, the Federal Communications Commission finally opened the official digital broadcasting proceedings, the first step toward setting a standard for the nation's next generation of television. The commissioners knew they faced no easy task. "These may be the most complicated questions about broadcasting in the history of the FCC," Hundt said from the dais as the meeting opened. "Everything will be different. The change is so extreme that many people have not grasped it."

The commissioners didn't actually make any decisions that day. Instead they laid out the big questions and gave interested parties several months to submit comments that the FCC would consider before deciding. Among the questions were these:

- Should the commission set any limits on how broadcasters use their second, digital channel?
- Should broadcasters be required to transmit a minimum number of hours of HDTV each week?
- How long should the transition period be; when would broadcasters be required to give one of the channels back?
- Given the cost of the transition, should smaller stations be given special consideration?
- How are existing laws, including the one requiring cable companies to carry broadcast television signals, affected by this?

Right away it became clear that the commissioners disagreed on some of the key questions. "Though admittedly consumer demand for HDTV is unknown," said Commissioner Andrew Barrett, who had been appointed to the FCC during the Sikes era, "I firmly believe that broadcasters' failure to use the 6 megahertz of transition spectrum at least some of the day for HDTV will ensure its demise."

Hundt vigorously disagreed. "That is not true at all," he said. "In fact, flexibility is the only thing that is going to guarantee high-definition television."

But Barrett was resolute. If there is no requirement to broadcast HDTV, he announced, "I will be forced to consider whether I would continue to support giving broadcasters the 6 megahertz of transition spectrum at no charge." And in the following days it became clear that several other commissioners agreed with Barrett, leaving Hundt in the minority. But while the FCC pondered these questions, the United States Congress threatened to scuttle the entire debate.

Months earlier, several members of Congress had talked vaguely about auctioning off the second channel that was to be set aside for HDTV. This had prompted the stunning reversal of view at the NAB convention in Las Vegas. Since then, Senate and House Republicans had laid out their goal to balance the federal budget in just seven years. Among the sources of money they had budgeted for this was more than $14 billion from auctions of unspecified spectrum.

Senator Larry Pressler, chairman of the Commerce Committee, decided that the second channel theoretically allotted to broadcasters was a wonderful place to raise that money. By one government estimate, auctioning the spectrum set aside for the second channel in just the nation's twenty largest cities would bring in $11 billion. Apparently Pressler had spoken to a broadcaster or two at home in South Dakota, and not surprisingly they had told him they weren't much interested in spending money for the new equipment they would need to offer digital TV. So if broadcasters don't really want this, Pressler asked, how can we explain to taxpayers why we are giving away such a valuable resource? Pressler was serious, and his committee staff was writing language to make the second-channel auction happen. Several other senators supported him, and lobbyists piled on, led by the group that started it all.

"It is absolutely ludicrous in the present political environment for broadcasters to suggest that they be granted free and exclusive radio spectrum that gives them the opportunity to provide commercial services in competition with entities that have paid literally billions of dollars for their spectrum!" complained a lobbyist for Land Mobile.

The broadcasters responded the only way they knew how: If you don't give us the second channel, they said once again, we won't be able to offer HDTV. And without HDTV, can we even survive?

The Association for Maximum Service Television reconvened its board to consider a new resolution, updating the one it had passed just before the NAB convention a few months earlier. "The MSTV board reaffirms its goal and commitment to broadcast high-definition television," the resolution said. "The board also reaffirms its commitment to the use by broadcasters of the advanced television channels substantially for HDTV. As part of the commitment to HDTV, broadcasters commit to broadcasting a reasonable minimum number of hours of high-definition television, as determined by FCC rules.

"The totality of the American public, which only local broadcasters serve, must have the opportunity to participate in the benefits of digital television and, most centrally, HDTV." The resolution was passed by unanimous vote.

Separately, the Broadcasters' Caucus, comprised of senior executives from the networks and related groups, took a vote on whether they should ask the FCC to *require* broadcasters to offer HDTV for at least a few hours a day. Everyone said yes, except Fox television and the National Association of Broadcasters. A few weeks later, however, as Senator Pressler's auction idea came up for a vote, the NAB lobbyists were up on Capitol Hill with everybody else taking the pledge: We *want* to broadcast HDTV, they were saying. We promise, we promise.

In the end, Pressler couldn't get the votes to pass the auction plan. Reluctantly he dropped it. But as long as the federal budget debate continued, the auction threat still hung in the air.

Now that the FCC was starting to consider the digital television rules, computer manufacturers began turning their attention toward the commissioners. The July 28 meeting had been a request for comments, and so the complaints from the computer manufacturers poured in. After all, Dick Wiley was barely paying attention to them anymore.

Early in July, an Apple Computer executive had written a screaming letter to Wiley. He'd asked the chairman to "halt the ACATS process at this time, pending input from the computer industry, which will form a task force to provide input to ACATS." Publicly this Apple officer said: The Grand Alliance will take the nation "into the future with a jaunty air and an anchor around our necks." His remarks were broadcast on the Internet, and his biggest complaint was, of course, interlaced pictures. To Wiley this complaint seemed so old, so stale. In any case, he would say, the Grand Alliance offers several transmission formats, and only one of them is interlaced. We've been through this so many times already. At the same time, broadcasters offered their own familiar rebuttals. Interlaced pictures are *better,* they would say. And besides, no one has even begun development of progressive-scan television cameras. When they do come along, they're certain to be ruinously expensive.

But in truth the situation had changed. By the summer of 1995, the Internet had grown to be a very visible part of American life. Just about every business had established a home page on the World Wide Web. Movie studios set up Web sites for each new release and published the Web addresses on billboards and in newspaper ads ("Preview SUDDEN DEATH on the Internet: http://www.mca.com"). In fact, all manner of companies were moving onto the Web. Ralph Lauren advertised a Web address for his new face cream: http://www.ralphlaurenfragrance.com. Even Kraft foods, a company not usually associated with high-tech marketing, ran magazine ads listing the Web site readers could visit for more information about its Boboli brand pizza crust: http://www.boboli.com.

By fall, Internet fever was sweeping the nation. Still, the hype surrounding it was spreading even faster. At the beginning of 1996, the percentage of Americans who used the Internet regularly was relatively small. The Electronic Industries Association reported that 40 percent of American homes had a computer, and fewer than half of these machines were equipped with a modem, the device that allows computers to communicate over telephone lines. Without a modem, Internet access was impossible. What's more, no one really knew how many computer-and-modem owners actually visited the Internet. The most generous estimate put the number at about 24 million. Other analysts said that that number was inflated, and a more realistic estimate was that about 10 million people—less than 5 percent of the public—regularly browsed the World Wide Web, a subset of the Internet.

Still, that number was growing fast, as was the relative importance of the Internet to the television industry. In December, NBC and Microsoft announced a deal to start a new all-news cable channel that would work in conjunction with Microsoft's Web site. And few in the industry could fail to note that when digital TV came along, viewers would not have to walk from the family room over to the den to call up the World Wide Web on the computer. They would be able to access the Web right there on the television screen, perhaps in a pop-up window activated with a remote control. In fact, early in 1996, computer makers and television manufacturers grew impatient and began offering television-capable computers and Internet-capable television sets using existing analog TV technology. These early devices were expensive, cumbersome compromises. Buyers usually wound up with a television that wasn't as good as others that cost less, and a sub-par computer as well.

But Internet mania was spreading so fast that the manufacturers didn't think they could afford to wait.

No one knew for sure where all of this might take the TV industry—or the computer business, for that matter. But surveys showed that the number of people who used the Internet had doubled in 1995, and no one expected that growth to slack off. Computer companies wanted to ensure that the new technology did not stand in the way of whatever came, for seismic changes affecting the entire nation appeared to be just over the horizon.

Sharks still circled; computer executives and congressmen still rumbled. The FCC still hadn't decided what to do. But by September 1995 Dick Wiley not only saw the end of his role in the drama, he scheduled it.

Laboratory testing of the Grand Alliance system had been completed on July 21, and now the equipment was in Charlotte for field testing. All the results so far convinced Wiley that no big problems had been found. So he wrote to all the other Advisory Committee members, saying, "Based on early reports from our technical experts, I am confident that the Grand Alliance system will prove to be worthy of the committee's recommendation to the Commission."

And so, eight years after he began, Wiley scheduled the final meeting of the Advisory Committee on Advanced Television Service for November 28, 1995. A year earlier, Wiley had worried that he might not even get a majority in the final vote, but now, Misener reported, "Dick is trying to get a unanimous vote." The chairman wanted this last meeting to go like all the others—every decision made, every vote counted, even before anyone arrived. A unanimous vote would be a fitting close to eight years of his stewardship, and it didn't seem impossible now. At the beginning, Dennis Patrick had stacked the Advisory Committee with broadcasters, and the industry's ever changing opinion had swung back in favor of HDTV. It seemed likely to hold there for at least a few weeks.

Lawyers in Dick Wiley's law firm, meanwhile, were at the Capitol lobbying their own issues, and one of them reported back that the NAB had used an interesting new argument to explain its change of view on HDTV. John Abel led us astray, they were saying.

Late in the summer, Abel left the NAB. High-definition television partisans were convinced he was forced out because the industry's position had changed, while Abel's had not. Abel insisted that was not so, and he was probably correct. He left to start a new company that would look for business opportunities in digital television. "This is my opportunity to implement a vision," he explained. As for his strong advocacy of digital flexibility instead of high-definition broadcasting, "I think it's clear that it was an important educational tool," he said. "Really, nobody was even talking about it before."

All along, none of the other broadcasters had been quite as willing to say exactly what they really thought. In the end, that got Abel into trouble. But John Abel started the drama. When he left, it seemed as if the show had lost its star.

Resolution finally came to Zenith. The company seemed unable to stanch its losses during the first half of 1995, even after bringing in one profitable quarter in 1994. For the first quarter of the new year, Zenith lost $24.3 million, double the deficit a year earlier. Soon after that, layoffs rolled through the Bunker once again. Ten percent of the company's remaining staff was let go.

But even that was not enough. Finally Zenith gave up. In November, the board of directors voted to sell a controlling interest to LG Electronics of Korea—Goldstar—for $351 million. Goldstar already owned a small portion of Zenith's stock, and the two manufacturers had cooperated on a number of projects, including the design of a digital VCR. Goldstar promised to keep Zenith as a separate company with its own name and management, though the board of directors would change to reflect Goldstar's 57.7 percent ownership share.

In the Bunker, the decision was greeted with both resignation and relief. No one welcomed foreign ownership, but at least the company was stable now. No more layoffs were likely. For eight long and difficult years, Zenith's leaders had hung on, hoping HDTV would save them. It turned out to be a bad bet. Technology evolved, moving in directions no one at Zenith (or anywhere else) had adequately anticipated. And the federal government, sponsor of the HDTV race, turned out to be an oh-so-fickle partner.

Nonetheless it was the allure of HDTV, in part, that attracted LG Electronics. Lee Hung Joo, the company's president, told a *New York*

Times reporter that Goldstar had first become involved with Zenith partly to get access to the company's high-definition television technology, adding, "if you want the technology of a company, you have to revitalize the company first." Did Lee expect anybody to protest the idea that a foreign company was buying America's last remaining television manufacturer? No, he said. "Don't you think they feel better than if Zenith were sold to a Japanese company?"

Jae Lim dropped out, too. In the fall of 1995, he took a sabbatical from his teaching job—to ponder his future, he explained. "I want to go back to teaching and writing," he said in October. "I've had enough of being involved in business." Of course, he didn't have to remain involved to get quite rich. When digital TVs went on sale, Lim stood to earn a lot of money.

Others at MIT continued as they always had, certain that they knew better than anyone else what the nation wanted and needed. HDTV "is one big yawn," Richard Solomon, one of the MIT guerrillas, posted in a general-circulation Internet memo in August. "Anyone with an eye in his head can see that interactive video will be on personal computers, not television sets." And in his 1995 book, *Being Digital,* Grand Vizier Nicholas Negroponte agreed, arguing once again that high-definition television "will be stillborn" since it's "just silly."

Keiichi Kubota, the NHK engineer, was leading a team back in Tokyo that was designing the network's own digital, high-definition television, even though most of the NHK's leaders still backed the old Muse system. By 1995, fewer than a hundred thousand Japanese had bought Muse sets. Prices were still ridiculously high.

As his team worked on a digital system, Kubota closely followed the tortured turns of the American effort, knowing that if the Grand Alliance system succeeded, the old guard at NHK would have to abandon Muse and take his team's research seriously. So he came back to Washington for each major Advisory Committee meeting to gauge the state of affairs. "I'm a little sad," he said as he showed up for the final Technical Subgroup meeting in October 1995. NHK had started all this thirty years before. Still, here he was watching the story's conclusion, "and we are not a part of it any longer."

A few months later, in the spring of 1996, the Japanese government quietly let it be known that Japan would switch to an all-digital system by the year 2000—a tacit admission, the newspaper *Asahi Shimbun* reported, that the government had made a mistake by sticking with analog technology while the rest of the world embraced digital.

Through the fall, Peter Fannon searched for a new job. No one expected the networks to continue funding the Advanced Television Test Center once Wiley's Advisory Committee shut down. But in November he stopped looking, at least momentarily. Westinghouse had bought CBS a few weeks before, and just after that sale the test center announced that Westinghouse/CBS had made an initial donation that would allow the test center to stay open at a much reduced level "to support the design and development of broadcast-related technologies for digital, high-definition television and other related television services," the announcement said. In other words, as the networks tried to figure out what to do with their new digital equipment, they could use the center to test their ideas.

The work didn't prove to be very rewarding in a lab with hardly any staff. Fannon even had to answer the phones. Then, early in 1996—as Congress and the FCC began giving serious consideration to digital television—consumer electronics manufacturers, labor unions, and a scattering of other organizations formed a new lobbying organization called Citizens for HDTV. Fannon was chosen to head it.

Woo Paik left General Instrument early in 1995. He'd been promoted into a job he didn't especially like, and he was wealthy enough that he didn't have to work at all. But Paik's real goal was to start a company of his own and produce new products based on the digital research he had done at GI. He worked out a business plan, and for a while he talked to Jerry Heller about joining him. Heller, too, had left GI as a wealthy man searching for a new project. But the collaboration idea didn't work out, and in the end Paik couldn't get financing to start his new company anyway.

So in November he took a senior engineering job with Qualcomm, another San Diego high-tech firm best known, at least to a segment of the public, for its popular Eudora E-mail software. Paik knew Qualcomm best for its chairman, Irwin Jacobs, the former MIT professor who had lured Jerry Heller out to California almost thirty years earlier to start Linkabit. This was also the company that had hired Paik, despite his uncertain immigration status, on a snowy winter morning more than fifteen years before. Having failed at starting his own company, Paik explained, "I decided to work in a corporate environment again, doing the same thing I intended to do. I'll be working on developing new products, just like I did at GI. I'll be trying to prove myself again in a new environment." But after a few months Paik decided he preferred his old environment. And with coaxing from Bob Rast, he came back to GI as an executive vice-president for technology, with permission to pursue just about any research project he wanted.

For his part, Rast remained a senior vice president at General Instrument. One of his new duties was to find business opportunities for the company in the digital, high-definition television world.

The FCC "notice of proposed rule making"—that list of questions published July 28—awoke a host of dormant players. In the fall of 1995, they began grousing about the plan to loan the broadcasters a second channel.

"The public owns these coveted airwaves, and for the government to give them away with no additional public interest safeguards and no compensation to the American public is scandalous," complained Andrew Blau, director of communications policy for the Benton Foundation, a member of an advocacy group coalition that had formed to fight the second-channel loan. Working in concert with him and others were lobbyists for Land Mobile and related business groups. All of them were fighting over the very same space on the airwaves that had been in dispute back in 1986, when John Abel launched his HDTV lobbying scheme. Now, nine years later, the process he started was crawling toward its conclusion, and Land Mobile was back. This time their arguments struck a chord, and soon newspapers across the nation began publishing not particularly well-informed editorials urging the FCC to auction the second channel.

"Don't Let the Broadcasters Off Free" cried the headline over an editorial in the *New York Times* on October 25. "Congress plans to take $270 billion out of Medicare, almost $200 million out of Medicaid and about $40 billion out of tax credits for low-paid workers. Yet it apparently has enough cash to send politically powerful broadcasters a gift worth perhaps $40 billion. The gift will come in the form of free licenses to use newly available frequencies to broadcast digital television programs and services.

"The broadcasters argue that they cannot afford to provide free, over-the-air television if they are forced to bid on the digital licenses. The argument is bogus." (Seven years before, the same editorial page, under different leadership, had declared, "For America not to compete in HDTV would be tantamount to abandoning a wide range of markets. But it may already be too late.")

Even with all this loud public support, Senator Pressler and the few other conservative Republicans who agreed with him continued to find that the rest of the Senate, as well as most of the House of Representatives, was uninterested in taking on the broadcasters. As the 1995 budget debate raged on through the fall, one auction proposal after another failed. Late in October, Senator John McCain of Arizona offered still another budget amendment requiring a full-blown auction of the second-channel spectrum. It lost by a vote of 64 to 25.

Nonetheless, in November Reed Hundt stepped into the fray and endorsed Senator Pressler's auction idea—lowering his standing among broadcasters and HDTV advocates even more, if that were possible. The auction proposal "has real merit," Hundt said. And then he made a showy visit to the Advanced Television Test Center, where he asked Peter Fannon's crew to give him a special demonstration of *standard-definition* digital television alongside HDTV.

For years, high-definition television advocates had been showing their product alongside conventional, *analog* TV. But neither the original contestants nor the Grand Alliance had ever offered a side-by-side show of high-definition beside standard-definition *digital* TV, which offered a cleaner picture even though the resolution was no higher. Political disagreements within the Grand Alliance had prevented it. Besides, the goal of Dick Wiley's race, and the Grand Alliance's raison d'être, had always been to give America *high-definition* television. By the fall of 1995, though, well over a million American homes had installed the new, small-dish, direct-satellite television service, and it provided standard-definition digital TV. So millions of people had seen it, and most of them liked it.

Hundt was convinced that standard-definition was good enough.

And in truth, digital standard-definition did look better than conventional, analog TV, even though the number of lines on the screen was no different. None of the defects that crept into the analog picture during the transmission process—ghosts, snow, interlace jitter—were present. With digital TV, every home got a "studio quality" picture or nothing at all. So as Hundt watched the demonstration Fannon staged for him, "if you looked hard you could tell the difference," *Broadcasting & Cable* magazine said, quoting one person who watched with Hundt. "But they both looked great." So, Fannon said later, Hundt "went away thinking: Standard definition looks just as good as HDTV, so what's the big deal?"

But Fannon didn't think the demonstration had been fair. The test center had been able to show Hundt only *uncompressed* standard-definition digital TV. The center had never acquired the equipment to run real-world tests of standard-definition TV that was heavily compressed so that it could be broadcast. Hundt hadn't seemed to care about that and "drew the wrong conclusion for the wrong reason," Fannon said. Bob Rast continued to argue that true high-definition TV crossed a visual threshold that standard-definition TV could never reach. "You're fooled into thinking you're looking at the real thing, not a picture," he said. "It's almost three-dimensional."

But Hundt wasn't listening. The demonstration sealed his view. "HDTV is a very marginal and possibly superfluous venture, because standard-definition digital is going to be just as exciting visually as high-definition," he told *Video* magazine a couple of weeks later. "[With the Grand Alliance system] you can have 18 million bits of information devoted exclusively to a 1,080-line image, which is called high definition, but there won't be any point to that for most people because standard definition is going to get just as nice a picture. So why not have three channels simultaneously at 480 lines, as opposed to one channel with 1,080 lines? They'll look the same. The notion of high-definition as a single or so-called second channel for broadcasters is defunct."

However, Hundt knew he could not impose his views all by himself. The chairman was at war with several of his fellow commissioners, who began loudly complaining that he was pushing proposals they did not endorse—including one that would require broadcasters to air several hours of educational television each week. Hundt had even suggested he would not loan them a second channel for digital TV unless they agreed.

Late in October, Commissioner Andrew Barrett exploded in anger, telling a group of reporters, "I am calling Reed Hundt a gutless, leaderless liar. I don't think the chairman has a high regard for the intellect or the judgment of his fellow commissioners. I don't know whether he's listening to us or not." Commissioner James Quello accused Hundt of "administrative extortion."

Hundt's response was philosophical. "Being chairman is very difficult," he said. "Your job as chairman is to get things done. But if you are a commissioner, your primary power comes from being able to block things."

When the *New York Times* published a long, prominently played story about all that, *Broadcasting & Cable* magazine came back the following week with an article anonymously quoting half-a-dozen presumably senior broadcasters who vented their own distrust of Hundt. "Nobody knows where he is coming from at any given moment," said one. "There is no respect in our industry for this man," said another. "No public official has been discredited as much as he has. It's the worst I've ever been around."

Not surprisingly, the primary focus of the broadcasters' rage was Hundt's advocacy of the second-channel auction. "While maintaining an official neutrality in the congressional debate," the magazine wrote, "the broadcasters say Hundt has worked against their [second-channel] plan. As they see it, Hundt tried to incite other industries against it, distributed estimates of the value of the spectrum that renewed questions about a broadcaster 'giveaway' and endorsed Senator Larry Pressler's plan [to promote auctions]."

So the television industry responded with its well-worn argument. As Edward Reilly, the president of McGraw-Hill Broadcasting, put it, auctioning the second channel "could kill free, local, and universal television service as we know it. I don't mean injure it around the edges, destroy the prospects of marginal stations, or delay the transition to digital high-definition. I mean kill the service!"

In early November, the four major television networks—ABC, CBS, NBC, and Fox—wrote a joint pleading to the FCC that asked the commission to *require* high-definition broadcasts. "The only way for the commission to assure that enough HDTV programs are in fact offered is for the commission to require each broadcaster to offer a minimum number of hours of HDTV," wrote ABC. "This type of programming will highlight the capabilities and attractiveness of HDTV to the consumer," added NBC. And CBS piped in to say, "The purpose of such a requirement would be to assure a fair marketplace test of the public acceptance of HDTV." Another purpose, of course, was to assure that the broadcasters got their second channel.

Finally, Hundt said he would leave the auction decision to Congress. Lobbying and debate continued into the new year, because the issue had been subsumed in the titanic budget fight between President Clinton and the Congress. As for whether or not the FCC would require high-definition broadcasts, Commissioner Barrett and at least one or two other commissioners wanted the broadcasters to air some high-definition programming. So the chairman knew that he could look forward to another fight when this question came up for a vote.

In the end, though, it wouldn't really matter whether the FCC required high-definition broadcasts. Even if the networks' promises to provide the service turned out to be wholly disingenuous, once they installed digital transmitters the networks would be able to broadcast high-definition programs at little extra cost. For decades, movies and prime-time action/adventure programming had been recorded on 35 millimeter film, a high-definition format. The networks already had vast libraries of this material. All they had to do was show it. And given the industry's highly competitive climate, they'd probably have little choice, since at least some cable and satellite TV services would certainly offer HDTV. High-definition movies would probably be available on videodiscs, too.

So there seemed little doubt that many TV stations would race to offer some HDTV, just as they had competed to provide color TV in the '50s and '60s and stereo television in the 1980s. In fact, at the end of 1995 a tiny item appeared in *Communications Daily* that lent support to this: "Dielectric Communications will design two HDTV antennas to be located atop the World Trade Center in New York City," it said. "Antennas are to be used by nine TV stations: WABC-TV, WCBS-TV, WNBC-TV, WNYW, WNYC-TV, WPIX, WWOR-TV, WNJU, WXTV." Among those stations were the affiliates owned and operated by the major networks. "One antenna tower will be used exclusively for HDTV; the other also will include NTSC antennas."

Here was a concrete indication that maybe, just maybe, the broadcasters were telling the truth. Finally the nation's consumers—the one group that had never been consulted during eight long years of regulation, research, and debate—would get a chance to see high-definition television and decide whether they liked it.

The final meeting of the Advisory Committee on Advanced Television Service was scheduled for November 28, 1995, and in the days leading up to it Wiley worked hard to prearrange every moment. More than anything else, he wanted his committee to issue a unanimous vote in favor of his recommendation to the FCC. At the last minute, though, Reed Hundt threw a handful of sand into the works. He appointed three new members to the Advisory Committee, and two of them were from the computer industry—one from Microsoft, the other from the Digital Equipment Corporation. Wiley had no problem figuring out what their concerns would be, and he spent the last weeks searching for a way to bring these new appointees into line. Of course, they came to every discussion programmed to complain about interlace scanning, but Wiley thought he had things under control. Meanwhile, his committee's leaders had received the final test results and declared that "the Grand Alliance system meets performance objectives and outperforms any of the four competing digital TV systems proposed earlier by Grand Alliance members. The system is superior to all known alternative systems." Wiley called the machine "an inspired technical triumph."

"It's the best of the best," Joe Flaherty chimed in.

Meanwhile the world's TV manufacturers began saying they wanted the FCC to set a digital TV standard *now,* so they could begin building and selling the new sets. The president of Sony Electronics in America publicly urged the FCC to hurry up and set the standard, adding that his company would work in conjunction with the Electronic Industries Association to speed things along. This group had been sending a letter to the FCC almost every week pushing the idea of an HDTV requirement. High-definition, the association still believed, was the service that would sell the sets.

At the same time, a consortium of more than fifty consumer electronics companies agreed on a standard for digital, high-definition videocassette recorders compatible with the Grand Alliance system. Conventional VCRs could not record a digital signal. The idea behind this agreement was to avert the VHS/Betamax format battles that had afflicted the VCR industry in its early years. The manufacturers knew full well that digital TVs would not sell if customers had no VCRs that could work with them.

Then on November 21, a week before the Advisory Committee meeting, Hundt gave a speech to the International Radio and Television Society in New York. The day before, the FCC press office called reporters and pitched the speech hard: This one's *important,* they said. "Digital TV: We Can Work It Out," the speech was titled. But for the designers of the nation's new digital TV system and the manufacturers who were so eager to build it, the speech might better have been titled "Digital TV: I Can Work You Over."

The networks' request for an HDTV requirement had landed on Hundt's desk just a few days earlier, and almost as soon as the chairman took the podium he remarked that "a lot of us were startled to see industry asking for an unprecedented level of regulatory micromanagement of digital broadcasts." Hundt declared that he was not interested in "force-marching the eminently successful American broadcasting industry and its hundred million home audience from analog to so-called high-definition television." But as Hundt went on, it became clear that he wasn't questioning simply the need for a high-definition requirement. The chairman was asking whether the public really wanted digital TV at all.

The Grand Alliance system was now complete after eight years of effort fostered and carefully directed by Hundt's own agency. Thousands of man-hours of work, costing about $500 million overall—all of it carried out with an explicit government promise of reward—had produced the world's most advanced television. It was a technological marvel that had forced Japan and Europe to abandon their own systems. Even Hundt frequently described the Grand Alliance's creation as "a wondrous genie in a bottle." Manufacturers were growing restive; *they* were willing to gamble that the public wanted these new televisions. But now, just one week before the conclusion of this great drama, Hundt was asking:

Is a government-mandated transition more of an expulsion from Analog Eden than a pilgrimage to the Digital Promised Land? Does the audience want to go on this journey? Can we be sure that consumers will purchase tens of millions of digital TVs that each cost between 15 percent and 50 percent more than analog TVs? Maybe instead consumers will regard the necessity of upgrading to digital reception as a multibillion dollar tax on what they have always thought was free television.... Zenith's Jerry Pearlman predicts that *a full decade from now only 20 percent of the country will have purchased a high-definition display. This isn't surprising since the average analog TV price today is about $300, and a high-definition home-theater display of the sort Jerry is talking about would run well north of $2,000.*

Hundt went on in that vein for several minutes, adding, as a final slap, "I believe the issue of expanding our auction authority to include, possibly, digital broadcast spectrum deserves serious, dispassionate study."

Almost as soon as he had finished, Pearlman faxed him a memo complaining that he had been misquoted. He'd never said any such thing. "While I always like the publicity when my name is spelled right," Pearlman wrote, "I like it even better if the quote is correct and in full context." As for the multibillion-dollar-tax idea, the equipment manufacturers believed that the prices for digital televisions would fall rapidly once they were introduced, just as prices had fallen for almost every other successful consumer electronics product of the last two decades. Pearlman, in his fax to Hundt, said Zenith expected to sell HDTVs for about $400 more than conventional, analog TVs by the fourth year of sales—roughly, 2001. And by the end of the ten- or fifteen-year digital-TV transition period, in the normal course of things most people would have replaced their televisions anyway. After that, anyone who still didn't want to buy a digital TV would be able to purchase an inexpensive converter box so that they could continue using their old analog TVs. Hitachi had already developed one of those. And the Clinton Administration had even begun talking about finding ways to subsidize the purchase of them.

As to the question of auctions—if Congress left the decision to the FCC, Hundt simply didn't have the votes. Lobbyists who followed the FCC day by day saw that the other four commissioners were firmly against the idea. Given the history and the current state of politics and technology, several important parts of Hundt's big speech just didn't make sense. But it certainly set the mood.

A week later, on the morning of the final Advisory Committee meeting, Bob Rast sat in General Instrument's Washington office and groused. "A year ago, Hundt was outraged because we were late. Now he says he inherited this problem from Al Sikes. But Hundt's been here two years now. He's an *owner* now. If he doesn't like what we've done, then

I say he led us down the garden path." Flying into town the previous evening, Rast had run into an Advisory Committee member on the airplane, who told him, "This is your big day. Congratulations!" But now, Rast said, "it just doesn't have the taste it should have."

Three hours later, just before two in the afternoon, Advisory Committee members, Grand Alliance partners, Federal Communications Commission officials, and scores of others began gathering in the commission's eighth-floor meeting room for the denouement. This was to be Wiley's carefully prearranged final meeting, the concluding moment when the committee would vote on its recommendation to the FCC and then dissolve. Everyone was there. Jim Carnes, Bob Rast, Jerry Pearlman, Jae Lim (who sat by himself, marking papers with a highlight pen while everyone else mingled), Peter Bingham of Philips, Bob Graves of AT&T, and many others. Only John Abel was missing from the room, where he had started all this nearly nine years earlier. Everyone else milled in the growing crowd, greeting each other, celebrating the completion of the Grand Alliance's work and the unambiguous success of the Advisory Committee—despite the pall that Reed Hundt had thrown over the event.

Ten minutes early, Wiley bustled in and began shaking hands and patting backs, working committee members to the end, pushing for the unanimous vote he so wanted. Hundt's speech was still on his mind, and Wiley adopted his sternest no-nonsense tone as he told a friend, "Hundt asked me what I thought of it, and I told him I didn't like it. We had a frank discussion. I told him I'm not going to throw stones, but I didn't like it."

A few feet away, Michael Sherlock, the NBC executive, was proclaiming the gospel; his network, Sherlock said once again, was "committed to HDTV." Next to him, Preston Padden, a Fox executive, smiled faintly but said nothing. A few months before, Fox had scornfully opposed the idea of offering HDTV. Asked about that now, Padden put up his hands and protested, "We've been good; we've been quiet. We're being good." Fox had signed the network resolution calling for an HDTV requirement but had added no ratifying language of its own. "We're scared," Padden added.

Behind him, Saul Shapiro, a senior aide to Hundt, was doing damage control. "We don't have the authority to do auctions," Shapiro was saying. His necktie was decorated with cockeyed TV-set caricatures, each of them sporting long rabbit ears. "Only Congress can give us that authority, and the chairman's not asking for it. He's not going up to the hill asking for it." That brought dubious grunts from his audience.

At two o'clock, the scheduled hour for the start of the meeting, Wiley pushed his way toward his chair, but stopped for a moment to whisper in the ear of one audience member.

"I want to share with you that I'm not going to get a unanimous vote," he confided. Two of Hundt's additions to the Advisory Committee—the representatives from Microsoft and Digital Equipment—weren't going along with the program, because the Grand Alliance system still offered one interlaced transmission mode. "There will be some abstentions," Wiley said with a small sigh.

Nearby, Jim McKinney was muttering, "If Microsoft opens his mouth with more complaints about how the computer industry was not included, I'm ready for him. I've got a fifteen-minute statement, and I'm going to read the whole thing!"

A moment later Reed Hundt slid into the room and took his seat next to Wiley at the head table, set up along the east wall. It faced three other long tables arranged to make a rough W shape that seated all thirty-six members of the Advisory Committee, who now settled into their chairs and glanced quickly around the room, trying to capture the historic moment in their memories. So much had already happened here.

Just behind Hundt's chair a 240-volt socket still poked out from the floorboard, installed nine years before for NHK's Muse equipment, which had been laid out on a table along this north wall. From the commissioners' dais a few yards to Hundt's right, where press photographers now sat, Chairman Dennis Patrick and the other commissioners had chartered the Advisory Committee eight years, one month, and twenty-eight days earlier and named Dick Wiley chairman. Five years later, Chairman Al Sikes and his commissioners had sat in the same chairs as they issued the rule that Hundt was now so intent on overturning: Every TV station will be loaned a second channel for the transition to HDTV.

And just eighteen months ago, Hundt had sat where he was sitting now, sandwiched between Wiley and Joe Flaherty as he addressed a Technical Subgroup meeting. It had been called so Hundt could utter the words of support that would scatter Wiley's enemies. Instead, however, Hundt had ignored the talking points that Joe Flaherty had so shrewdly prepared and given a speech that sounded as if John Abel had written it.

Now, addressing the Advisory Committee's final meeting, Hundt kept away from the contentious areas he raised in his speech a week earlier. "Today is an historic moment in the annals of American broadcasting," he said as he looked around the room. "At the outset I want to salute the members of the Advisory Committee and the more than one thousand volunteers who contributed over the years to its development for producing a remarkably capable, high-performance, flexible, and fundamentally mysterious system that far exceeds the visions and goals of the committee's early days in 1987. The Grand Alliance system is a prime example of what private enterprise, unfettered by strict government guidelines or instructions, can and will provide in a technologically competitive world."

Finally the chairman promised, "We will not let you down in terms of moving this forward in an expeditious manner. But let's not worry for just a few moments about the work ahead. Let's celebrate our accomplishments so far instead."

"Thank you for those meaningful remarks and for your presence here," Wiley said when Hundt finished. Many in the room listened carefully to see if a droll tone had crept into Wiley's voice. It was hard to tell.

Looking at his committee, Wiley immediately attacked his unanimous-vote problem once again. "I want to say that there are legitimate and sincere differences among members of the Advisory Committee and members of industry. But we've made an earnest effort to find a solution. There are policy differences, too, but we will reserve them for a later day to be determined by Chairman Hundt and his colleagues."

After that came a few official reports; most of them were deserved self-congratulations for the long job now complete. Robert Hopkins, the former Special Panel chairman who was the executive director of the Advanced Television Systems Committee, the broadcast industry technical group, described how his organization had "documented" the Grand Alliance system, producing a paper that set out all the system's technical details so that manufacturers could duplicate it. That "ATSC standard" was to be the basis for the recommendation today.

"And if you'd like to get a copy," Hopkins said, "it's available on the Internet." Craig Mundie, the Microsoft representative, smiled at that.

A few minutes later Wiley came to the comment period. Maybe the FCC was going to make the policy decisions at a later time. But Wiley intended to take a last shot here. So he called on Thomas Murphy, the president and chairman of Capital Cities/ABC. "Tom, I understand you have something you'd like to say," Wiley called out, affecting a tone suggesting that he had no idea what Murphy had on his mind.

"I'd like to take this occasion," Murphy said, looking down at a sheet of paper, "to reaffirm my company's commitment to the implementation of high-definition television." Cable television is going to offer it, Murphy added, "so we must offer it over the air to remain competitive. In the short run, it's unlikely to give us any market gain. But assuming the spectrum is assigned, it is our intention to do it to meet our public interest obligations."

Wiley nodded with a smile and then turned to Ervin Duggan, the former FCC commissioner, who was now the president of PBS and the third of Hundt's Advisory Committee appointees. "Ervin, I understand you have something you want to say?"

Duggan looked up at Wiley with a slightly quizzical expression and asked, "Dick, did you want me to speak on the recommendation?" Wiley stuttered a bit as he answered, "Uh, you should say whatever you want to say."

"Well," Duggan said, "I'd like to say that PBS is fully committed to the full implementation of HDTV."

Michael Sherlock, the NBC vice president, was the last voice in Wiley's Greek chorus. "As you know," he said, "NBC is straightforward in our support of HDTV. We certainly hope to get it implemented as soon as possible."

After that Wiley opened the floor to comments from other committee members—remarks that had not been prearranged. First up, a cable TV representative complained (as Hundt had warned months earlier) that a *standard-definition* digital system had never been tested. That set off a round of pursed-lipped assurances: We know it will work. No big deal. Don't worry about it. Then Craig Mundie, the Microsoft representative, raised his hand. Wiley called on him and hunkered down in his chair.

"It's obvious that during the progression of this process, the personal computer was launched," Mundie said. "I find it interesting that if you want to get a copy of the standard document, you can get it on the Internet. Even a year ago, everyone was less aware that the convergence between TVs and computers was taking place. This standard is a compelling improvement over analog, but there is a realization in the computer industry of the effects some choices that have been made will have." His complaint was veiled, but his meaning was clear: *Interlace!*

Jim McKinney sat up like a rifle that had been cocked. Wiley called on him, and McKinney read his fifteen-minute statement, starting with a list he'd drawn up of endorsements, contacts, and meetings involving the computer industry over the last couple of years. McKinney didn't mention the first one, which had taken place three years earlier, when Michael Liebhold, the Apple functionary, had pressed his complaints on Al Sikes. McKinney had been there then, too, sitting not far from where he sat now, and he'd barked at Liebhold, "How dare you bother the chairman with this!" But Sikes had stopped McKinney short, saying, "You are being rude."

Now McKinney was saying, "Our attention has been constantly focused on the needs and interests of the computer industry."

None of this swayed the other computer representative, from Digital Equipment, who looked up as she said, "We need to strive to get to an all-proscan format as soon as possible if we are to remain competitive."

"MIT supports the statement of Digital Equipment Corporation," Jae Lim squawked, and his Grand Alliance partners rolled their eyes.

The discussion was over. At last it was time for a vote, and it looked like the outcome was going to be messy after all. Wiley's pre-printed press releases announcing the vote yet to come already lay on a table. The Advisory Committee "overwhelmingly" approved the recommendation, they said. That was the best Wiley thought he'd be able to do. No unanimous vote. Somebody read the motion: "The Advisory Committee on Advanced Television Service recommends that the Federal Communications Commission adopt the ATSC digital-television standard as the U.S. standard for advanced television broadcasts."

Several people, including Michael Sherlock of NBC, waved wildly to be on record as seconding the motion. Around the room then, people settled back in their chairs, pulled out pens, and prepared to mark their committee rosters with the votes: Yes, No, or Abstain.

But then Wiley sat up in his chair, and with no hesitation he bellowed, "Will all in favor say Aye."

Aye! an unknowable number of voices around the tables called out.

"All opposed say Nay," Wiley shouted immediately after. No one spoke up, and about two seconds later the chairman shouted, "Thank you very much!" and gaveled the meeting to a close. Around the room, committee members and others stood up, shaking their heads with bemused, startled smiles on their faces. Wiley had offered no opportunity for abstentions!

In the hall moments later, Wiley bubbled happily, "It all worked out. It was a unanimous vote!"

Did everyone on the committee know he was going to call a voice vote?

"Well, I don't know if they did or not," Wiley said, grinning. "But my momma didn't bring me up stupid."

Nobody complained in the hours ahead. Unanimous or not, the motion certainly would have passed "overwhelmingly," no matter how the vote had been taken. Right after the vote, Reed Hundt took the entire Advisory Committee back to his office for a group portrait; they all stood in front of a large American flag that hung on the back wall. For the last shot, Hundt and Wiley posed in front of the group shaking hands, false smiles on both faces.

Moments later, Hundt stood at his office door shaking everyone's hand as they left. Jerry Pearlman of Zenith was one of the last out, and he patted Hundt on the back as he chastised the chairman. "Now that part's done," Pearlman said. "One thing at a time. Now we have to get you over some hurdles." Hundt said nothing.

From there, the group moved to the Sheraton Washington hotel for a reception honoring Wiley. Advisory Committee members and Grand Alliance partners read tributes and gave him plaques, including one adorned with a large brass rat. They pressed him to tell the now legendary brass rat joke once again, and he reluctantly complied. The punch line—"What I want to know is, do you have any brass lawyers?"—brought gales of laughter.

After that, Wiley spoke briefly. "The real payoff for all this work," he said, "will be when we actually get this available to the American public." And with a laugh he added, "I hope my reward will be that you give me the first digital TV."

In the following days, Wiley decided he was free to speak his mind at last, now that his Advisory Committee work was done. So he began writing op-ed columns, giving interviews. "The numerous technical experts who looked at standard-definition and high-definition using our system did not see standard-definition as being as good as high-definition," he told *Broadcasting & Cable* in answer to a question about Reed Hundt's stated view. The magazine had put Wiley's picture on the cover. "There's no way you could transition away from analog" if there were spectrum auctions, he added. And in a low-key way he blamed the broadcasters for their own troubles. "When you begin to put an emphasis, as did John Abel and others, on standard-definition and only standard-definition," then you opened the door for questions about whether the broadcasting industry really needed a second channel.

In the end, Wiley said: "People aren't going to go out and buy these sets just to get more of the same. You've got to look for a quantitative leap forward—a dazzlingly clear picture on a large screen."

Then the magazine asked Wiley one final question: In your long career, you've done so much. You've been the chairman of the FCC. You've established the leading communications law firm in Washington—and much more. Where does your Advisory Committee work rank among all of that?

Wiley hesitated not a moment.

"First," he said in a clear, certain voice. "Without a doubt. It crowns it all."

22 *Limping to the finish line*

For almost a year, Congress debated a new version of the telecommunications bill, introduced early in 1995, after the first one unexpectedly died. Angry arguments raged around the bill month after month, including fights over broadcast ownership rules and the parameters of competition between phone companies. Through all this, however, the broadcasters' little "flexibility amendment"—the one giving them the right to use their second channel however they chose—slumbered unnoticed deep in the bill's text. This innocuous-sounding amendment, written for very different reasons almost two years before, now promised to save the industry. The amendment would virtually guarantee the broadcasters a second channel for digital TV, cutting off any talk of auctions. No high-definition broadcasts would be required, and that had seemed like a grave problem when the amendment was first introduced in 1994. But now, given the far larger threats that hung in the air, it seemed almost beside the point. Finally at the end of 1995, Congress and the White House announced that they had worked out the last wrinkle. At last everyone seemed happy with the bill. It was ready for passage, until ...

On December 29, Bob Dole, the Senate Majority Leader and presidential candidate, was talking about the bill during a news conference when out of the blue he said he was troubled by the broadcasters' amendment. It was a "giveaway" of valuable spectrum, he complained, suggesting that he wanted this problem solved before the bill was passed. Not once in the previous months had anyone even mentioned this amendment. Now all of a sudden it had snagged the entire telecommunications bill, and groans could be heard all over town. "If we get into a controversy over the spectrum again, the bill is going to die," moaned Senator Larry Pressler. He had been a leading advocate of auctions before, but now even Pressler wanted to move on. Still, Dole's comment found many sympathetic ears.

"Stop the Giveaway," *New York Times* columnist William Safire wrote. "The ripoff is on a scale vaster than dreamed by yesterday's robber barons." Other editorials echoed that one, and lobbyists from competing industries joined the chorus. So Dole decided he had found a winning issue. Most other members of Congress were loath to take on the broadcasters; they needed to be on TV to get reelected. But Dole, the presumed Republican presidential nominee, must have figured that he was going to get all the free television time he wanted regardless. In mid-January 1996, he delivered a florid speech on the Senate floor in which he condemned the second-channel plan, calling it "a big, big corporate welfare project. Here we're cutting Medicaid and doing all the painful things while we lend them spectrum for twelve years. Why shouldn't they pay for it?" Dole made it clear he would block passage of the telecommunications bill until this issue was resolved to his satisfaction.

Over the next several weeks, few senators or representatives of either party came forward in support of Dole's point of view. All of them had been debating this issue for months; they'd already made up their minds. So in February, Dole agreed to a compromise. He would allow the telecommunications bill to pass. But in exchange, the FCC would promise not to begin issuing second-channel licenses until the end of 1996—after Congress had been given a chance to discuss Dole's concerns and consider new legislation authorizing auctions. With that, the telecommunications bill passed—including the broadcasters' amendment.

Within days, the NAB launched a lobbying attack on Dole's auction idea. "We'll spend whatever it takes," promised Eddie Fritts, the NAB president. Thirty-second advertisements began appearing on stations in Washington and a few other cities. One showed a multicolored kite drifting effortlessly in the sky as an announcer intoned: "Air is a wonderful thing. Free air lets us send you all the shows you love and local news, sports, and weather. Now Congress has a new idea. They tax everything else. Why not tax the airwaves?" Then a toll-free number appeared on the screen. The NAB said more than forty thousand callers reached operators who then passed them on to the offices of their congressmen.

The connection between the ads' allegations and the actual political situation was faint at best, and congressmen began yelping that the spots were unfair. Broadcasters "are bullying Congress," Dole complained. Opponents from other industries threatened to sue the NAB. But no matter. Congress held a few dyspeptic hearings and lined up the usual advocates and complainers. ("It's time to fold the Grand Alliance," Grand Vizier Negroponte groused at one of them.) In the end, however, the TV ads and Congress's natural inclination to stay friendly with the broadcasting industry did the trick. Dole had turned the auction idea into a campaign issue, so the White House reflexively swatted it away. Vice President Gore declared: "We have opposed every suggestion from the Gingrich–Dole Congress that we should immediately auction the digital spectrum and let the winners do whatever they want with it." By spring it seemed clear that there would be no auctions. When Dole left the Senate to run for president in mid-May, the broadcasters jubilantly proclaimed victory.

The 1996 National Assocation of Broadcasters convention that spring was a high-definition television festival. "1,000 lines—No waiting!" declared banners hanging from the ceiling of the Las Vegas convention hall. A small plane circled the city, towing a trailer that read: "HDTV: Television for the 21st Century." In the convention center exhibit hall, the Westinghouse Electric Corporation, CBS's new owner, was showing a new product: a solid-state, digital, high-definition television transmitter. Westinghouse chairman Michael Jordan stood in front of the prototype for publicity photos, hand resting on the cabinet. And during the Association for Maximum Service Television's annual meeting (the scene of lawyer Jonathan Blake's embarrassed 180-degree turnaround a year earlier) Jordan told the station owners, potential customers for his new product, to start the transition to digital TV as soon as possible. "At CBS we're looking at that right now," he said, adding: "One thing that is very important is adoption of a standard."

Meantime, Dick Wiley was urging the broadcasters to stop waiting around for the FCC to set a standard. The television industry, he said, "ought to just go ahead and start building the stations, get going with it." That's just what some of them did. An alliance of broadcasters and TV equipment manufacturers announced in April that they planned to build a model high-definition television station. It would allow real-world tests of new digital TV equipment. The sponsors could also use the facility for HDTV demonstrations. They decided they would choose a Washington-area TV station as the host, "to give our nation's lawmakers and regulators a front-row seat," explained Eddie Fritts.

Just about every station in Washington applied. Certainly they were interested in the $6 million worth of equipment that manufacturers planned to donate, but several also said they wanted the bragging rights. They were all eager to become the nation's first high-definition television station. Jim McKinney, the former FCC official who had more recently served as the head of a broadcast-industry technical group, was chosen as the project director, and he promised that the new station would begin broadcasting by fall. Meanwhile, at least one other commercial station—WRAL in Raleigh, North Carolina—also announced plans to put an HDTV station on the air as soon as possible. So did several public television stations. None of these stations would actually broadcast programming to anyone's home right away. The first digital televisions weren't expected to go on sale until sometime in 1998. The goal, as the public broadcasters put it, was to demonstrate high-definition television, stimulate public interest, and work with manufacturers to design new digital equipment while also developing marketing and promotion plans. In other words, they just wanted to get ready for the digital revolution.

Even then, the FCC was preparing its next Notice of Proposed Rulemaking, the second of three proposed regulations setting the digital TV standard. This one was to ask the question: Should the FCC adopt the Grand Alliance system as the new television standard for America? Not surprisingly, Reed Hundt was saying he didn't want to do it. Maybe something better will come along, he argued. If it does and we have locked ourselves into using the Grand Alliance standard, the nation will be stuck with an obsolete system.

In an interview just over a year earlier, Hundt had marveled at the flexibility of the Grand Alliance system, calling it "a wondrous genie in a bottle" that could accommodate any technological advance anyone might imagine. "Any new application you want to provide, anything anyone ever wants to invent in the future, you can just add it to the mix and put it through the door into this system," Hundt had said. "Anything, I don't know. This is stuff for other people to think of. But the commission asked them to build us a better bicycle, and they came back with a Ferrari. I think it's great for the country."

Now, however, Hundt was saying: "How can we balance the need for certainty about the standard with the goal of encouraging future advances?" So Hundt wanted the FCC to merely "authorize" use of the

Grand Alliance standard without mandating it. Under that scheme, anyone who came up with another idea would be free to market it. The Grand Alliance system might be used in New York, as a result, while stations in Silicon Valley could use a system put together by the computer industry.

TV manufacturers and broadcasters hated this proposal. Hundt would turn the digital television transition into a Betamax-VHS style commercial battle, they argued. Why, they asked, should TV stations spend millions of dollars buying Grand Alliance HDTV equipment to produce and broadcast HDTV programs—and television manufacturers millions more designing high-definition television sets for sale—if none of them could be sure the system would prevail? And how many consumers would spend money on an HDTV set if they feared they'd have to buy a new one when they moved from Philadelphia to Detroit?

Once again, however, Hundt was all by himself. FCC commissioner Susan Ness, often an ally, spoke for all of the other commissioners on this issue when she said: "Absent a mandated standard, investment and manufacturing decisions could be stalled, thwarting the ability to convert rapidly and smoothly to digital broadcasting." When Hundt realized that the other commissioners would vote against him, he backed down. And on May 9, the Federal Communications Commission issued its Fifth Notice of Proposed Rulemaking in the Matter of Advanced Television Systems. (The first was issued when Dennis Patrick was chairman.) It said:

We believe [the Grand Alliance standard] embodies the world's best digital television technology and promises to permit striking improvements to today's television picture and sound; to permit the provision of additional services and programs; to permit integration of future substantial improvements while maintaining compatibility with initial receivers; and to permit interoperability with computers and other digital equipment associated with the national information initiative. It was developed and tested with the unparalleled cooperation of industry experts. . . .

As a gesture to Hundt, the notice also offered his arguments against adopting a standard. The chairman had changed his position a bit. Hundt wanted the digital television standard to have a sunset provision; after a certain date it would expire. But even as the FCC notice proposed this idea, it described the reactions from industry groups: "Parties addressing the issue of adoption of a standard for a limited duration were uniformly in opposition."

The vote in favor of proposing the rule was unanimous. But the notice, like the previous ones, was just a statement of intentions. Approval of it opened a several-month comment period. After all the comments had been received and reviewed, the commission would begin discussion of adopting the actual rule. The commissioners said they planned to adopt the new standard by the end of 1996.

A couple of weeks later, the Senate selected a new majority leader to replace Bob Dole, and the choice—Senator Trent Lott, a Republican from Mississippi—left lobbyists at the NAB smiling with contentment. It turned out that Lott and Eddie Fritts were close friends; they had been roommates at the University of Mississippi. The value of that became apparent just one month later, when a letter on Congress of the United States letterhead arrived on Reed Hundt's desk, saying:

As you know, during Congress's consideration of the Telecommunications Act of 1996, questions were raised regarding the assignment of a second channel to television broadcasters for the transition to advanced television services. In deference to the concerns raised by the distinguished former Senate Majority Leader and others, several of us asked the Commission to postpone awarding such licenses prior to the end of this Congress in order that the issue may be further examined by the Committees of jurisdiction . . .

The House and Senate Commerce Committees have held hearings on this issue, and the consensus view of the Commerce Committees and the Congress is, in our view, that the Commission should move forward as expeditiously as possible on its current plan to award a second channel to television broadcasters for the transition to advanced television services. . . . Thus we recommend that the Commission complete all actions necessary to prescribe rules to permit the deployment of over-the-air digital broadcasting no later than April 1, 1997.

Finally, we would note that the Commission does not need any additional statutory authority to proceed with the assignment of digital licenses. We would, therefore, expect the Commission to proceed with bringing this exciting new technology to the American people without further delay.

Lott, House Speaker Newt Gingrich, and three other congressional leaders signed the letter. Although Lott grumbled that it had been made public before he wanted, he said he agreed with everything in the letter, nonetheless. With that, it seemed, all debate about the second-channel auctions was over.

But of course, Reed Hundt did not agree. For months he had been promising that the FCC would publish the new rules by the end of 1996, formally opening the digital television era. But when he got the Lott-Gingrich letter, once again Hundt grew a brand-new opinion. In separate interviews with two *New York Times* reporters, Hundt uttered disparaging and, in one case, profane assessments of the broadcast industry's lobbying strategies. He threw the Grand Alliance into the same basket, telling one of the reporters: "The Grand Alliance was a creation of the broadcasting industry, the primary purpose of which was to make sure they could get the spectrum [for free]. It's not widely reported that way, but that's my opinion." (And it was certainly a curious opinion, considering the open hostility between the Grand Alliance and the NAB over the years.)

Then, for a *Times* story published on July 1, another reporter interviewed Hundt and wrote that he showed "no signs of rolling over" to Lott and Gingrich. "Seizing on a passage in the letter that gives him until April 1997 to turn over the channels," the story went on to say, "he said he planned to use that time to lobby for several provisions that are almost certain to offend the broadcasting industry." Namely, Hundt said, he would wait until the new Congress took office in January 1997 and urge its leaders to impose new requirements on the broadcasters in exchange for the second channels—for instance, mandatory public-service programming and auctions of some broadcast spectrum. Ignoring his earlier promises to conclude the digital proceedings by the end of 1996, Hundt said he wanted to reopen battles he had already lost—even though the *present* Congress was telling him to hurry up and finish the digital TV work.

In fact, however, a few weeks later the White House and the broadcasting industry came to agreement on a long-debated rule requiring television stations to air three hours of children's programming every week. That seemed to satisfy Hundt's request for public-service programming. And in mid-July, the FCC began preparing the last of its Notices of Proposed Rulemaking for digital television. This one set out plans for assigning second channels, and it proposed to put none of the digital channels at the top of the broadcast-television band, in channels 60 to 69. These were the least valuable broadcast channels; the higher the channel, the more power was necessary to broadcast a signal. As a result, few TV stations used these channels.

The commission's plan was to hold those channels back and auction them. In mid-July the White House endorsed the idea. The broadcasters, of course, complained loudly. But some of them also realized that the new children's TV agreement and the partial auction plan gave Hundt most of what he wanted and might smooth the way toward quick approval of digital TV rules. That seemed even more important when, late in the summer, Bob Dole announced a plan to cut taxes by 15 percent if he were elected president. One source of money to pay for this, Dole said, was broadcast-spectrum auctions.

On July 25, the FCC approved the final Notice of Proposed Rulemaking for digital television. But even as Hundt voted in favor of the notice, he swatted at the broadcasters once again. The second channels stood as "the biggest single gift of public property to any industry in this century," the chairman grumbled. Commissioner Quello immediately argued with him. It's not a gift, Quello said. It's a loan.

A few weeks later, President Clinton offered his own view for the first time, telling *Broadcasting & Cable* magazine, "I think digital television will be of tremendous benefit to the American public, and I fully support the transition. In addition, by waiting [until after the transition is complete] to auction the spectrum that is returned by broadcasters we will be able to auction contiguous spectrum blocks, which are far more valuable than scattered pieces, [and ensure] a smooth transition to digital technology for over-the-air television."

With all of the proposals published and everyone's position clearly stated, all that remained was the vote. Or so it seemed.

Back in March, the broadcasters and television-equipment manufacturers had formed a lobbying organization called Citizens for HDTV. Peter Fannon, the former test center director, was its leader; and it had been quite active. The computer industry had no similar group. The industry's lobbyists worked more or less independently, and they seemed to be having little effect. They were, of course, flailing away at the interlace dragon, but in early July, Jim Burger, Apple Computer's chief lobbyist in Washington, admitted, "I feel like Don Quixote."

Then in mid-July the computer industry was given important new ammunition. The Polaroid corporation demonstrated a new digital, *progressive-scan* television camera that provided a stunningly clear picture, even though it was capable of transmitting only about 720 lines—360 fewer lines than the proposed interlaced high-definition standard. For as long as anyone could remember, the broadcasters had been arguing that it would be *years* before anyone was ready to build marketable, affordable progressive cameras. But now Polaroid had done it, and when the company demonstrated the new camera to the FCC in July, Jim Burger crooned, "The Polaroid camera has them scared to death."

About the same time, ABC television issued a surprise announcement. Preston Davis, a senior ABC official, told *Communications Daily* that the network would begin transmitting high-definition programming in early 1998 and was leaning heavily toward using progressive scanning. "It's clear that producing programming in progressive will provide better pictures and be cheaper," Davis said. The other networks said nothing, but the NAB was still saying that interlaced pictures looked better, though the arguments were growing strained. A long *Barrons* article on the subject in July included this passage:

> *[Broadcasters] argue that interlace is better for TV viewers and remains the most efficient way to transmit picture data. Of course, sticking with interlace effectively keeps computer firms out of the competition. Lynn Claudy, senior vice president for science and technology at the National Association of Broadcasters, says interlace offers viewers a better spatial rendition of life, while progressive offers superior motion shots: "Do you want to see beads of sweat on the quarterback's face or the stitches on the ball as it hurtles through the air?" Claudy would rather see the quarterback's perspiration.*
>
> *In Washington, interlace has pretty much dominated the discussion, in part because the computer industry, which uses the progressive-scan approach, had not gotten actively involved until recently.*

Finally, Reed Hundt betrayed his true sympathies. He called Bill Gates, Microsoft's president, and urged him to get into the game lest he be rolled over by the broadcasters. And sure enough, a short time later, the computer and entertainment industries announced the formation of a high-powered lobbying coalition to compete with Citizens for HDTV. This one was called Americans for Better Digital TV, and among its members were leading computer and motion-picture companies. (Both groups had adopted the lobbying industry's latest deceptive ploy: Give a new lobbying group a name that makes it sound like the members are housewives from Kansas.)

"The standard being considered by the FCC would stand in the way of future innovation and force consumers to use inferior technology over time," Gates said in the new group's first press release. "If approved, this standard will become a roadblock to the convergence of the television and the PC." Filmmaker Steven Spielberg added: "As we move into the next century, it is important that the standards for advanced television give the public the opportunity to see the images of film with progressive scanning, without interlace . . ." (As it happened, the Grand Alliance system was specifically designed to transmit Hollywood films only in progressive format; that was possible because 35-millimeter films are produced with a lower number of frames per second, allowing more compression.)

The broadcasters' statements and the press releases from the new lobbying group helped draw an even clearer picture of the players' true motivations. Despite the lofty statements of principle, all of them were simply jockeying for commercial advantage as the FCC prepared to set the rules for the most important new telecommunications service launched in half a century. The computer lobby proposed its own digital transmission standard, and it happened to be directly compatible with the VGA video standard that dominated the computer world. The computer people wanted to abandon high-definition television altogether; it held no particular advantages for them. About the same time, the cable television industry, after working with the broadcasters for many years as partners in the Advanced Television Test Center, suddenly reversed course and began lobbying against the Grand Alliance standard it had helped create. The Cable Mongols didn't want their greatest commercial competitors to be given important new digital powers.

Buried in the Americans for Better Digital TV press release was this statement: "Audio industry leaders warn that the proposed Grand Alliance standard would be a step backward, essentially locking consumers into outdated technology [because it includes] a mandated, exclusive audio standard that is already technologically obsolete." Here was a new argument. Before now, the public debate had never included any particular criticism of Dolby's AC-3 digital sound system. In fact, AC-3 (now known commercially as Dolby Digital) had recently been chosen as the audio standard for the new DVD digital movie-disc format that was to be introduced soon.

It happened, though, that one member of this new coalition was a company named Digital Theater Systems, a direct competitor of Dolby. DTS, as the company was known, had developed a sophisticated, multichannel digital audio system used in more than three thousand movie theaters nationwide. DTS had wanted to be chosen as the audio standard for the new digital discs, but the company lost to Dolby. At a hi-fi show sponsored by *Stereophile* magazine in New York City in May, DTS announced that it was entering the audio and home-theater markets—taking on Dolby once again in another field that Dolby dominated. Now that DTS had joined this new coalition, it was throwing mud at its greatest competitor under the cloak of high-minded concern about the nation's technological future.

Some of Hundt's assistants had little patience for the arguments from the computer industry and its allies. "I have great sympathy for the idea of interoperability with computers," Saul Shapiro, Hundt's aide, said in July. "But the broadcasters are committing themselves to spending millions of dollars on all this new digital equipment. At the end of the day, the computer people don't have to put up one penny."

The other FCC commissioners, too, seemed tired of the debate—even as Microsoft stepped up its criticism in October. Speaking at an industry luncheon, Commissioner Rachelle Chong noted that the Grand Alliance standard already included compromises made on behalf of the computer industry. "I'm a little confused about why this can't work," she said. She and Commissioner James Quello wanted to approve the Grand Alliance standard just as it was and perhaps set a date "sometime in the future to relook at the standard," as Ms. Chong put it.

Commissioner Susan Ness began pushing the computer people and the broadcasters to meet, at least, and discuss possible compromises—though both sides said they were all too familiar with each other's point of view. But Ms. Ness was also saying she wanted to approve the standard within a few months one way or the other. Andrew Barrett, the fifth commissioner, had resigned a few months earlier. That left Reed Hundt the only other vote on the commission, and as usual he was being enigmatic. In a speech, he started talking about spectrum auctions once again, and he proposed requiring each new digital TV station to dedicate 5 percent of its programming time to "educational TV, free time for political debate, and the like." As for the computer industry's complaints, Hundt said Wiley's Advisory Committee "worked in good faith to produce a consensus standard. Unfortunately they did not succeed. I'm still hoping for a consensus to emerge." In late September, he declared that the warring parties should be locked into a room until they reached a compromise and that if they could reach "a de facto understanding, we might consider sprinkling holy water on it."

But while these old arguments lumbered along, digital high-definition television was *already* becoming a commercial reality.

On June 17, 1996, WRAL television, a family-owned CBS affiliate in Raleigh, North Carolina, applied for a license to operate an experimental high-definition television station using the Grand Alliance standard. The FCC granted the request, and WRAL quickly began assembling the equipment. In Washington, meanwhile, Jim McKinney chose WRC as the host for the industry's model HDTV station. WRC was an NBC station, owned and operated by the network. It had been one of General Sarnoff's most important properties, and now engineers at the Sarnoff Shrine at Princeton were selected to help design and install the Grand Alliance equipment. The race was on to become the nation's first active high-definition television station.

Strictly speaking, the Raleigh station won. On July 23 WRAL began transmitting digital *data* on channel 32. The station hadn't been able to get a Grand Alliance decoder. There were only two in the country, leaving WRAL to receive its transmissions only on a spectrum analyzer. Still, a press release boasted: "WRAL-HD is the first commercial station in the country to broadcast the Federal Communication Commission's new digital television standard." Jim Goodmon, a warm, voluble southerner who was the station's CEO, said: "We're not going to be whipped in the technology area. My notion about this has always been: Be first; get out there with the new technology."

In Washington, the WRC-Sarnoff team wasn't happy. These were General Sarnoff's men, after all, heirs of RCA's traditions. Keepers of the legends. First with radio, first with television, first with color TV . . . and they'd been beaten by a little station down in *Raleigh*? But WRC was having trouble getting its new equipment in place. For one thing, the truck driver bringing the new digital-television antenna got lost on the drive down from Connecticut. He arrived three days late. Finally, however, just after 9:00 P.M. on July 30, WRC was ready for the first broadcast on its new digital channel: WHD-TV, channel 34. And this one wasn't going to be just a data stream. WHD *did* have one of the Grand Alliance decoders, and so the station was going to go on the air with a full, live high-definition television picture. Really, the General's heirs were saying, this would be the first *real* high-definition television broadcast. "We're proud, once again, to be contributing to history," said Alan Horlick, the station's president and general manager.

At 9:45 P.M., the station engineers and Sarnoff's men were on post, and one of them threw a switch. The digital signal coursed from a high-definition camera set up there in the equipment room, through the Grand Alliance decoders and transmitters, up to the new antenna on WRC's tower out back, then over the air across the city, stretching to the far suburbs in Columbia, Maryland, and Burke, Virginia. The only digital television set up to receive it, however, sat on a table in WHD's equipment room, and the first live broadcast from a commercial station using the Grand Alliance standard filled the HDTV screen bright, sharp, and crystal clear. The picture was prosaic—except for one thing. The high-definition camera was pointed at the Grand Alliance equipment racks—just aluminum panels and faceplates studded with switches, knobs, dials, and twinkling lights. But as the Sarnoff engineers looked at the TV, they couldn't help but smile. There in the center of the high-definition television screen was their own special symbol, a curious emblem tying this moment to the great legends they carried with them.

Hanging from one equipment rack was a hand-painted bunch of blue bananas.

In the fall, Paul Misener resigned from Wiley's law firm and took a new job. He became a senior lawyer-lobbyist in the Intel Corporation's Washington office, and his decision to move turned out to be a transforming moment. Intel, of course, was a leading member of the computer-industry lobby, and Misener became the company's chief digital-television strategist.

"I want to get everybody to meet," he said shortly after he moved into his new office. Susan Ness, the FCC commissioner, had been saying essentially the same thing. Over the preceding months, Ness had emerged as the commission's mediator—the one who tried to clean up the messes caused by Hundt's bombast. So Misener suggested that she write a letter to all of the parties—broadcasters, TV manufacturers, computer and software makers, filmmakers—urging them to meet and settle their differences. On October 24, the letter went out. "I believe there is significant common ground," she wrote. "I would like to see the standard issues resolved as swiftly as possible—at the outside, by mid-December. I therefore ask you to agree to meet with other interested parties, to report to me by October 30th with a schedule, and then again, by November 25th, with the resolution of your discussions. Let us aim to have a recommended solution by Thanksgiving."

Within a few days, all sides had agreed to meet, but that didn't stop them from hurling more bricks across the divide. The Grand Alliance standard "will cost consumers $91 billion and stifle growth of the country's most important industry," the computer lobby argued in printed ads. The Silicon Valley lobbyists came up with that number by taking the most inflated estimate they could find for a digital television's introductory price and multiplying it by the number of TVs Americans would presumbly buy over the following fifteen or twenty years. Filmmakers—grafted to the computer-industry lobby so it would appear to have more heft—continued to wail about their favored issue, the shape of the HDTV screens. "We will no longer tolerate the mutilation of films when they are shown on TV," complained Martin Scorsese, the film director who was serving as vice president of the Artist Rights Foundation.

Not to be outdone, the broadcasters and filmmakers held a "rally" and press conference in Washington at the end of October. "We will not trade away the integrity of the standard," bellowed Eddie Fritts. Peter Lund, a CBS executive, argued that the computer industry's "only aim is further delay in adopting any standard other than the one that suits their own narrow business interests."

And so it went, even as the combatants agreed to a series of meetings in November to see if, despite all the acrimony, they could find some common ground. To say the talks were opening with bad blood on both sides would be a vast understatement. Neither side intended to give up much of anything. The broadcasters were convinced that they only had to show up and pretend to work toward compromise. They believed they had the votes on the FCC to approve the standard, just as it was, as long as it looked like they had made a good-faith effort. The computer industry, led by Misener, put out a four-point plan for a settlement: Drop the 1,080-line interlaced format from the standard, denuding it of the capability to offer the highest-level high-definition images. Include modifications so that television signals could more easily carry data as well as pictures. Change one of the standard-definition formats so that it had computer-friendly square pixels. And give the entire standard a sunset date: After ten years it would expire.

The broadcasters offered no reaction. "They have not taken it seriously or even really read it," observed Jon Blake, a broadcast industry lawyer-lobbyist. "There's still a lot of bluster out there, a lot of breast beating. I don't know if they can get past it." But there was no shortage of bluster among computer executives, either. At the Comdex show, the computer industry's annual convention in Las Vegas, Andrew Grove, head of Intel, declared that his industry was entering a war with the broadcasters and TV makers. Soon, he predicted, Americans would spend more time in front of their personal computers watching digital-television broadcasts and "other interactive, lifelike experiences. And in this war, he who captures the most eyeballs wins."

In that atmosphere, the interindustry negotiations opened in Washington on November 4. On one side were executives from Hollywood and Silicon Valley; on the other, broadcasters, Grand Alliance representatives, and leaders of the consumer-electronics industry. They quickly agreed to three ground rules: The talks would be held in complete secrecy: no leaks. The combatants would announce either a complete settlement or an impasse: no partial solutions. And, successful or not, the talks would be completed by Thanksgiving: no extensions.

"We're saying nothing," Sherlock, the NBC vice president, said as the meetings opened. Blake added: "People are freer to explore common ground if there isn't concern about views being publicized." Rast said he took that to mean "that the good guys will keep their mouths shut and the bad guys will leak to the press to shape issues their way." But in fact there were no leaks; the talks proceeded in total secrecy—demonstrating tremendous self-control, considering what was going on inside.

The first meeting, the one on November 4, was devoted to organizing. Then, during the second session a few days later, both sides laid down their cards. The broadcasters and their television-industry allies liked the Grand Alliance standard just as it was. They saw no reason to change it. Meanwhile, by the time the computer industry and their Hollywood allies offered their position, that four-point plan of a few weeks earlier had swelled. The filmmakers loaded on their demands, chief among them that all televised movies be shown in their original aspect ratio—the shape of the screen—whatever that might be. "Pan-and-scan"—the practice used to make a wide-screen movie fit on a square-screen set—would be banned. Movies would be letterboxed when necessary. And their final, unlikely directive was that TV manufacturers would be forbidden to offer any advanced features with their sets, allowing consumers to alter the aspect ratios of the programming they watched on their own TVs.

At the same time, the computer makers announced that they wanted to strip the standard of *all* of its 1,080-line high-definition formats—no matter whether they used interlaced or progressive scanning. And for the first time, in the privacy of the closed meeting, they explained the real reason for this: They didn't want to pay for it. To display high definition, a receiver had to include a certain kind of memory chip costing $12, $20, $30, or $50—the price estimation depended on the political position of the speaker offering it. The computer companies said they were unwilling to absorb this expense because, they announced, they intended to start building digital television receivers into every personal computer starting in 1998. And they wanted to keep the cost of those new PCs below $2,000. By including 1,080-line images, particularly in interlaced versions, "you add costs to get a lower quality," remarked Robert Stearns, a senior vice president with Compaq Computer who, the other side quickly saw, possessed an arrogance that offered no observable bounds.

All this meant that the computer industry wanted to challenge the television industry. And as the computer industry leaders entered this new market, they were proposing to strip the TV industry of its greatest marketing advantage—HDTV—so they could offer competing products that were more cost competitive. Meanwhile, the decade of effort to produce a world-beating digital high-definition television system would have been for naught. The American public would never see it.

For the broadcasters and set makers, this idea was dead on arrival. To fight it, the consumer-electronics industry lobbyists tried to split the filmmakers away from their computer brethren, figuring Hollywood wanted televised movies shown at the highest resolution. But that didn't work, so the negotiators came up with a strategy to evade the HDTV debate altogether: Enact the standard just as it was and establish a sunset date after which it would expire.

The meeting was adjourned for several days to consider this, and a penultimate two-day session was scheduled at the Hyatt Regency in Denver for Friday and Saturday, November 15 and 16. There, the broadcaster consortium offered its sunset proposal: The Grand Alliance standard would be adopted as written, with an eleven-year expiration.

But for fifteen years after that, television receivers would still be built with "backward compatibility." In other words, no TV receiver could be made that did not comply with the standard until the year 2023. Misener called that "not so much a sunset as a sunspot." He said computer makers might be willing to consider a sunset of one or two years, perhaps even three. But no more than that.

The two sides bickered back and forth but made no headway. After half a day of this, according to one participant's transcription of the discussion, Stearns said: We've been all over the world on this problem, and we are not going to agree. Craig Mundie, the Microsoft executive, was serving as the lead spokesman for his side. Unlike some others from his side, Mundie had a history with this issue. He'd been at the table, representing Microsoft, when Dick Wiley rolled right over him with his "unanimous" vote a year earlier. Now Mundie said: The net result of a year of discussion is that there has been no yielding on the computer industry's two largest economic concerns—1,080-line progressive and 1,080-line interlaced.

Joe Flaherty, representing CBS, responded: The cable and satellite television industries both use interlaced transmissions. You can't disallow it only from broadcast television. Mundie shot back: It's not at all obvious that they will continue to use interlace. But in any case, why don't we just take the display formats out of the standard, not mandate them?

Here was a new idea. The standard listed eighteen possible display formats—high definition, standard definition, interlaced, progressive, and more than a dozen other related alternatives. With that on the table, the two sides withdrew to private rooms for strategy discussions. The broadcasters and TV makers immediately realized that if they stripped the display formats out of the standard, leaving the rest of it intact, they could still agree among themselves to restore them in an informal fashion. The computer industry could do whatever it wanted. But the broadcasters would be able to choose from among the eighteen formats for their programs, and TV makers would build digital sets that could receive any of them.

Back in the main meeting room, the broadcasters' side broached the idea of ending the discussions with a partial solution—all the secondary points that had already been settled—while leaving the two intractable issues, high-definition formats and aspect ratios, for the FCC to decide. But the computer industry insisted that they stick to the original agreement: A complete agreement or nothing at all. Mundie spoke up to say:

We are disappointed. The two key issues are economic, and you have made no concessions. I believe the process has ended. We will go back to the processes we used before.

In other words, both sides would return to loud, deceptive lobbying, misleading advertising, and crude arm twisting. The meeting adjourned. Both sides agreed to think things over before making any decisions about how next to proceed. Almost two weeks remained before Thanksgiving, the deadline.

That night, some of the consumer-electronics and Grand Alliance negotiators discussed the situation over dinner. They concluded that "the best solution would be to agree that formats not be included in the FCC-mandated standard," one of the diners wrote in a briefing memo for his colleagues. "That is because of a belief that it is important to get the standard done quickly. We could separately agree between the broadcast and consumer-electronics industries on formats, presumably including 1,080-P [progressive] and 1,080-I [interlacing]. In such a case, it would appear useful to come up with, and establish with consumers, a symbol of compliance, which would appear on products."

The symbol would certify that that TV receiver was equipped to receive all eighteen display formats set out in the original Grand Alliance standard. And if the computer industry carried through with its threat not to support HDTV, their PC-TVs would not be eligible to carry it.

In a series of conference calls over the next few days, the broadcasters and consumer-electronics manufacturers agreed privately among themselves to accept exactly that plan. After that, they scheduled a final, last-ditch meeting in Washington, Sunday, November 24. From the moment that meeting opened, the group haggled and complained. They adjourned Sunday evening and returned Monday morning. Finally, Monday evening they agreed: No formats. Immediately—before anyone could backpedal—they drafted a statement and carried it over to FCC commissioner Ness. Trying in her own way to make sure that no one had second thoughts, she quickly released a statement: "I'm delighted by the resolution of this controversy. This deal will provide the American consumer a future rich with digital broadcasting and computer-friendly programming."

The deal also set up a mammoth competition for control of the American living room. Computer-industry executives started telling reporters that they now had the tools they needed to enter the television business. They boasted that they would have millions of digital-television receivers in people's homes, conforming to *their* standard, before the television industry could even get out of the gate. "We'll ship 15 million of these things in 1998," Mundie boasted.

Over the following days, industry analysts began warning of a new format war between these two mammoth industries—similar in some ways to the Beta-versus-VHS battle in the early days of the VCR.

"I hope we're VHS," Rast said with a chuckle. But Gary Shapiro, head of the Consumer Electronics Manufacturers Association, was more confident. "Frankly," he said, "I think the TV industry will be a bigger threat to the computer world than vice versa." He promised that the first high-definition television sets would go on display at the Consumer Electronics Show in January 1998. And John Taylor, the Zenith spokesperson, predicted that they would cost between $1,000 and $1,500 more than an equivalent conventional television. Bruce Allan, a Thomson executive, agreed, and said the first sets would appear in the stores by mid-1998. He and everyone else in his industry announced that they planned to make HDTVs equipped to receive all eighteen display formats—as if the negotiations with the computer industry had never taken place.

Reed Hundt lauded the agreement, too, and said the commission intended to keep its promise to vote on the standard before the end of the year. The only party to the negotiations that seemed unhappy was the Hollywood group. As soon as the computer-industry leaders had seen that they could cut a deal, they dropped the Hollywood team. Left out in the cold, the filmmakers stomped their feet and whined for a few weeks. "We have not ruled out anything, whether a legal challenge or a legislative challenge," warned Henry Goldberg, a lawyer representing the Coalition of Film Makers. But in the end they did nothing; gradually they faded out of the picture.

Announcement of the agreement was the lead story in the *New York Times* and appeared on the front pages of newspapers nationwide. Not surprisingly, some of the nation's local broadcasters read about it with something less than mirth and glee.

"Why would anybody be enthusiastic about this?" asked John Larkin, general manager of KTZV, the NBC affiliate in Bend, Oregon. He was interviewed in the industry journal *Electronic Media*. "To play in this new world," he added, "is going to be very expensive."

Van Vannelli, station manager at WHIZ-TV in Zanesville, Ohio, grumbled: "We are about as raring to go for this thing as we are to jump from a high building into a pit of flaming gasoline."

At the FCC, staff members modified the proposed Grand Alliance standard by striking out the display formats—just one half page in the 64-page document. Every other part of the standard remained intact. The document was then passed around the commission, and one by one the commissioners voted to approve it. The last vote was cast, and the standard for the next generation of American television was formally approved by unanimous vote midday on December 24.

Asked that afternoon how he felt as he presided over this historic moment, Reed Hundt tried to frame his legacy: "What we did was kill the idea of handing on a silver platter to the Grand Alliance the old idea that all this was about high-resolution pictures, pretty pictures," he said. "That was the Japanese idea. I have been battling against this idea for the last three years. The most important thing, the most essential thing we did today, was to liberate this technology from the Grand Alliance deal, because the computer industry is where the machines will be in the twenty-first century."

23 HDTV is real

By January 1997, the model high-definition television stations in Raleigh and Washington had been on the air for six months. Two more had started up in Seattle. And from all this operational experience, station managers were beginning to accrue some important lessons about broadcasting in the new digital age. To their surprise, they were finding that under the withering stare of an HDTV camera, some things actually looked worse.

A few months after the start of operations, engineers with WHD-TV, the model station in the basement of the NBC-owned station in Washington, had carried a high-definition camera upstairs to an active studio so they could broadcast the local news. But the viewers who clustered around the high-definition monitor were greeted by an unpleasant surprise. For years, the station's evening news team had been doodling on their felt-covered desktop while they waited to go on the air, defacing it with drawings, caricatures, and short quips. None of that had been visible on conventional televisions. But when the WHD-TV crew turned the high-definition camera on the set, "you could read everything they had ever drawn—right there on the TV," Jim McKinney, WHD's station manager, said with a laugh. The viewers could also see all the dirt and dust that had accumulated on ledges and in the cheap shag carpeting on the platform's floor.

But the problem with the sets ran far deeper than graffiti and dirt. With conventional television, it didn't really matter what studio sets were made of—mahogany or cardboard, chrome or duct tape—or whether they were faded, dinged, and battered. The resolution of conventional TV was so poor that viewers couldn't tell the difference. So in studios across the country, bookcases were actually crude paintings. Pillars and columns typically were painted cardboard tubes. Through a high-definition camera, however, all of it looked like just what it was: pathetic fakery.

Late in 1996, NBC decided to design a new set for *Meet the Press*, the Sunday morning talk show that was produced in Washington. The network hired Jim Fenhagen, a leading set designer, to build it. After visiting WRC and WHD-TV, he decided he had better make the new set future proof because, he realized, "with the old TV, you can get away with murder, but in high-definition it is appalling." So on the new set, Fenhagen used real wood veneer instead of painted cardboard, real pewter instead of gray duct tape. A piece of wood painted to simulate brass in the NBC peacock had to be replaced with real brass. Even with all of that, some old habits died hard. For the first broadcast on the sharp new *Meet the Press* set early in 1997, the shelves held real books. Still, small groups of them were held together the old way—with duct tape wrapped around the backs. On one shelf of legal books, the tape was peeling off a copy of *Corpus Juris Secundum*. That was a high-definition no-no; it was clearly visible on the station's HDTV monitor.

If the old sets looked bad, the people looked worse. For fifty years, television personalities, both men and women, had applied heavy powder and thick pancake makeup to cover wrinkles, five o'clock shadows, and other facial imperfections. Though the makeup was far from subtle, on regular TV it looked just fine. Even for problems that makeup couldn't easily hide, conventional television's low resolution usually smoothed the rough edges. Not so with high-definition TV. "Where'd that mole come from?" Willard Scott, NBC's longtime weatherman, asked rhetorically the first time he saw his face on a WHD high-definition monitor.

All of this held a sobering message for the 1,600 TV stations nationwide that were preparing to make the transition to digital broadcasting. Not only would they have to buy new cameras, transmitters, recorders, switchers, and every manner of production equipment, but now they realized that they'd also have to renovate or replace their sets and retrain their makeup artists. When a story on this appeared in the *New York Times*,[1] executives from several CBS affiliate stations erupted.

---

[1] Early in 1997, the *Times* asked me to spend some time as a reporter writing about the transition to digital television. This interim assignment actually lasted more than a year. References to *New York Times* stories from this point forward refer to my work. This chapter draws in part from my work for the *Times*.

CBS executives had spent months reassuring these stations that the price of the transition would be manageable. Now here was a huge new cost no one had ever mentioned.

At a breakfast for several hundred network engineers in the spring, Charles Cappleman, a senior vice president for CBS, tried to assure them that they didn't have to worry about their sets; the ones they had would look just fine. "You may need to wipe fingerprints off the desk, do some dusting, vacuum the carpet," he said. "But in our opinion you won't have to change the sets. Studio production will not cost one cent more."

The reason, as Cappleman explained it: For most shots, the photographic depth of field is quite narrow. In other words it wouldn't matter if the sets looked fake. Usually they would be out of focus, anyway.

In the days after the FCC approved the Grand Alliance system as the nation's new television standard, broadcasters and television manufacturers cooed with satisfaction. They believed they now had the certainty needed to move forward. But within a few weeks their tune changed. They forgot about the battle just won and focused on the next one: the assignment of a second channel for every TV station in the nation. Now the two industries were saying: We can't do *anything* until those new channels are assigned. And Reed Hundt was making it clear that he had no intention of giving those broadcasters he so disliked broadcast spectrum worth billions of dollars unless he got some clear commitments in return. Specifically, he wanted the networks to promise that they would have a significant number of digital stations on the air in just one year.

Well, the broadcasters squealed like aggrieved children. "It's completely unrealistic to think we can be on the air in a year," Michael Sherlock, the NBC vice president, complained in March. "It took us eight years to develop this system, and it sat around for more than a year at the FCC. Until just three weeks ago, we had been planning on a six-year rollout." (That was the schedule under the old rules written by Al Sikes's FCC.) "There has to be time to develop the equipment, establish capital-spending plans. In some cities we have to build new towers. Even a two-year period is completely unreasonable in certain circumstances."

NBC, Sherlock offered, could have WHD on the air in a year. That wasn't hard since it was already broadcasting. A year or two later, maybe the NBC-owned station in Philadelphia could start up its digital channel.

The other networks offered timetables that were no more aggressive, in some cases rolling back earlier promises that had been made while the industry was still trying to convince the FCC to approve the Grand Alliance standard.

As an example, a few months earlier, Preston Davis, president of broadcast operations and engineering at ABC, had said his network would begin transmitting high-definition television signals "in a limited way in 1998." Then in the spring of 1997, he changed his tune, saying: "Our current target is to begin a limited rollout in the fourth quarter of 1999." Speaking for the entire industry, the NAB sent the FCC a proposal that offered to have exactly one digital station on the air in the first year. Eleven other stations in nine cities would go on the air in the second year, a handful more in year three, and the rest sometime in the following four years—most of them toward the tail end of this proposed timetable, roughly 2004.

"This thing languished at the FCC for so long," noted Lynn Claudy, an NAB official, "and now Hundt wants to put the FCC on a fast track?" He shook his head.

TV manufacturers yelped in complaint and said they could not start selling HDTVs in 1998 if broadcasters delayed the start of the new service. "If there's no signal, then there's no sales," said Bruce Allan, a vice president with Thomson. And Hundt summarily rejected the broadcasters' proposal. "They want to slow-roll digital television into oblivion," he opined. "It just doesn't compute. For the networks, the cost of this transition is pocket change, *pocket change*! The satellite folks spent hundreds of millions to get started, and they had to hope that the rocket worked. But this . . . it's so cheap I can't believe it." He and others at the agency said they believed this proved beyond all doubt that broadcasters had never been interested in HDTV; all they cared about was "warehousing the spectrum," as Hundt put it.

"I say if the broadcasters don't want to do this," Hundt added, "they ought to sell their licenses to someone who does. There are thousands of people in Silicon Valley who would take this up in a *nanosecond*!"

Hundt's interest in moving the transition along rapidly wasn't born simply of a public-spirited desire to bring digital television to the American people as rapidly as possible. No, as it turned out, the Clinton Administration was counting on a rapid transition to help balance the federal budget. In fact, the new balanced-budget bill wending its way through Congress called for an auction of the analog channels in 2002!

The purchasers wouldn't take possession of the channels until 2006. But by holding the auction earlier, the Clinton Administration could include the paper profits in its plan to have a balanced budget within seven years.

So it was no wonder that the FCC was unwilling to accept the broadcasters' slow-paced transition plan. If the industry had its way, digital broadcasting would have just begun at the time the analog stations were sold. As a result, only a handful of households would have digital TVs. All along, the government had promised a fifteen-year transition —time enough for everyone to buy a new digital set. Under the Clinton plan, though, the transition was compressed to just eight years, leading opponents of the administration—and even some friends—to call the whole thing a silly fiction. House Commerce Committee Chairman Thomas Bliley said the spectrum-auction plan had "emerged like a snake-oil salesman at a local carnival." But not surprisingly, the Democratic commissioners on the FCC, including Hundt, were mouthing support for the Administration's budget fantasies.

"I don't think this will even be an issue anymore in 2006," Hundt explained. "It will be like saying we need to make provisions for people who have Betamax VCRs."

As usual, Hundt's strident demands had polarized the debate over the broadcasters' implementation plans. Commissioner Ness had come to be the mediator in these situations, and she began trying to devise a compromise. The FCC had promised to award the new licenses by April 1, the date laid out in that letter from Lott and Gingrich the previous year. And slowly through March, proposals and counterproposals were faxed back and forth between the NAB and Commissioner Ness's office. Both sides postured and threatened in the newspapers; with each new offer, the NAB said the FCC was drawing more blood from its members. "You have to understand, this is an extremely aggressive schedule for us," Claudy, the NAB official, said with one new offering in late March.

The FCC meeting was scheduled for the morning of April 3. By April 1, the broadcasters still had not offered a schedule aggressive enough to satisfy Hundt. He wanted a significant number of stations on the air by the fall of 1998, to stimulate sales of the first HDTVs for that Christmas season, when most televisions are sold. He and some of the other commissioners were saying they did not want to put the second-channel awards on the agenda until broadcasters sweetened their offer. If the vote was delayed, months could pass before the FCC took up the issue again. Meanwhile, the spectrum-auction advocates in Congress would have powerful new ammunition and plenty of time to make trouble.

Throughout the day on April 2, proposals flew back and forth until, finally, late in the day, the broadcasters made their last offer: Network-owned television stations in the ten largest cities—twenty-six stations in all—would begin digital broadcasting by November 1, 1998. On that date, about 14 percent of the nation's households would be able to receive at least three digital signals. More stations would go on the air in the spring of 1999, raising the number of households receiving at least three digital signals to 30 percent. And by Christmas of 1999, the number would rise to 53 percent. By 2003, under the new proposal, every station in the nation would be on the air.

Michael Sherlock said: "We have made a major commitment that involves faith on our part that things will go smoothly. But we have made a commitment, and we intend to stick with it."

Over at the FCC, Ness said she was "comfortable with these commitments." Hundt agreed. And so, late in the evening, the second-channel allotments were added to the meeting's agenda.

When the meeting opened the next morning, the business was pro forma; the hard work had already been done. Outside the commission office building, however, a handful of protesters allied with Ralph Nader organizations marched and shouted about "the great spectrum giveaway." One of these people sat in the audience and shouted his objections during a public comment period, prompting Commissioner Rachelle Chong to reply, "I do not agree with those who say this is a free giveaway to broadcasters. This is a technology transfer. You just can't shut off analog TV and start digital broadcasting without a transition period." Aside from that brief moment of rancor, the commission worked through a choreographed script of prepared, preprinted statements and prearranged votes. Even Reed Hundt behaved. He had vented his recitations of personal accomplishment during interviews before the meeting, saying: "This is terrific. We have completely changed the entire policy that was thrown into my lap by Dick Wiley, who told me I would just have to get out of the way."

A little later, however, as Hundt sat behind the dais before a packed meeting room, a tangled bank of TV cameras gazing at him from the back wall, he said simply, "Today we're reinventing analog TV, and we're making it a digital business for the twenty-first century." Commissioner Quello added, "This is a historic moment for all of us. The possibilities are endless." After every commissioner had taken the opportunity to speak, the commission voted unanimously to loan the second channels to all 1,600 television stations, legally ratifying former chairman Al Sikes's grand plan first sketched out in this same room so many years earlier. Now the nation's television stations need only apply to get their second channels, and agree to accept the fast-paced implantation schedule negotiated over the previous weeks. Finally, under the new rule, the original channels would have to be returned by 2006.

"Broadcasters have been waiting for this day for ten years," Bob Wright, NBC's president, crooned as the meeting ended. Sherlock of NBC added, "Eighteen months from now, we will bring network television in high definition to the nation." And Gary Shapiro, head of the Consumer Electronics Manufacturers Association, saw good times ahead for his industry, too. "HDTV sets will be the highly sought-after 'Tickle Me Elmo' gift of the 1998 holiday season," he said. Shapiro had already craftily boxed the members of his trade group, the television manufacturers, into moving faster than they ordinarily might. Back in January, without consulting his membership first, Shapiro had announced that the first HDTVs would go on display at the annual Consumer Electronics Show, or CES, in January 1998—just one year later. He'd repeated that promise over and over again, and by time of the FCC meeting in April, the TV manufacturers were beginning to make the same promise as if it were their own.

"We'll have HDTVs to show at the 1998 CES," Bruce Allan, a Thomson executive, said after the meeting. Zenith, Philips, and other TV makers echoed that.

Even NAB President Eddie Fritts, whose opinions on HDTV had been as variable as the trade winds over the last decade, waxed enthusiastic. HDTV broadcasts by his members might even convince people to hook up their television antennas again, he suggested. "This new high-definition programming is so superior that this premise may hold true for broadcasters," he said. And what a gift Fritts was taking to his members, who were gathering for the NAB convention just three days later.

As they assembled in Las Vegas, the nation's broadcasters were ascendant. In the last few months, they'd won approval of the Grand Alliance standard with only a small modification, and they were free to broadcast interlaced programming, or anything else.

The Supreme Court had just ruled in the broadcasters' favor in an important battle with the cable industry. The justices had said cable companies nationwide were required to carry local-broadcast television programming if the stations asked to be on the system. And now, on the eve of the convention, the industry had beaten back fierce opposition to win the greatest gift of all: those second channels for digital broadcasting. At NAB '98, all seemed right with the world. As a result, it was an odd moment, and place, for the computer industry to lay down its formal challenge. But then, for all of the industry's other laudable accomplishments, no one was surprised that when it came to politics, computer executives were once again utterly tone-deaf.

Microsoft, the world's leading manufacturer of software; Compaq, the largest manufacturer of personal computers; and Intel, the largest maker of computer processor chips, had joined forces in what they called the DTV Team. Their representatives, three senior vice presidents, took out ads in broadcast-industry publications to publicize that they would be making a big announcement during the NAB convention. And sure enough, on Monday, the team staged a series of conferences for the press and show attendees, at the Treasure Island hotel.

Before the first event, Craig Mundie, the Microsoft executive, explained: "Since the beginning of the year, we have been working together to produce a recommendation that we could offer to the broadcast world. And what we have come up with is a recommendation to use layered formats for the introduction of digital television." The first format, which the DTV Team was calling HD-zero, would be used for a few years. Then when technology improved at some point in the future, the nation would move to HD-one, allowing for higher-resolution transmissions. And finally at some shining point far over the horizon, the computer industry would allow the nation to move to the final stage, HD-two, and the highest-resolution transmissions—roughly equal to the ones that the broadcasters were planning to put on the air in the fall of 1998.

And what was HD-zero, the format to be used for an indefinite time? Well, it was none other than 480-line progressive-scan television, with occasional use of 720-line progressive. This was simply standard-definition television—the same old dog the computer industry had been beating for more than a year, though now he had an apt new name: HD-zero.

"By adopting HD-zero," Bob Stearns, the Compaq representative, said, "we will bring literally millions of PC-TV products into the home, big screen and little screen, and speed up the transition to digital TV." Ron Whittier, the Intel executive on the team, added, "What we are talking about is accelerating the deployment of receivers." Though none of them were actually saying it directly, the three of them were still talking about abandoning HDTV—all so computer makers could bring less expensive digital-television receivers to market.

Outside the hotel conference room, where these pronouncements were being made to successive audiences of astonished listeners, the DTV team had set up a demonstration of 480-line progressive television. As usual, it looked clean but soft; there wasn't even one pixel of additional detail in the picture. "Doesn't that look good," Stearns intoned as he pointed at the monitor.

For a while, the team tried hard to seem oh-so-reasonable as they made their underwhelming offers. But their true nature soon slipped out. "I think we are doing these guys a favor," Mundie offered. "It's not clear to me that, without this, terrestrial broadcasting is going to be a viable business much longer."

Stearns said, "The broadcast industry has walled itself off from where the action is." A few weeks earlier, he had remarked: "What we're trying to say is that if you don't listen, you're going to be a buggy whip." Now he and the others were telling the broadcasters: We are going to start building digital-television receivers into personal computers starting next year—15 million or 20 million of them a year. They will be equipped to receive *only* the HD-zero signals, not high-definition programming. If you broadcasters go ahead with your plans to broadcast 1,080-line HDTV, then these computers will go dark. "We would not be able to receive your pictures," Mundie said. "If I were an advertiser, I think I'd be saying that's a lot of eyeballs I don't want to overlook."

Not surprisingly, the broadcasters were appalled. At one of the conferences, a broadcaster stood up to say: "I am a little puzzled. What you are saying is that to get these treats you are offering us, we have to throw away all the work we've done over the last ten years and start again." Some others in the audience cheered.

Mundie retorted: "Those business and political decisions didn't conform with an awareness of developments in the software and computer industry."

No matter. The broadcasters were feeling flush. They just laughed at the computer proposal. "The computer industry must think broadcasters are just plain stupid," snapped Robert Graves, the former AT&T representative for the Grand Alliance.

"Call me myopic," added Lynn Claudy, an NAB officer, as he waved his hand toward the showroom floor a little later, "but I see the computer industry proposal not as a threat but as a loser. Just look at the trends at the convention center. You see all these wonderful HDTV programs, and you see lots of digital equipment designed for interlaced high-definition programming. If you embrace the computer industry proposal, you'll see none of this programming in their plan."

Stearns left town convinced, he said, that "we are striking up a good relationship with them." He could not have been more wrong. The DTV team had simply antagonized an entire industry, polarizing the debate. More than one broadcaster caustically joked that the computer executives must have recruited Reed Hundt as an advisor.

The broadcasting industry firmly rejected the computer industry proposal, but few broadcasters were being clear about their own plans. They'd been given a gift, a second channel with which they could do whatever they pleased. The law said only that they had to be on the air with a digital signal by a certain date. As a Texas broadcaster liked to quip, his station could fulfill the requirement by broadcasting digital noise from a laboratory signal generator.

In a general sense, all three major networks were promising to offer some high-definition programming. CBS was the most specific of them; in January the network announced that, beginning with its fall schedule in 1998, CBS would broadcast HDTV programming at the highest-level format: 1,080 vertical lines by 1,920 horizontal lines—interlaced. But no one at CBS was saying how much of the network's schedule would be high-definition, or what CBS would air the rest of the time.

The other networks were even less specific. During the final Advisory Committee meeting more than a year earlier, Dick Wiley had wrested a pledge from all of them except Fox to broadcast HDTV. And all three of them were repeating those promises now. But how much HDTV, and what technical formats... no one was willing to say. The only consensus seemed to be that programming made in Hollywood and recorded on 35-millimeter film—about three-quarters of the networks' prime-time schedules—would most likely be broadcast in high-definition. This film was already in a high-definition format, so the networks could broadcast it without buying the expensive new high-definition cameras, recorders, and other equipment that would be needed to produce original high-definition shows. Local broadcasters, meanwhile, generally indicated they would pass through whatever programming the networks sent them. And they offered a full range of different opinions on what they would air the rest of the time.

Still, it seemed clear that buyers of the first HDTVs the following year would find at least a small sampling of high-definition television shows in the early months—if they were among the minority of Americans who got their TV programming from a television antenna. But the fact was, about 65 percent of the nation's homes were connected to a cable system. And to the Cable Mongols, the broadcasters were no longer the most important threat. They were losing customers to DirecTV and the other direct-broadcast satellite companies that had signed up 5 million customers so far. That was the root of a new problem. Nearly all of the new satellite customers were former cable subscribers. These people had switched to one of the satellite companies for a mix of reasons. The satellite services boasted of an improved, digital picture and freedom from the cable companies that many Americans had come to loathe. But most important, in the eyes of the Cable Mongols, satellite TV offered up to 175 channels.

Most cable companies, meanwhile, had been slow to upgrade their systems to fiber optics and other advanced technologies. So they were able to offer only thirty-five or forty channels on average, and in many cases the picture quality was terrible—particularly when the subscribers' homes were far from the cable office. As a result, some of the major cable companies were placing orders for new digital equipment with a completely different motivation. They wanted digital boxes that would allow them to increase the number of channels they offered—and, not incidentally, reduce the quality of the pictures they transmitted. None of the major cable operators were making plans to provide high-definition programming. Quite the opposite, in fact.

Broadcasters were going to use digital compression to squeeze one high-definition show into the space previously filled by one analog, standard-definition program. Cable companies, meanwhile, were planning to use digital compression to squeeze six or eight or twelve *lower-*

resolution digital programs into the space formerly occupied by one analog show. "Some programmers are going for volume instead of quality," remarked Mike Hayashi, a vice president with Time Warner Cable, the nation's second-largest cable system. "We are planning for three-quarters resolution or better." But Tele-Communications Inc., the nation's largest cable company with 14 million subscribers, had the most extreme plan. TCI, as the company was known, had not spent money to upgrade most of its systems around the country. The antiquated equipment would not allow the company to offer more channels and compete effectively with the satellite services. But with digital compression, TCI planned what David Beddow, a senior vice president with the company, called "an electronic rebuild," using "very high levels of compression to increase our channel capacity." The result would be half-resolution images—"VHS-quality" pictures, as they are known in the industry. When a consumer records a television show on a VHS videotape, the taping process degrades the image quality, leaving a picture only about half as clear as the original. And since millions of Americans watch VHS videotapes, Beddow said his company had decided that VHS-level resolution would be good enough.

None of the production companies that supplied cable programming were talking about high-definition, either. At HBO, for example, senior vice president Robert Zitter said he didn't think there was going to be any real demand for HDTV. So HBO had no immediate plans to provide it.

When word of this began circulating in the news media, some of the Cable Mongols grew embarrassed. Television set manufacturers, of course, complained of betrayal. But Gary Shapiro of the Consumer Electronics Manufacturers Assocation was philosophical. "I think it will just increase the exodus of customers from cable," he said. Worried, perhaps, that Shapiro might be right, the Cable Mongols came up with a new strategy in the spring of 1997. The industry had a well-established reputation for talking out of both sides of its mouth, and cable leaders set up a new opportunity to demonstrate this talent.

In May and June, cable-industry executives began calling HBO and asking its leaders to transmit some high-definition programming in 1998. "Our customers came to us and asked us to do this," said Jeffrey Bewkes, HBO's chairman. The company agreed. Privately, HBO officers explained that they had changed their minds because John Malone, chairman of TCI; Ted Turner, vice chairman of Time Warner; and other cable titans—their most important customers—all had put in personal requests. So Bewkes announced that HBO would offer two channels of high-definition programming starting in the summer of 1998.

Not surprisingly, TCI, Time Warner, Cox Cable, and other leaders of the industry lauded the announcement. But there was a catch: HBO could send out whatever it wanted. But not one viewer would see the high-definition shows unless the local cable companies agreed to carry them. And here the cable industry's double-talk reached a level of high art. TCI put out a statement saying the company's leaders were happy about HBO's decision. That was hardly a surprise; they had asked HBO to do it. Still, TCI said, "we have not decided if we will carry the HDTV signal." Even Time Warner Cable, part of the company that owned HBO, was equivocal. "High definition is very much on the horizon, and we'll try to make room for it." That was all Joe Collins, chairman of Time Warner Cable, was willing to say.

Now the cable industry comfortably had it both ways. If anyone complained, cable leaders could accurately say: Yes, of course we are doing HDTV. HBO is going to broadcast *two channels* of it. At the same time, however, not one cable company anywhere was promising to air it.

Network executives gloried in this; anything that showed their longtime competitors in a bad light made them giddy. Meanwhile, however, they were involved in some lowbrow maneuvering of their own. As Congress debated mammoth budget bills in the spring, NAB lobbyists quietly worked provisions into senate and house versions that would allow them to evade parts of the FCC's recent digital-television decisions that they most disliked. Specifically, one amendment would scrap the requirements for putting digital signals on the air by the agreed deadlines. Another would allow TV stations to keep their second channels more or less forever.

The circumstances were a mirror of the quiet work NAB lobbyists had sneakily carried out a couple of years earlier, when they worked the "HDTV killer" amendment into the telecommunications bill. This time, while the ever vulnerable Congress was preoccupied with the far larger issue of balancing the federal budget for the first time in thirty years, lobbyists from the NAB and allied groups snuck the proposals into senate and house bills through friends of the industry, including Representative Billy Tauzin, the Louisiana Republican who was chairman of the Telecommunications Subcommittee, and Senator Conrad Burns, a Republican from Montana who was a member of the Senate Commerce Committee. Tauzin's amendment would allow a broadcaster to keep his second channel if at least 5 percent of the homes in his area were still watching analog television—a situation likely to prevail for decades. Five percent "is a low number," Tauzin acknowledged, "but it's designed that way on purpose so that, as we enter this new world, we don't leave some people behind."

Burns's amendment softened the rollout schedule for digital broadcasting; the FCC's requirements became merely suggestions. "I'm an old broadcaster," the senator explained, "and I know that you can run into a lot of difficulties in the construction phase of these digital stations. And I think we need to give the broadcasters flexibility to deal with unforeseen and unknown circumstances." One unforeseen circumstance was press coverage of these two amendments. The Burns amendment died after the *New York Times* published a story about it. The Tauzin proposal stayed in the budget bill, but it was softened before becoming law. Under a new, complex formula that changed the 5 percent requirement to 15 percent, broadcasters could conceivably be required to return their second channels at some point in the future—a significant change from the original wording.

Still, no one had really expected the broadcasters to return those channels in 2006 in any case. It seemed impossible to believe that the entire nation would have bought a digital TV by then, and no congressman would want to be accused of turning off his constituents' television service. The 2006 date was simply a budget gimmick. Given the political realities of the transition, the channels would be returned when the broadcasters decided they didn't want them anymore. And that day might come sooner than the NAB lobbyists generally supposed, because operating two television stations was a very expensive proposition. The electric-power bill for running a transmitter was several hundred thousand dollars a year—higher if the channel was in the UHF band, as most of the new digital channels were. And during the transition, stations were going to have to operate two stations, pay two power bills. Then what would happen in 2007, 2008, 2009, when most of the nation had made the switch to the new digital channels and the stations' old analog equipment started wearing out? Would TV stations be happy about buying a new analog transmitter costing half a million dollars to send out old-style analog signals that few people were watching any longer? Economic realities were certain to answer questions that legislators and lobbyists seemed unable to resolve.

By late summer, the broadcast networks still had not made any decisions about what they planned to put on their digital channels—even though the first broadcasts were to begin in just over a year. Even the general commitment to broadcast HDTV was beginning to soften. CBS was unwavering in its promise to air 1,080-line interlaced programming, and NBC was still making general promises to put on some HDTV shows; but Fox was dismissive of HDTV. Though there were no public announcements, Fox executives said privately that they intended to broadcast 480-line progressive-scan signals.

Preston Padden had been a senior executive at Fox for several years, and he had apparently infected Fox with this view. Back while the Advisory Committee was still active, Padden had been the Washington lobbyist for Fox, and he had regularly browbeaten Wiley, urging him to abandon HDTV in favor of multichannel broadcasting, saying something like: You have to evolve your thinking; you can't stick with these rigid formulations. Wiley's response had been unvarying. He'd say: Preston, you know what the program is. I'm just gonna tell you, I'm not going that route. I may be the last body alive, but I am going to stay the course. It's HDTV. Within the Fox network, however, Padden had a more willing audience. But then, in May of 1997, Padden got a new job. He was appointed president of the ABC Television Network. And in time, he would roil the debate over digital broadcasting as no one had before.

ABC, of course, had already pledged to broadcast HDTV, just like the other major networks. But when Padden stepped into the job, ABC announced that all bets were off. The network would reexamine everything with no prejudgments about what the answers ought to be. Everyone knew, however, that Padden came to the debate with a strong point of view.

For ABC and the other networks, there was no easy answer to the debate over what should be done with the digital channels. The terms of the discussion had changed because the resolution of this question had moved out of the engineering labs and into the executive suites. As

Michael Jordan, the chairman of CBS, put it: "This whole digital transition has been left to the engineers until just about six months ago. All of a sudden we got this thing approved, and nobody has a clue what they are going to do." The primary consideration was no longer determining what was technologically possible. Now it was: How can we earn a profit? For years broadcasters had been saying they saw no way to make money from broadcasting high-definition television; advertisers weren't going to pay more just because the products showed up better. At the same time, the latest thinking in the industry was that there was no easy way to make money from multichannel broadcasting, either. Networks were already struggling to find appealing programming for the single channels they already had. What good would it do to fragment the audience over four or five different channels—while also trying to sell four or five times as many advertisements to smaller audiences? Local stations faced the same dilemma. As some of them were saying: If there is so much appealing programming out there to fill channels 4A, 4B, 4C, and 4D, why aren't those shows already on the air? With all those great programs just waiting to be broadcast, how come the weakest channels in many cities are simply airing home-shopping services?

So at ABC and all the other networks, both points of view had advocates and detractors. During internal meetings, these issues were debated endlessly, with no happy resolution. At CBS, the network with the clearest commitment to HDTV, Jordan, the chairman, liked to say: "When you have a superior product, customers will always move to it. The networks are already losing their audiences as more channels are added to cable and satellite services, just as Campbell's soup lost market share when Progresso came onto the market, and more again when Goya started selling soup.

"We have to assume the current trends will continue," Jordan went on. "But to me high definition with sports and movies and major entertainment is a way to arrest that decline. The History Channel is not going to do high-definition. Only the major players are going to do it. And it's going to help us; it gives us an advantage."

The multichannel advocates had their own reasoning. Some of them talked about banding several of a city's TV stations so that, together, they had fifteen or twenty channels. They could then offer a miniature "cable" system that provided only the most popular programs, allowing these broadcasters to take an important share of the audience away from the Cable Mongols.

Through all this debate, television-set and broadcast-equipment manufacturers were nagging the networks to make up their minds so these companies could get down to the business of making the equipment that matched the networks' broadcasts. Of course, they were most interested in HDTV. That, in their view, was the only service that would sell the new sets.

"We are very focused on the digital transition; it's the only way we are going to survive," said Jim Meyer, executive vice president of Thomson, the nation's largest television manufacturer. "But if the broadcasters don't choose to offer products that take advantage of this, then that's another thing." But the broadcasters were complaining, too. "Manufacturers have not locked in well-communicated marketing plans," said Bob Wright, president of NBC. "I don't know what they are doing at Sony, Thomson, Panasonic."

Trying to break this chicken-or-egg stalemate, Panasonic, among others, began staging demonstrations of its new convertor box. It was capable of showing broadcast signals at several different levels of resolution—1,080-line interlaced, 720-line progressive, 480 progressive and 480 interlaced. The idea, said Jukka Hamalainen, president of Panasonic American Laboratories, was "to let the broadcasters see for themselves so they can make informed decisions."

One audience was Preston Padden and other senior executives at ABC. Patrick Griffis, a senior Panasonic officer, narrated the demonstration, staged in August at ABC corporate headquarters in midtown Manhattan. And the outcome seemed preordained. Padden had formed his opinions years earlier; and at about the same time as the demonstration, ABC announced layoffs and early retirements among its senior executives. That was certain to focus the thinking of the people who remained. In fact, Padden liked to say that his staff was so worried about the general decline of the television networks that they were spending "all of their time clinging to the arms of their chairs like they're on a sinking ship." With that attitude, they didn't seem likely to offer points of view that were strikingly different from Padden's.

Griffis put on Panasonic's show, and as Padden described it later, ABC executives and engineers "were getting down on the floor to point out tiny differences" between high-definition and standard-definition progressive images. "You could barely tell." As Griffis recalled it, "Padden was mostly interested in seeing the difference between 480-interlaced and 480-progressive."

In the weeks following that demonstration, Joe Flaherty and others over at CBS loudly complained that the presentations were flawed. Panasonic had not taped or shown the material properly, degrading the high-definition images, they claimed. CBS even implicitly threatened to withdraw a huge order for Panasonic HDTV equipment if the company did not stop putting on these shows. But no matter; the damage had already been done.

In mid-August the Paul Kagan Company put on another seminar for broadcasters at the Park Lane Hotel in New York City and invited Preston Padden to speak. Padden accepted, and word leaked out that he would finally break the stalemate and announce what ABC intended to do with its digital channel. When the morning came, the room was packed with broadcasters—and press. Padden was the first speaker, and virtually the first words out of his mouth were that he was not there to make any official announcements. But then he articulated the most effective economic justification for abandoning HDTV in favor of multichannel broadcasting that any broadcaster had offered in public until then. Maybe it was not a formal network statement issued on ABC stationery. But everyone in the audience took it as an announcement.

Year after year, the networks continue to lose their share of the television audience, Padden noted. "And our share of the viewing audience will continue to erode as long as we remain a single channel in an expanding multichannel universe." The solution, Padden said, might well to be offer "multiple streams of high-quality pictures—multiple channels that each constitute a significant improvement in both picture and sound, compared to what consumers receive today." He was talking about the 480-line progressive images that Panasonic had shown him a few weeks before. Americans already paid $30 billion a year for various pay-TV services, Padden noted, and ABC wanted a share of that. "One way to finance continued, free, over-the-air service," he said, was to charge fees for some of the channels.

So there it was, ABC's plan: Broadcast one free channel, as the law required, plus three or four additional channels of pay-TV programming, as well. Under the new FCC rules, this idea was perfectly legal. ABC would have to pay a fee to the government if it aired pay-TV programs. Otherwise, nothing seemed to stand in Padden's way—not even the pledges his predecessors had made as part of their effort to win the channels in the first place.

Of course, the request a decade earlier from all of the broadcasters, including ABC, had led to the appointment of the Advisory Committee and everything that followed. Back then, the industry had said that if broadcasters were not given the ability to offer HDTV, "there is strong reason to fear that they will become a second-class service." And at the Advisory Committee's final meeting in 1995, Thomas Murphy, chairman of Capital Cities/ABC, had joined Dick Wiley's Greek chorus to say: "I'd like to take this opportunity to reaffirm my company's commitment to the implementation of high-definition television."

Then, a few months later, ABC had joined the other three networks in a petition formally asking the Government to *require* broadcasters to offer high-definition programming. As ABC put it in that pleading: "The only way for the commission to assure that enough HDTV programs are in fact offered is for the commission to require each broadcaster to offer a minimum number of hours of HDTV." Now, apparently, Padden was intent on proving that the advice was apt.

Padden wasn't alone in his view. At the same Kagan seminar, David Smith, president of Sinclair Broadcasting Group, which owned or provided programming for twenty-nine stations nationwide, announced that his company, too, would forgo HDTV and offer several channels of pay-TV instead.

"I have to ask: Where is the money in this?" Smith said. "We have yet to see how anyone makes money as an HDTV broadcaster." He and several other broadcasters offered a perverse explanation for their new view. It went like this: All of the discussion about broadcasting in high definition was dreamed up by the engineers downstairs in the labs. Upstairs in the executive suites, we had no idea what they were doing. If only we had known, we would have stopped it before things went too far.

At the seminar later in the day, Jay Fine, a senior vice president at CBS, offered the countervailing view. "I don't know where all the extra content for multichannel broadcasting is going to come from," he observed. "Stations are struggling to fill the one channel they have now." And a few days later, Robert Wright, president of NBC, chimed in. "We have made our decision," he said. "Broadcasters have to embrace high-definition television. We cannot afford to have anything but the best picture. If HBO is going to do it, then I'm going to do it. That's my starting and my ending point."

In the days following the Kagan seminar, computer-industry officers and their allies cheered. At least one major network was going to air 480-line progressive programming! But television-set manufacturers bitterly complained. Who on earth would be willing to spend thousands of dollars for a digital TV that offered nothing better than 480-line progressive images? But then complaints also began coming in, quietly at first, from an unexpected quarter: Congress. The House and Senate had stood by quietly while the FCC set the digital-television rules earlier in the year. But now some members were unhappy.

Padden's remarks "are so disillusioning," said Senator John McCain, chairman of the Commerce Committee. "It is a clear revocation of a commitment that was irrefutably made."

Over the House, Representative Michael Oxley, a member of the Telecommunications Subcommittee, said, "If this was a trial balloon sent up by ABC, I would suggest that it's made of pure lead, if my discussions with my colleagues are any indication." Subcommittee chairman Billy Tauzin had recently sponsored that bill to let broadcasters keep their second channels more or less forever. But now he was upset with Padden and like-minded broadcasters. "The whole idea was that they would exchange one channel for another channel to broadcast HDTV," he said. "We gave them the spectrum for a new service, for HDTV. I don't think Congress will let the decision not to use HDTV pass without some serious debate and discussion. There will be a quid pro quo. If there is no HDTV, then the question becomes: How much spectrum do you need to do a digital broadcast? We could take back the rest and auction it." It's true, Tauzin noted, that he had sided with the broadcasters on other issues. "But I think they need to hear from me now because I am a friend."

Facing that criticism, Padden began to backpedal a bit. "I never said I wasn't going to do any HDTV," he said in his office one afternoon, clinging to a technical accuracy. "All I was trying to do was reassure people here who think there's no future in network television. I think you'll see ABC doing the same things that the other networks do."

In Baltimore, home of the Sinclair Broadcasting Group, the retreat was even more pronounced. "We are prepared to commit that we are going to do some HDTV," said Nat Ostroff, a Sinclair vice president. "We have to find out if the public wants it." But none of that was enough for Senator McCain. He decided to call a hearing, and he summoned Padden and Smith as witnesses.

The Park Lane Hotel was crowded for Padden's original speech, but the line waiting for admittance to the Senate hearing in September extended from the hearing-room door, along the hall in the Senate office building, down two flights of steps, and along another hall. Everyone with the tiniest interest in digital television was there. Senator McCain opened the hearing with an attack on the network executives who had said they were blindsided by their engineers. To suggest, he said, "that this unprecedented effort to make digital TV a reality was really some sort of skunk-works engineering project that caught the business guys by surprise, is hard to accept." What's more, he added, the nation lost the income from auctioning the channels because broadcasters insisted that they needed the spectrum for HDTV. Now, it seems, "consumers will lose twice." No auction revenues, no HDTV, "and I will not accept that.

"I have no desire to penalize broadcasters for misstatements to date. But I hope the broadcast-industry witnesses here today will use the occasion to describe how the industry *actually* intends to proceed with digital television and HDTV."

A few moments later, Senator Ernest Hollings, a Democrat from South Carolina, read congressional testimony from broadcasters ranging back over the last decade, and his tone dripped with scorn as he said, "For years, HDTV was what they were asking for—the entire industry. Since we were in on the deal, we let the deal ensue just as they said it would." Senator McCain, Senator John D. Rockefeller, and others joined in with angry, mocking remarks, all in the same vein. And Padden finally told them "we do not have a plan" to broadcast pay-TV programming and "have not yet come up with a plan" for multichannel broadcasting. ABC, he added, "remains committed to broadcasting some HDTV programming."

Smith, the president of Sinclair, changed his mind, too, though he added a caveat. "We are going to launch the new service making the assumption that the public wants HDTV," he said, "and we are going to offer as much HDTV as the public wants. If the public doesn't accept it, we won't do it."

Senator Burns snapped back, "I see no risk that your ratings will decline if you provide a high-definition service." In the end, however, the senators seemed satisfied with what they heard. And they clearly enjoyed toying with one of the hearing's additional witnesses—Reed Hundt, who had recently announced that he would soon resign from the FCC. The members of the Senate Commerce Committee didn't like Hundt, and they particularly disliked his approach to setting rules for the deployment of digital TV. Still, during Hundt's forty-five minutes on the stand, he managed to keep his temper through a bipartisan tongue-lashing.

"If you're trying to change the rules and set a new policy, you'd better get a new Congress," Hollings told Hundt at one point. "This was always about HDTV, and the whole thing was on track until you muddied the waters." Speaking softly, Hundt replied, "I think high definition turned out to be more of a lobbying idea than a business strategy." But he was speaking into a gale; nothing he said really mattered. When the extraordinary hearing ended, Hundt, Padden, and Smith slunk out of the room.

Piling on, the leaders of CBS and NBC sent written testimonials to the committee, reaffirming their commitment to HDTV. "Our core belief is that consumers will demand HDTV pictures, particularly in the premium time period," said Bob Wright of NBC. "I believe our pay-television competitors will offer HDTV to their customers. I am convinced that broadcasters must respond to this marketplace demand by offering HDTV." Jordan of CBS said: "I am writing to clarify CBS's plans for digital television and HDTV. CBS fully intends to live up to the spirit and letter of commitments made and understandings conveyed to the FCC and the Congress in our rollout of digital television and HDTV."

Fox television was silent, as usual. Later that very week the network had been planning to announce its digital-television strategy, and Fox executives had been saying privately for weeks that the network intended to pursue multichannel broadcasting, not HDTV. A senior Fox executive had been telling broadcasting colleagues that Padden, a former Fox executive, had gotten in trouble simply because he had presented his ideas badly. "We know Preston, and he can do it better," he said. But after the hearing, the network decided not to announce its strategy after all. "What, do you think we're crazy?" the executive quipped.

While the networks politicked, many local stations were facing different pressures. In Dallas, for example, station WFAA, the market-leading ABC affiliate, was quite excited about HDTV and had started planning over the previous summer to put on a high-definition demonstration at the Texas state fair in the fall. The station contracted with a local production company to provide high-definition material, rented space at the fairgrounds, and plunged enthusiastically into the rest of the planning—convinced they would score a competitive coup against their archrivals over at KXAS, the NBC affiliate.

By August, planning had swung into high gear—then WFAA suddenly heard that, in New York, Padden was saying the network would probably abandon HDTV. The network feed for WFAA would likely be several channels of pay-TV programming instead. The station's leaders were furious. Actually, "that understates it," said Beaven Els, WFAA's engineering director. Ward Huey, president of the parent company's broadcast division, stopped by to see Padden while he was in New York a few days later and told him exactly how bad an idea he thought ABC's plan really was.

But over at rival KXAS, Doug Adams, the station's president, gloated. "Great! Terrific!" he said. All this angst in Dallas was the result of circumstances that no one in Washington or New York had anticipated as they fought their political battles. In Dallas and the rest of the country, local stations were intensely competitive. Digital television was coming, very soon, and nobody wanted to be the last to offer it.

"I'm really excited," Adams said. "By putting a superior picture into people's living rooms, we can build on our audience. And there's a real advantage to being first." At WFAA, Huey's view was similar. "High definition is undeniably attractive," he said, "and we want to be there first. I cannot imagine how we can be advantaged by being late to the party." Across the country, hundreds of TV-station executives were offering similar opinions.

"They all want to be first," said Mark Richer. When he worked for PBS, he had monitored system testing for Wiley's advisory committee. Now he was general manager of Comark Digital Services, which was helping numerous stations across the country, including KXAS, make the transition to digital broadcasting. "The most important thing for them," he added, "is the competition with other stations."

Despite Padden's remarks, WFAA did stage its high-definition demonstration at the state fair. And it was a big hit. Even the competitors over at KXAS had to admit it. "The response was phenomenal," grumbled George Csahanin, director of engineering for KXAS. "Everybody was talking about that damned thing."

"We had 150,000 people go through to see it," said Bob Turner of WFAA, adding with a dry tone: "I guess they were kind of demoralized at KXAS."

And then, a few months later, WFAA beat its rival once again by becoming the first Dallas-area station to begin broadcasting a high-definition signal on its new digital channel. But, as with all things related to digital television, this inaugural broadcast was a bittersweet victory. The same day, at a hospital across town, all of the heart monitors in the coronary-care unit suddenly stopped working. As it turned out, these wireless devices had been sending data to the nurses' stations over the same "channel" that WFAA had been given for digital broadcasting. No one was hurt, but WFAA had to take its signal off the air just a few hours after the broadcast began while the hospital looked for heart monitors that used a different frequency.

A short time later, the U.S. Food and Drug Administration sent a warning to hospitals nationwide: Beware! Digital television is coming. Check to be sure your wireless monitors are not using a frequency that has now been assigned to a local television station. And at an industry conference, Joe Flaherty, a CBS executive, quipped: "We've made plans to shut the system off anytime any of us are in the hospital."

In early December 1997, the Intel Corporation called reporters to its Santa Clara, California, offices for a series of new-product announcements. But in the newspapers the next day, only one of these declarations attracted any attention. Intel was backing out of the format war it had been fighting with the other members of the DTV Team, Compaq and Microsoft.

"Our earlier proposal was a smashing failure," acknowledged Ron Whittier, Intel's representative on the team. "Our objective now is to remove barriers between us and the broadcasters. The format issue was an unfortunate discussion that sidetracked us from making investments and getting on with implementation."

This wasn't just a principled change of heart. For one thing, none of the nation's other personal-computer manufacturers had shown any interest in getting into the DTV Team's little war with the broadcasters. They saw no benefit in adding digital-television receivers to their personal computers. In fact, executives with IBM, Dell, Packard Bell, Hewlett-Packard, Gateway 2000, and Sony all said they had no such plans. "It's very much not a priority for us," T. R. Reid, a spokesman for Dell, said after discussing the matter with his company's leaders. "Our customers are much more interested in mainstream computing." And at Hewlett-Packard, Lawrence Sennett said, "We collect enormous amounts of consumer data, and in recent months we have asked people, flat out, basically, if watching TV on a PC was a big thing for them. And the answer we got back was no." At year's end, Mundie acknowledged that he knew of not one computer company, other than Compaq, that had plans to build digital-TV receivers into their PCs.

Faced with that information, Intel had made a technological breakthrough. At the Santa Clara press conference, Whittier and other Intel officers demonstrated a format convertor that would enable personal computers to receive 1,080-line interlaced signals and convert them to 480-line progressive or anything else. And within a year or so, Whittier predicted, processing power would advance far enough that the computers would be able to display a 1,080-line HDTV signal.

Actually, the Hitachi Corporation had designed the convertor. The company's New Jersey lab had been working on it for several years. Over the summer, Intel had asked to have a look at it, and then Intel engineers managed to create a software version of the convertor box. Now Whittier was saying that by the time HDTVs came onto the market, Intel would be able to add this technology to personal computers at little or no cost. As a result, Intel was dropping out of the format war. Technology had made it irrelevant. The broadcasters "had it right, and we had a naive notion," admitted Serge Rutman, a senior Intel researcher.

Intel's partners on the DTV Team were not nearly so magnanimous. Steve Goldberg, director of corporate development for Compaq, said his company was still hoping broadcasters would come around to Compaq's point of view. As for Intel's new software, "it's unclear right now what this means. These things are still in the lab, and we are evaluating a number of different approaches." At Microsoft, Craig Mundie refused to comment. But a few weeks later he made it clear that his company was not changing its position at all. The whole world should adopt 480-line progressive television and nothing more. At a conference in January, Microsoft chairman Bill Gates and one of his employees, Steven Guggenheimer, a manager of the company's digital-television strategy group, stood on a theater stage in front of nearly 2,000 show attendees and pointed at a large television monitor demonstrating a "480-line progressive high-definition signal," as Guggenheimer put it. Standing next to him, Gates added, "That sure looks good, that higher resolution."

By this point, the television manufacturers simply laughed at Microsoft and Compaq. The 1998 Consumer Electronics Show was opening. They had worked hard to meet Gary Shapiro's challenge. Most of them were setting up show-floor displays of the first digital high-definition television sets intended for consumers. All of the sets were designed to display 1,080-line interlaced images. Reality had already marched past the Microsoft-Compaq plans.

At the show, CBS, PBS, and Gary Rebo, a longtime HDTV producer, were going to feed 1,080-line high-definition programming around the show floor for display on these prototype HDTV sets. And the Consumer Electronics Manufacturers Association, managers of the show, were ceaselessly promoting the debut of the first sets. "Sixteen manufacturers to show HDTVs!" screamed one release. "Consumers want HDTV," said another.

Las Vegas was uncharacteristically cool as the show opened, encouraging the ninety thousand attendees to stay inside. That was hardly unpleasant. Just as promised, almost every one of the world's TV manufacturers had an HDTV on display—products that were real, or nearly so. Over the previous year, these companies had kept their plans to themselves, for competitive reasons and because of many of them really had had no idea what they should offer, given the many uncertainties that still hung in the air. But in the end Gary Shapiro's gambit had worked. All of the companies believed they had better be there—or they would find they were the only ones that weren't. So they scrambled to assemble prototype products, and all of them ended up with almost the same thing: big, boxy rear-projection televisions as large as a double-door refrigerator turned on its side—all with wide screens using the 16-to-9 aspect ratio. Some of them looked magnificent, while some others clearly had been pulled out of the lab too soon. And, by and large, these were the only sets the manufacturers planned to put on sale in the fall. A larger array of additional products would go on sale in 1999. But for now, only one company, Sony, said it intended to offer a traditional TV with a picture tube, known in the trade as a direct-view set. Over the previous months, the entire industry had come to the realization that they just didn't know how to make a full high-definition direct-view HDTV set, even though more than 90 percent of all TVs sold were direct-view models. And at the show, industry executives sprang this nasty little secret on the world.

"Right now, we don't have any way to do 1,080-line-by-1,920-line direct-view sets for consumers," said James Newbrough, a senior vice president for Philips. All of the direct-view sets that had been used for demonstrations over the last ten years had been professional monitors costing $10,000 or more. The problem was, it just cost too much to make a picture tube with more than 2 million tiny pixels (compared to only about 350,000 in a conventional set). "We intend to sell direct-view HDTV sets in 1999, and our feeling is that anything with 1 million pixels or more is high-definition," said Bruce Babcock, a Thomson vice president. In fact, a Sony executive explained, the company's direct-view set would offer only 1,080 vertical lines by about 1,200 horizontal lines—or about 1.3 million pixels.

Rear-projection sets were easier to manufacture; they could be made to offer the full resolution. Still, they were going to be quite expensive. Opening prices quoted on the show floor ranged from about $6,000 to $10,000. (Those numbers required quite a stretch of imagination to fit into the promise, made a year earlier, that the first HDTVs would cost $1,000 to $1,500 more than equivalent conventional sets.) But even as salespeople uttered the figures, many of them were also listening over their shoulders to hear what the competitors were saying. Given the traditions of the industry, the prices were certain to drop before fall. Nonetheless, nobody thought huge numbers of these sets were going to sell. "Something this expensive and big doesn't exactly fit into everybody's living room," acknowledged John Taylor, spokesman for Zenith. "It's ideal for showing HDTV to the masses. I think you'll see it in sports bars, shopping malls, and showrooms." Others were a bit more optimistic. "Yes, they're expensive," said Jeff Cove, a Panasonic vice president. "But we already sell a lot of sets in this price range." And Meyer, a Thomson executive, agreed. "Last year nearly 100,000 consumers spent more than $3,000 for a TV, and 50,000 people spent more than $4,000," he said. "I think there is a sizable market for this kind of luxury product. It's a time of great wealth in this country, and I think this will appeal to the Lexus crowd."

At the same time, Meyer acknowledged: "Not everyone will buy an HDTV this year, but I believe everyone will want to see HDTV. That's a tremendous opportunity for our retailers. It'll turn stores into HDTV theaters."

But in many ways it didn't really matter whether these first, refrigerator-sized TVs sold in great numbers. The first color TV that went on sale in 1954 had had a tiny 12.5-inch screen, a crummy picture, and a price tag of $1,000 (about $6,000 in 1998 dollars, when adjusted for inflation). Hardly any of them sold. But over the years the quality of the color sets had improved dramatically, while prices had fallen. The average price for a color TV in 1998 was under $300 (less than $50 in 1954 dollars), with a screen size of, about 20 inches.

"These are the worst HDTVs we'll ever see," Joe Flaherty said as he wandered the show floor. And some of them did look pretty bad. For many showgoers, however, it didn't seem to matter. Mitsubishi, for example, was showing a huge prototype with a 73-inch screen. A receiver box sat atop the 7-foot tall cabinet; the internal circuitry had not been finalized. And the picture, frankly, just wasn't very good. In addition, the booth's lighting was terrible; a spotlight shone right on the screen, giving the picture a washed-out look. Still, as a wandering group stopped and looked up at it, salespeople's mouths dropped open and one of them said, "Boy, I want one; you can't beat that!" Scenes like that were repeated across the show floor. For many people, it seemed, as long they were told they were watching HDTV, it didn't really matter what it looked like.

On the show's opening day, the president of DirecTV, the nation's largest direct-broadcast satellite company, announced that starting in the fall his company would broadcast two channels of HDTV programming. Not to be outdone, his direct competitor, EchoStar, reacted to the DirecTV announcement the next day. Charles Ergen, head of EchoStar, said, "Anything anybody can do with HDTV, we think we can do better and more of. We can do dozens of channels." All of that caught the attention of the cable industry, and a few days later Time Warner Cable said it would begin carrying HBO's HDTV shows in the fall, and any other high-definition signals it received.

More announcements like that followed, including the networks' decisions on what they planned to do with their digital channels. NBC chose 1,080-line interlaced as its high-definition format and said *The Tonight Show*, certain prime-time programs recorded on film, and big movies—including *Titanic* and *Men in Black*—would be broadcast in HDTV.

"Our issue is, we want to make sure we offer the best picture available," said Bob Wright, NBC's president. "I don't ever want to be in a position where we don't have that." ABC chose 720-line progressive as its high-definition format, and Preston Padden, ABC's president, staged a boisterous press conference to trumpet his choice. Among the invited guests were nearly a dozen of the academic and computer-industry zealots who'd been campaigning for progressive scanning over the last decade. Each of them was given a moment at the podium to extol the wonders of progressive-scan broadcasts, and Padden pronounced: "We believe 720-P offers the highest picture quality of any available format." ABC was not ready to say how much high-definition programming would be aired. Padden said only: "We want to be the leader."

CBS, meanwhile, stuck with its plan to use 1,080-line interlaced as its high-definition format and announced that it would air at least five hours of HDTV programming each week, plus some football and basketball games. "By providing the highest-quality HDTV programming, we will gauge the reaction of consumers of HDTV and the response of set manufacturers to the new technology," said Michael Jordan, chairman of the network.

Few people were surprised that Fox, alone, decided to forgo HDTV and broadcast in the 480-line progressive format. The higher-level formats "are wasteful of spectrum," said Andrew Setos, a senior vice president for Fox. "And these different formats are in the standard to give people the ability to choose." Fox had cleverly invited Representative Billy Tauzin and Senator John McCain to California for demonstrations of 480-line progressive, and for meeting with Rupert Murdoch, the company's chairman. Tauzin and McCain, chairmen of important committees, were the only people who were intent on keeping the broadcasters to their promise of airing high-definition programming. And if they could be swayed, no one remained to hold the broadcasters to their word.

After the meetings and demonstrations, both Tauzin and McCain made remarks that made it sound as if they had been convinced that 480-line progressive was good enough. That emboldened Fox to announce its strategy. But right away, Mr. Tauzin remarked: "Frankly, [given] additional information, which I have since come to acquire," the meetings with Fox "would have been persuasive. But to say today that they have turned me away from high definition is not correct." He and Senator McCain both warned of sanctions against Fox if the network did not offer some high-definition programming. The next day, Lawrence Jacobson, president of the Fox network, announced that "our high-definition format is 720-P, and we will also broadcast some of our programs in 480-P. We are committed to HDTV." Setos added, "We will be using both of these formats in our broadcasts."

All of the networks said these were only their initial choices. Each of them reserved the option to change its mind if the public showed a preference for something else. Still, for the people who had spent the last ten years fighting the high-definition television wars, everything finally seemed to be falling into place.

Certainly there'd be other battles. If the past ten years had proved nothing else, that, at least, was a certainty. Nonetheless, when digital broadcasting began in the fall of 1998, the public would see a fair sampling of HDTV from broadcasters, satellite providers, cable operators, and perhaps others. Viewers would also be offered some 480-line progressive programming and some multichannel broadcasting, too. They'd be shown some interactive and data services. Then, as digital broadcasting matured in 1999 and beyond, clever programmers in some cities would begin offering new formats and services no one could even conceive of at the start. Viewers would have a look at all of it and finally get to decide for themselves what they liked. Just as it should be.

Toward the end of the Consumer Electronics Show, one of the longtime HDTV warriors, Jim McKinney, sat in the HDTV information booth at the front of the convention hall, answering showgoers' questions about digital television. McKinney, of course, was a former FCC officer who had attended the first Muse demonstration at FCC headquarters in 1987. He then became a principal in Wiley's Advisory Committee and director of the model high-definition station, WHD-TV. Here in the CES booth, he sat directly in front of Zenith's first HDTV set, a 64-inch rear-projector, playing high-definition feed from PBS. It looked absolutely magnificent, the best of the show, and McKinney was determined to have one in his home in the fall.

Watching all the activity around him that late afternoon as the show began to wind down, McKinney leaned back in his chair, hands clasped behind his head, as a self-satisfied smile settled over his face. "I'll tell you," he said. "I had no idea this show would turn out the way it did. People now know. HDTV is real."

Appendix How to buy a digital television set

Until now, buying a television has been a relatively simple affair. Almost anyone could do it without fear of intimidation because the choices were few. A customer had to choose a screen size, look over the remote control, decide whether he wanted picture-in-picture (often purchased, seldom used), and perhaps look at the inputs on the back.

With digital TV, all of that is changing. A digital set is closer to a computer than a television. The number of options, display formats, and related capabilities can be numerous and complex. What's more, the options are almost certain to change over time. The lessons you learn today may be of little help when you go back to buy a new set a few years from now. You might feel like the people who bought a personal computer loaded with DOS programs in the 1980s, then returned to the store in the 1990s to find that the entire industry had switched to Windows.

Still, some basic advice should hold up for a while.

Should I buy an HDTV now?

Typically a small group of people known in the industry as "early adopters" buy hot new products as soon as they come onto the market. Usually, that's about 1 percent of the market for a moderately priced product. The percentage is smaller for a truly expensive item, and the first-generation high-definition television sets are very expensive. They cost $5,000 or more. Most people will wait, and for those who do, prices will fall quickly while the capabilities of the sets will rise dramatically.

In truth, the TV industry does not see the sets they will sell in 1998 and early 1999 as the actual product launch. Speaking privately, they call this the demonstration period. Sure, some people will buy the first-generation digital televisions—huge rear-projection sets. But for the rest of the nation, the industry hopes that a look at these TVs will whet consumers' appetites for cheaper, more appealing models that will come onto the market later in 1999 or 2000. As James Newbrough, a senior vice president for Philips, put it in early 1998: "What we have to tell people is, this is just a start. In reality it will be better to wait until 1999."

How fast will prices fall? James Meyer, a senior executive for Thomson, said prices for consumer-electronics products usually fall by one third every two or three years. By that calculation, an HDTV costing $6,000 in the fall of 1998 will cost $4,000 in 2000 or 2001. But the prices will probably drop more rapidly than that. For one thing, the costs of the first-generation TVs are unusually high because the sets are almost handcrafted devices. In 1998 mass-production lines for digital television sets had not been built, and only a handful of companies were supplying microprocessor chips at the start. Over a couple of years' time, however, all of that will change. Almost every manufacturer of televisions in the world will be making digital TVs. The competition to begin efficient manufacturing will be intense. And the television industry, more than any other, is known for cutthroat price competition.

Should you buy an HDTV at the launch of the new service? If you're an early adopter who lives in a city where broadcasters are offering some high-definition programming, you will almost certainly enjoy your new TV. At the same time, though, those who wait will be rewarded. Prices will fall quite fast. And manufacturers—who in 1998 were just figuring out how to build an HDTV that *worked*—will begin selling a far greater variety of sets that look and work much better.

If I wait, does it make sense to buy a new analog set in the meantime?

If you plan to buy an ordinary set in 1998 or 1999, the kind that costs less than $1,000, go ahead. It will be at least several years before high-definition televisions are sold at that price. If you're going to buy a fancy home-theater model costing several thousand dollars, however, the considerations are different. By the turn of the century, HDTVs will almost certainly be available at comparable prices, and you could feel foolish for having spent all that money on a standard-definition set when, with a bit of patience, you could have had a high-definition television for roughly the same price.

If you do buy an analog set in the early years of the digital service, you might want to be sure it is equipped to work well with convertor boxes. Manufacturers will sell convertors that allow viewers to watch high-definition programming on their analog sets—though not in high definition, of course. But these boxes will work best if plugged into an advanced input on the back of your set. These inputs come in two styles: "S-video" and "component video." Most TVs can be bought with an S-video jack, a little round plug with four pins. An S-video connection separates the video signal into two parts, reducing iterference, and some manufacturers hope to use those jacks as the input for their convertor boxes.

The more advanced form of input called "component video" is an input with three separate jacks. A component input separates the video signal into three different parts, and high-end convertor boxes will take advantage of this. S-video jacks are quite common; most inexpensive TVs can be bought with one of these for a few dollars more. But component inputs are generally available only on expensive TVs. It's not necessary to hold out for that. Most manufacturers are likely to build their convertors to work with the far more common S-video inputs.

Suppose I do want to buy a high-definition television set. How do I know if the set I'm looking at really is an HDTV?

"High-definition" is a subjective term. Back in 1935 when British government officials were working to establish a standard for the first commercial television broadcasts, they declared that they would consider "nothing less than high-definition television" with 240 lines of vertical resolution—double the number in test broadcasts of the time. In the United States a few years later, *Broadcasting* magazine announced the beginning of "this country's first regular schedule of high-definition television broadcasts," with 441 lines.

Today, *high-definition* is generally described as a television picture with at least twice the resolution of conventional TV. And resolution is most easily determined by counting the number of lines on the screen, vertical and horizontal. Conventional television offers 480 vertical lines of resolution and 720 horizontal lines. So by that definition, the highest-level format in the Grand Alliance standard—1,080 vertical lines and 1,920 horizontal lines—easily qualifies as high definition. But what about all the intermediate formats, including 720-line progressive TV?

In the fall of 1997, the Consumer Electronics Manufacturers Association polled its membership, TV manufacturers and retailers, to reach consensus on what constituted high-definition television. The purpose was to set out enforceable definitions and prevent misleading advertising—by television or computer salespeople. To qualify as an HDTV under the industry's definition, the association said a digital television must offer at least:

—Vertical resolution of 1,080 interlaced lines or 720 progressive lines
—A wide screen with a 16-to-9 aspect ratio
—An audio system capable of reproducing a Dolby AC-3 surround-sound signal
—A receiver capable of receiving all eighteen possible display formats set out in the Grand Alliance standard

Even within that definition, however, there's still wiggle room. Only vertical resolution was stipulated, not horizontal. The highest-level HDTV signal is 1,080 vertical lines by 1,920 horizontal lines. But in early 1998, DirecTV, the direct-broadcast satellite company, announced that the service would broadcast a signal that has 1,080 vertical lines but only 1,280 horizontal lines. Under the industry's definition, that still qualifies as HDTV, and the picture will look far better than conventional TV. But it won't look as good as a full-resolution signal.

This becomes important for consumers buying direct-view sets, the ones with picture tubes. At the start of the new service, manufacturers were saying they didn't know how to make an affordable direct-view set that offered the highest-level signal. So Thomson (makers of RCA sets) and Sony both said they would build direct sets with horizontal resolution levels comparable to the DirecTV model. "Our feeling is that high definition starts at about that level," said Bruce Babcock, the Thomson executive.

If you want a direct-view set, there may be no choice, at least in the beginning. But ask about both vertical and horizontal resolution capabilities so you know what you're getting before you take it home.

What about progressive and interlace; should I worry about that?

Probably not, but once again you should know. Progressive-scan receivers are far better for viewing text, and many engineers think progressive scan offers better television pictures, too. But interlace scan has its own defenders and positive attributes. The debate has raged for a decade and may never be resolved. To evade it, the Grand Alliance engineers designed their system so that it can receive progressive or interlace signals. That's one of the reasons the system has eighteen different formats.

But there's a difference between being able to *receive* eighteen formats and being able to *display* all of them. Receiving all those signals is the work of a microprocessor chip. But building a picture tube that can switch automatically among eighteen different formats is extraordinarily expensive. As a result, most manufacturers are building high-definition televisions that display only one or two signal types, called the "native-display formats." The TVs are able to *receive* any of the eighteen signal types. But all of those signals are electronically converted inside the set to the television's native-display format. Early on, most manufacturers are choosing 1,080-line interlaced as the native-display format. That seems to make a lot of sense since many broadcasters seem to be favoring this format. But that could change.

Find out the native-display format of any digital television you want to buy. The wisest course is to make sure that the format matches that of the television programming you watch most often, since converting a signal will degrade it. If, however, you intend to use your digital television for computer applications, then look for one with a progressive-scan native-display format. All the 1,080-line interlaced signals will be converted to that.

With all those different transmission formats, how will I know that my digital TV will be able to receive all of them?

Originally, designers of the Grand Alliance system hoped that the government would *require* televisions to be built so they could receive all eighteen transmission formats, just as conventional TVs are required to receive NTSC programming. But after the negotiations between the computer industry and the broadcasters in 1996, that requirement was removed. At that time, a Grand Alliance officer wrote in a briefing memo for his colleagues: "It would appear useful to come up with, and establish with consumers, a symbol of compliance which would appear on products." And a year later, the industry did.

Digital televisions capable of receiving all of the eighteen different formats will carry a special seal. It shows a 16-to-9 aspect-ratio box with the letters DTV inside and the phrase "ATSC-certified digital television." ATSC stands for Advanced Television Systems Committee. Any new television that does not carry this seal stands the risk of going dark if an unusual signal comes in.

Since only a minority of the programming will be in high definition at first, is the rest of it going to look terrible in comparison?

Billy Byers, an engineer with Thomson, was issuing a warning to broadcasters who intended to ignore HDTV when he said, "There's an immediate psychological reaction to high definition. And what I think broadcasters don't realize is that, in the new broadcast environment, high definition is always going to be just a channel change away." But the same sentiment can apply to viewers. How are they going to feel, after watching a magnificent wide-screen high-definition movie on their new HDTV, to find a fuzzy letterboxed conventional show on the very next channel. Will people still want to watch when the picture quality is so much worse than the HDTV program's? There's a corollary problem for broadcasters: How will advertisers who have not yet adapted to the new age feel if their low-definition advertisements are dropped into the middle of a high-definition program?

Anticipating these problems, almost every segment of the television industry is planning to deploy technologies that home-theater enthusiasts have used for years to improve the quality of television images. Only now these devices, called "line doublers," will be used to make conventional programming look better in the high-definition world. A line doubler converts a TV signal from interlace to progressive, immediately improving it. Some "upconvertors," as the products are generally known, also create additional lines of resolution by adding new information to the signal derived from the signal information that is already there.

CBS, among others, intends to "upconvert" all of its standard-definition programming before it is broadcast on the digital channel. And several set manufacturers, including Sony and Mitsubishi, intend to build upconvertors into their digital-television sets, for use with shows that are not upconverted at the source. An upconverted signal is not high definition, but it looks quite a lot better than conventional television. So it makes sense to buy an HDTV with a line doubler or upconvertor built in.

The aspect-ratio program is trickier. High-definition televisions will have wide 16-to-9 screens, and high-definition programming will generally be broadcast in that format. But all of the television programming of the previous fifty years was filmed or taped in the conventional format, with an aspect ratio of 4-to-3. Several companies sell convertors that stretch 4-to-3 images to 16-to-9, but most broadcasters don't like the technology. The picture actually *looks* stretched. Unless that view changes, conventional shows watched on an HDTV will appear in vertical letterboxing, with broad black bars on the left and right. And high-definition shows watched on a conventional set with a convertor box will appear in horizontal letterboxing, with a black bar at the top and bottom of the screen. That, it appears, is simply an unavoidable feature of the transition.

What if I want an improved, digital picture—but I don't want to spend the money for a full high-definition set?

Most of the first-generation digital televisions, perhaps even all of them, will be full HDTVs. The television industry wants the public to see these first. But in time, a range of models will be offered. RCA, for example, likes to talk about selling three levels of digital TVs: good, better, and best. The principal difference among these different models will be the native-display formats. The "good" TVs will probably feature lower-resolution displays, and they will cost less. In fact, most manufacturers say they will sell a broad range of digital models with different prices and capabilities—everything from a standard-definition digital TV to a full HDTV. Some will offer progressive-scan displays, line doublers, modems, additional memory, or built-in direct-broadcast satellite receivers.

Most of those options should be relatively straightforward. But for many people, the most confusing problem may be understanding the difference between what kinds of signals the television can *receive* and the kinds it can *display*. Just keep in mind the notion of the native-display format. It isn't determined by what kind of signal comes into the television. It's the highest level image the set can display.

In the new age, asking about the screen size and the shape of the remote control will no longer tell you everything you need to know about your new digital television set. But with a bit of education, buying a high-definition television set does not have to be a consumer's nightmare.

Sources and acknowledgments

I owe thanks to many—most of all to the contestants and managers of the HDTV race who so generously shared their lives with me. In fact, the most important source for almost everything in this account is the players themselves. During more than four years of research, I interviewed almost everyone in this book over and over again. I visited all the contestants' labs several times, held roundtable talks with scores of engineers, watched their machines take form, and sat in on uncounted meetings and discussions.

Descriptions of conversations that I did not witness are based on the participants' own accounts of what occurred, or printed records of what they said. Similarly, accounts of someone's thoughts are based on what that person told me he was thinking at the time. I have taken one reportorial liberty, however. If someone described a statement in the past tense, in many cases I moved it into the present tense as I placed it in the narrative. ("I told him I didn't like it" became "I don't like it.")

I should caution that the players' memories often were far from clear, even for events that took place a relatively short time earlier. For some meetings and other events, I interviewed everyone involved—six or eight people in several cases—and found that recollections differed, sometimes radically. In those cases I selected the common elements of their recollections and offer here the closest possible consensus view. Still, readers should be cautioned that memories often are colored by experiences after the event.

Documentary sources filled in facts when memories failed. Dick Wiley and Paul Misener provided invaluable assistance; they gave me unrestricted access to Paul's meticulous, exhaustive files of documents and correspondence from the Advisory Committee's activities. Similarly, Peter Fannon was unstinting in his willingness to help me understand

and observe the important work at his Advanced Television Test Center. John Taylor at Zenith was zealous in his efforts to help me learn about his company. So were Cynthia Gray, formerly of Sarnoff, and Bob Ford, formerly at AT&T. Bob Rast could not have been a more helpful advocate, as a representative of both General Instrument and the Grand Alliance.

At the same time, I owe a debt to some journalists and writers whose interest in this subject preceded mine. Tom Lewis's book *Empire of the Air: The Men Who Made Radio* was a valuable source on the early years of broadcasting. *Communications Daily* chronicled the HDTV race in detail, and its account of events up to the spring of 1990 was published in two books—*HDTV: The Gold Rush Begins* (1989) and *HDTV: The Rush to Reality* (1990). Those books—catalogs of clippings, really—were of great use to me. They were out of print when I started my work, but Keiichi Kubota dug copies out of his files in Tokyo. That was but one of Keiichi's many kindnesses. *Broadcasting & Cable* magazine provided thorough, important coverage of the race, particularly in the later years. *Electronic Engineering Times* and the newsletter *HDTV Report* were also valuable sources.

I am indebted to several others as well, including the editors of the *New York Times Magazine,* who sponsored my initial research—though in the end no magazine article resulted from it.

Andrew Pollack, Tokyo correspondent for the *Times,* carried out a difficult but critical interview for me in Japan when I could not get there. Thank you.

My brother Alan, an accomplished historian and author, offered valuable advice on the near-final manuscript. My friend Roberta Baskin, one of the few broadcast journalists I openly admire, also gave much-appreciated suggestions and encouragement. My editor, Walter Bode, gently guided me through a revision that markedly improved the book. And my wife, Sabra Chartrand, also a talented journalist and author, advised me through several rough patches and provided encouragement when many other spouses might have resented the time and attention this project consumed.

About the author

Joel Brinkley has been a reporter, editor and Pulitzer Prize–winning foreign correspondent for the *New York Times* since 1983.

He is a native of Washington, D.C., and a 1975 graduate of the University of North Carolina at Chapel Hill. He worked for the Associated Press, the Richmond (Va.) *News Leader* and the Louisville *Courier-Journal* in the 1970s and early 1980s.

In 1979, he was dispatched to Southeast Asia to cover the Vietnamese invasion of Cambodia, the Pol Pot genocide, and the resulting refugee crisis. For that he won the Pulitzer Prize for international reporting in 1980.

After joining the Washington bureau of the *Times* in 1983, he covered the American involvement in Lebanon, international drug trafficking, and the contra war in Nicaragua, among other issues. He was Washington editor of the paper's coverage of the Iran-contra affair and White House correspondent during the final years of the Reagan Administration.

In March of 1988, he was named chief of the *Times* bureau in Jerusalem, Israel, where he served from the beginning of the Palestinian uprising through the Gulf War. In 1991 he returned to the *Times* Washington bureau to become project editor. And in 1995 he moved to New York, where he held several reporting and editing positions, including political editor and editor in charge of the paper's coverage of the crash of TWA Flight 800. Through 1997 and part of 1998, he served as a reporter in charge of covering the nation's transition to digital television. In 1998 he returned to the *Times* Washington bureau.

Over the last fifteen years, Mr. Brinkley has won more than a dozen national journalism and writing awards. He and his wife live in Chevy Chase, Maryland.